Aerodynamic Theory

A General Review of Progress

Under a Grant of the Guggenheim Fund
for the Promotion of Aeronautics

William Frederick Durand
Editor-in-Chief

Volume III

Div. F · The Theory of Single Burbling · C. Witoszyński and M. J. Thompson
Div. G · The Mechanics of Viscous Fluids · L. Prandtl
Div. H · The Mechanics of Compressible Fluids · G. I. Taylor and J. W. Maccoll
Div. I · Experimental Methods—Wind Tunnels · A. Toussaint and E. Jacobs

With 167 Figures
and 6 Plates

Berlin · Julius Springer · 1935

ISBN-13: 978-3-642-89629-3 e-ISBN-13: 978-3-642-91486-7
DOI: 10.1007/978-3-642-91486-7

GENERAL PREFACE

During the active life of the Guggenheim Fund for the Promotion of Aeronautics, provision was made for the preparation of a series of monographs on the general subject of Aerodynamic Theory. It was recognized that in its highly specialized form, as developed during the past twenty-five years, there was nowhere to be found a fairly comprehensive exposition of this theory, both general and in its more important applications to the problems of aeronautic design. The preparation and publication of a series of monographs on the various phases of this subject seemed, therefore, a timely undertaking, representing, as it is intended to do, a general review of progress during the past quarter century, and thus covering substantially the period since flight in heavier than air machines became an assured fact.

Such a present taking of stock should also be of value and of interest as furnishing a point of departure from which progress during coming decades may be measured.

But the chief purpose held in view in this project has been to provide for the student and for the aeronautic designer a reasonably adequate presentation of background theory. No attempt has been made to cover the domains of design itself or of construction. Important as these are, they lie quite aside from the purpose of the present work.

In order the better to suit the work to this main purpose, the first volume is largely taken up with material dealing with special mathematical topics and with fluid mechanics. The purpose of this material is to furnish, close at hand, brief treatments of special mathematical topics which, as a rule, are not usually included in the curricula of engineering and technical courses and thus to furnish to the reader, at least some elementary notions of various mathematical methods and resources, of which much use is made in the development of aerodynamic theory. The same material should also be acceptable to many who from long disuse may have lost facility in such methods and who may thus, close at hand, find the means of refreshing the memory regarding these various matters.

The treatment of the subject of Fluid Mechanics has been developed in relatively extended form since the texts usually available to the technical student are lacking in the developments more especially of interest to the student of aerodynamic theory. The more elementary treatment by the General Editor is intended to be read easily by the average technical graduate with some help from the topics comprised in Division A. The more advanced treatment by Dr. Munk will call for some familiarity with space vector analysis and with more advanced mathematical methods, but will commend itself to more advanced students by the elegance of such methods and by the generality and importance of the results reached through this generalized three-dimensional treatment.

In order to place in its proper setting this entire development during the past quarter century, a historical sketch has been prepared by Professor Giacomelli whose careful and extended researches have resulted in a historical document which will especially interest and commend itself to the study of all those who are interested in the story of the gradual evolution of the ideas which have finally culminated in the developments which furnish the main material for the present work.

The remaining volumes of the work are intended to include the general subjects of: The aerodynamics of perfect fluids; The modifications due to viscosity and compressibility; Experiment and research, equipment and methods; Applied airfoil theory with analysis and discussion of the most important experimental results; The non-lifting system of the airplane; The air propeller; Influence of the propeller on the remainder of the structure; The dynamics of the airplane; Performance, prediction and analysis; General view of airplane as comprising four interacting and related systems; Airships, aerodynamics and performance; Hydrodynamics of boats and floats; and the Aerodynamics of cooling.

Individual reference will be made to these various divisions of the work, each in its place, and they need not, therefore, be referred to in detail at this point.

Certain general features of the work editorially may be noted as follows:

1. Symbols. No attempt has been made to maintain, in the treatment of the various Divisions and topics, an absolutely uniform system of notation. This was found to be quite impracticable.

Notation, to a large extent, is peculiar to the special subject under treatment and must be adjusted thereto. Furthermore, beyond a few symbols, there is no generally accepted system of notation even in any one country. For the few important items covered by the recommendations of the National Advisory Committee for Aeronautics, symbols have been employed accordingly. Otherwise, each author has developed his system of symbols in accordance with his peculiar needs.

At the head of each Division, however, will be found a table giving the most frequently employed symbols with their meaning. Symbols in general are explained or defined when first introduced.

2. General Plan of Construction. The work as a whole is made up of *Divisions*, each one dealing with a special topic or phase of the general subject. These are designated by letters of the alphabet in accordance with the table on a following page.

The Divisions are then divided into chapters and the chapters into sections and occasionally subsections. The Chapters are designated by Roman numerals and the Sections by numbers in bold face.

The Chapter is made the unit for the numbering of sections and the section for the numbering of equations. The latter are given a double number in parenthesis, thus (13.6) of which the number at the left of the point designates the section and that on the right the serial number of the equation in that section.

Each page carries at the top, the chapter and section numbers.

W. F. Durand

Stanford University, California
January, 1934.

GENERAL LIST OF DIVISIONS WITH AUTHORS

Volume I.

A. Mathematical Aids

W. F. DURAND — Professor (Emeritus) of Mechanical Engineering, Stanford University, Calif., Member of the National Advisory Committee for Aeronautics.

B. Fluid Mechanics, Part I

W. F. DURAND

C. Fluid Mechanics, Part II

MAX M. MUNK — Lecturer in Aerodynamics at the Catholic University of America, Washington, D. C., and Technical Editor of the "Aero Digest".

D. Historical Sketch

R. GIACOMELLI — Lecturer in History of Mechanics at the University of Rome, Italy, and Editor of "L'Aerotecnica".

with the collaboration of

E. PISTOLESI — Professor of Mechanics at the Royal School of Engineering at Pisa, Italy, and Editor-in-Chief of "L'Aerotecnica".

Volume II.

E. General Aerodynamic Theory—Perfect Fluids

TH. VON KÁRMÁN — Director of the Guggenheim Aeronautics Laboratory, California Institute of Technology, Pasadena, Calif., and formerly Director of the Aerodynamic Institute, Aachen, Germany.

J. M. BURGERS — Professor of Aero- and Hydrodynamics at the Technische Hoogeschool at Delft, Holland.

Volume III.

F. The Theory of Single Burbling

C. WITOSZYŃSKI — Professor of Aerodynamics at the Warsaw Polytechnical School and Director of the Warsaw Aerodynamic Institute, Poland.

M. J. THOMPSON — Assistant Professor of Aeronautical Engineering at the University of Michigan, Ann Arbor, Mich.

G. The Mechanics of Viscous Fluids

L. PRANDTL — Professor in Applied Mechanics at the University of Göttingen, Germany, and Director of the Kaiser Wilhelm Institute for Fluid Research.

H. The Mechanics of Compressible Fluids

G. I. TAYLOR — Yarrow Research Professor of the Royal Society, Fellow of Trinity College, Cambridge, England.

J. W. MACCOLL — Research Officer, Department of External Ballistics, Ordnance Committee, Woolwich, England.

I. Experimental Methods—Wind Tunnels

A. TOUSSAINT — Director of the Aerodynamic Laboratory, Saint-Cyr-l'École, France.

E. JACOBS — Associate Aeronautical Engineer, in charge of the National Advisory Committee for Aeronautics' variable-density wind tunnel, Langley Field, Virginia.

Volume IV.

J. Applied Airfoil Theory

A. Betz — Professor at the University and Director of the Aerodynamic Research Institute at Göttingen, Germany.

K. Airplane Body (Non-Lifting System) Drag and Influence on Lifting System

C. Wieselsberger — Professor of Aerodynamics and Director of the Aerodynamic Institute, Technische Hochschule, Aachen, Germany.

L. Airplane Propellers

H. Glauert [1] — Past Fellow of Trinity College, Cambridge, England; Principal Scientific Officer at the Royal Aircraft Establishment, Farnborough.

M. Influence of the Propeller on other Parts of the Airplane Structure

C. Koning — Rijks-Studiedienst voor de Luchtvaart, Amsterdam, Holland.

Volume V.

N. Dynamics of the Airplane

B. Melvill Jones — Professor of Aeronautical Engineering in the University of Cambridge, England, Member of the Aeronautical Research Committee of Great Britain.

O. Airplane Performance

L. V. Kerber — Former Chief Aerodynamics Branch Materiel Division, U. S. Army Air Corps, and former Chief, Engineering Section Aeronautics Branch, Department of Commerce.

Volume VI.

P. Airplane as a Whole—General View of Mutual Interactions Among Constituent Systems

W. F. Durand [2] — Professor (Emeritus) of Mechanical Engineering, Stanford University, Calif., Member of the National Advisory Committee for Aeronautics.

Q. Aerodynamics of Airships

Max M. Munk — Lecturer in Aerodynamics at the Catholic University of America, Washington, D. C., and Technical Editor of the "Aero Digest".

R. Performance of Airships

K. Arnstein — Chief Engineer of the Goodyear Zeppelin Company, Akron, Ohio.

W. Klemperer — Research Engineer of the Goodyear Zeppelin Company, Akron, Ohio.

S. Hydrodynamics of Boats and Floats

E. G. Barrillon — Director of the Naval Experimental Tank, Paris, France.

T. Aerodynamics of Cooling

H. L. Dryden — Physicist in the United States Bureau of Standards, Chief of the Aerodynamics Section, Washington, D. C.

---

[1] Deceased August 4, 1934.

[2] In the original plan, it was expected that this Division would be prepared by Professor M. Panetti of the R. Scuola di Ingegneria of Turin. Unfortunately, at the last, Professor Panetti found himshelf unable to give the needed time for this work, and in order not to delay publication, the General Editor has undertaken to prepare a brief treatment of the subject.

CONTENTS

DIVISION F

THE THEORY OF SINGLE BURBLING

By **C. Witoszyński,**

Professor of Aerodynamics at the Warsaw Polytechnical School and Director of the Warsaw Aerodynamic Institute, Poland

and **M. J. Thompson,**

Assistant Professor of Aeronautical Engineering at the University of Michigan, Ann Arbor, Mich.

DIVISION G

THE MECHANICS OF VISCOUS FLUIDS

By **L. Prandtl,**

Professor in Applied Mechanics at the University of Göttingen, Germany, and Director of the Kaiser Wilhelm Institute for Fluid Research

DIVISION H

THE MECHANICS OF COMPRESSIBLE FLUIDS

By **G. I. Taylor,**

Yarrow Research Professor of the Royal Society,
Fellow of Trinity College, Cambridge, England

and **J. W. Maccoll,**

Research Officer, Department of External Ballistics,
Ordnance Committee, Woolwich, England

DIVISION I

EXPERIMENTAL METHODS—WIND TUNNELS

PART 1

By **A. Toussaint,**

Director of the Aerodynamic Laboratory, Saint-Cyr-l'École, France.

PART 2

By **Eastman Jacobs,**

Associate Aeronautical Engineer, in charge of the National Advisory Committee for Aeronautics' variable-density wind tunnel, Langley Field, Virginia.

NOTATION

The following table comprises a list of the principal notations employed in the present Volume. Notations not listed are either so well understood as to render mention unnecessary, or are only rarely employed and are explained as introduced. Where occasionally a symbol is employed with more than one meaning, the local context will make the significance clear.

DIVISION F

x, y	Coordinates along axes X, Y	
Z, z	Complex $(x + iy)$	
ξ, η	Elliptical coordinates	
ζ	Special complex variable $\tau + i\sigma$,	**2**
s, S	Length of arc of a curved profile	
α	Angle of incidence	
θ	Angle, usually of inclination to axis of x	
V, v	Velocity in general	
u	Velocity along X	
u	Velocity of origin in case of moving axes,	**6**
w, v	Special velocities,	**6**
Ω	Induced velocity,	**7**
ω	Angular velocity	
C	Circulation,	**7**
f	$f(z)$	
φ	Potential	
ψ	Stream function	
P	Force in general	
P_y	Lift	
P_x	Drag	
M	Moment,	**5**
N	Moment,	**5**
g	Absolute acceleration,	**6**
i	$\sqrt{-1}$	
$\mathfrak{J}$	Imaginary part of complex quantity	
$\mathfrak{R}$	Real part of complex quantity	
ϱ	Density	

DIVISION G

X, Y, Z	Coordinate axes	
x, y, z	Coordinates along X, Y, Z	
ξ	Special coordinate along x,	**10**
	Also non-dimensional length along x,	**13**
η	Non-dimensional length along y,	**13**
b	Dimension, usually width	
δ	Thickness of boundary layer,	**14**
δ^*	Special length corresponding to the lateral displacement,	**14**
δ^{**}	Special length corresponding to the momentum loss,	**17**

ε	Linear extension, **5**
k	Diameter of grain of roughness, **22**
l	Length in general
r	Radius
S	Area
α, β, γ	Inclinations to axes X, Y, Z
γ	Angular deformation, **5**
u, v, w	Component velocities along X, Y, Z
U	Basic velocity, **26**
u', v'	Disturbing velocities, **26**
v_*	Special velocity deduced from the shearing stress, **21**
ω	Angular velocity, **6**
Γ	Circulation, also special parameter, **24**
Ψ	Stream function
φ, χ, ψ	Special functional symbols
D	Resistance or drag
D_i	Induced drag
E	Kinetic energy
M	Momentum
p	Stress or pressure
P	Basic pressure, **26**
p'	Disturbing pressure, **26**
σ	Normal stress
τ	Shearing stress
	Also special time variable, **13**
c	Special coefficient $l = cx$, **25**
C_D	Coefficient of resistance or drag
C_f	Coefficient of frictional resistance
k	Coefficient or factor
μ	Coefficient of viscosity
ν	Kinematic coefficient of viscosity
R	Reynolds number
T, t	Time
ϱ	Density

DIVISION H

X, Y	Coordinate axes
x, y	Coordinates along X, Y
x, y, z	Used for pressure ratios p_2/p_0, p_1/p_0, p_3/p_0, IV **2**
T	Thickness of layer of transition; also temperature
S	Area
α, β	Special angles, IV **2**
M, m	Mach angle, IV **1, 5**
θ	Angle in general
a	Velocity of sound
q	Resultant velocity at a local point in the stream
u, v	Velocities usually along X, Y, sometimes along and at right angles to radius vector, III (3.2), IV **5**
U	Velocity of flow far from any obstacle or disturbance
K	Circulation
φ	Potential
L	Lift
γ	Ratio of specific heats of a gas at constant pressure and at constant volume
J	Joule's equivalent, I **6**

$\varkappa$	Coefficient of heat conductivity
μ	Coefficient of viscosity
∇^2	$\partial^2/\partial r^2 + (1/r)\,\partial/\partial r + (1/r^2)\,\partial^2/\partial \theta^2$ (polar coordinates), III (2.1)
M, m	Mass
p	Pressure
ϱ	Density
R	Gas constant

DIVISION I

x	In general, distance in direction of flow
y	In general, distance transverse to flow or along span of wing
z	Used for complex $(x + i\,y)$
b	Half span
c	Chord
D	Diameter of circular tunnel
R	Radius of circular tunnel
a	Interval between successive vortices in vortex wake
h	Width between trains of vortices in vortex wake
h	Width of tunnel of rectangular section
k	Height of tunnel of rectangular section
L	Length in general
l	Major axis of oval, III **6** b)
t	Minor axis of oval, III **6** b)
S	Area of wing
S_0	Area of cross section of tunnel
α	Angle of incidence
α	Total angle of conical diffuser
V, v	Velocity in general
U_0	Basic air velocity at point of observation
u, v, w	Supplementary velocities due to vortex systems
w	Used for complex $(u - i\,v)$
Γ	Vorticity
φ	Functional symbol
φ	Potential
ψ	Stream function
F	Potential function $(\varphi + i\,\psi)$
E	Energy
F	Force
P_m	Power absorbed by fan
P_u	Useful power (time rate of energy) at experimental section of tunnel
C_L	Lift coefficient
C_D	Drag coefficient
C_t	Tunnel coefficient
η	Efficiency
R	Reynolds number
m	Mass
n	Direction along normal
μ	Coefficient of viscosity
ν	Coefficient of kinematic viscosity
ϱ	Density

DIVISION F

THE THEORY OF SINGLE BURBLING[1]

By

C. Witoszyński and **M. J. Thompson**
Warsaw Ann Arbor, Mich.

EDITOR'S PREFACE

The aerodynamics of the Lifting-Wing lies naturally at the center of progress in aviation. In the mathematical investigation of the details of the flow of the air about an airplane wing, and with special reference to the discontinuity of flow which may appear at the trailing edge, the use of a discontinuous form of potential, complementary to the classic form, as developed by Professor Witoszyński and his collaborators, offers especial interest. By the use of these forms of potential Professor Witoszyński has succeeded in developing for profile forms derivable from the circle by the usual methods of conformal transformation (and hence with close approximation for standard forms) expressions for lift, moment and profile drag (exclusive of friction). The latter, in particular cannot be developed without the use of something representing discontinuous flow and in this respect the work of Professor Witoszyński and of his collaborators furnishes an interesting and important addition to the general body of aerodynamic theory.

The Editor desires to express his acknowledgments to Professor Thompson, co-author with Professor Witoszyński, for the translation of the original manuscript from the Polish, and for valuable aid in checking formulae, preparation of diagrams and in proof reading the work through the press.

W. F. Durand.

---

1. The Discontinuous Potential. The circulation theory of Joukowski which is generally accepted in the study of two-dimensional fluid motion as the basis for the calculation of aerodynamic forces, gives too large values for the lift, even for very pronounced aspect ratios. The method which follows consists in completing the classic velocity potential by a discontinuous term which permits the annulment of the velocity at the trailing edge of the profile in the same way as Joukowski's

[1] In the present Division, this term is used in the sense of a single point or line of separation.

complementary term, but which at the same time gives the values of the lift closer to those obtained in experiments.

Let us suppose that we have placed a profile provided with a sharp trailing edge in a position such that the lift (component normal to the wind) is equal to zero. It is clear that the flow around this profile may

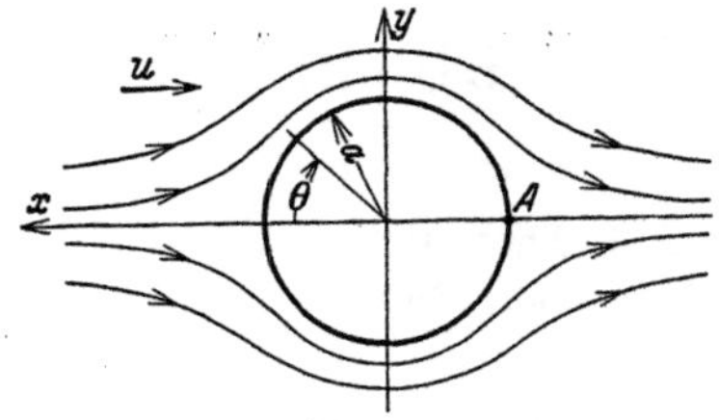

Fig. 1a.

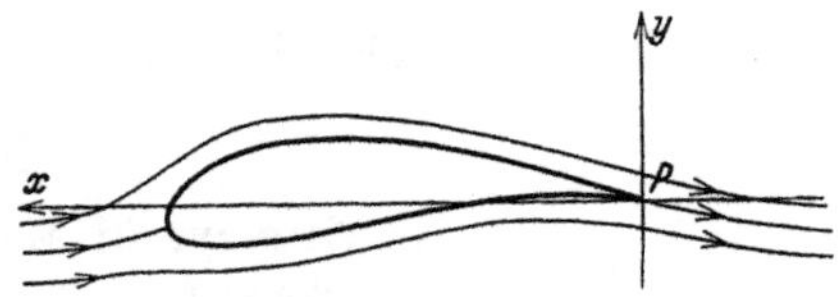

Fig. 1b.

be determined by the method of conformal transformation, in taking as a basis the classic flow around a circle, to which there corresponds the complex potential:

$$\Phi + i\Psi = -u\left(Z + \frac{a^2}{Z}\right) \tag{1.1}$$

In (1.1) u represents the velocity of the wind, which at infinity is in the direction of the negative X-axis, $Z = X + iY = re^{i\theta}$ and a is the radius of the primitive circular profile. In this case the trailing edge P corresponds to the point A of the circular profile, where the velocity becomes zero, see Figs. 1a and 1b. But if the direction of the wind at infinity is changed so that it forms an angle $-\alpha$ with the x-axis instead of zero, the complex potential takes the form:

$$-u\left(Z e^{i\alpha} + \frac{a^2 e^{-i\alpha}}{Z}\right) \tag{1.2}$$

The point P, see Figs. 2a and 2b, corresponds to the point A as before but the velocity at A is no longer zero, the calculations giving an infinite velocity at P, which does not conform with reality.

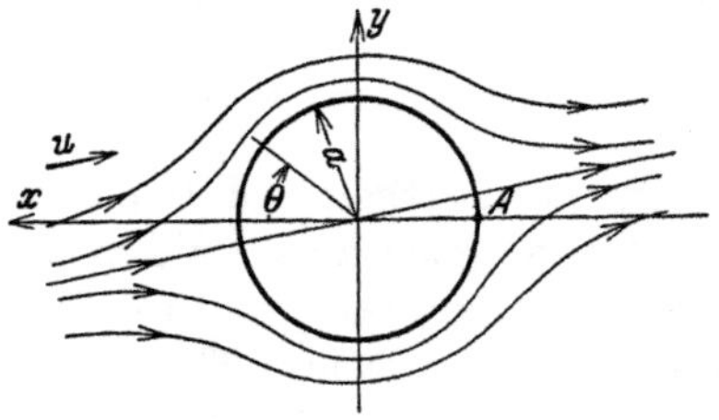

Fig. 2a.

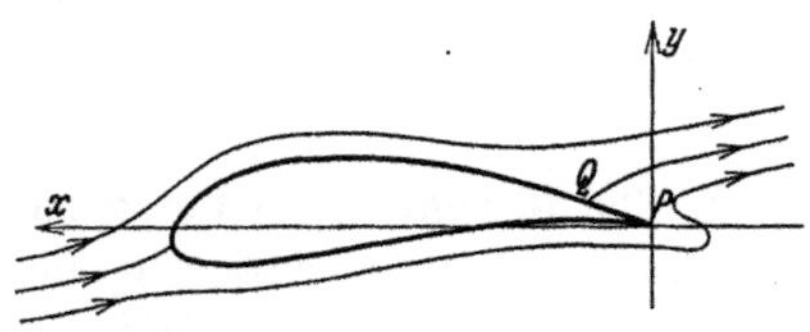

Fig. 2b.

Experimental observations show that the fluid filament surrounding the profile on the lower surface does not prolong its path along PQ (Fig. 2b) but detaches itself from the profile (burbles) at the trailing edge P. In this way between the point P and the point Q, which also changes its position, a "dead" space is formed bounded on two sides by liquid filaments possessing different velocities. Thus a sort of layer of discontinuity is obtained which begins in the neighborhood of the trailing edge of the profile. In order to fix our ideas, let us suppose that the points P and Q coincide at the trailing edge, that is to say that the layer of discontinuity has a thickness equal to zero at its point of origination. We shall, in what follows, call this type of flow—flow with *single burbling*. This type of flow is illustrated in Fig. 3. In the case of the flow around a wing at a high angle of attack or around a bluff object, we have *double burbling* since the fluid filaments close to the body detach themselves at two points in its rear.

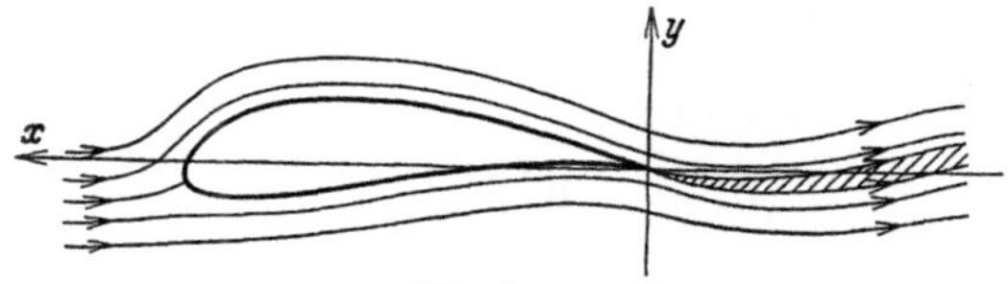

Fig. 3.

Now it goes without saying that the expression (1.2) cannot be used as the complex potential for the circular profile which may serve as the point of departure for the determination of the flow shown in Fig. 3. In order to obtain such a potential, it is necessary to complete the expression (1.2) with a term corresponding to the discontinuity which exists in the rear of the profile. Let us choose for this discontinuous complement referred to the circular profile, a function of the form:

$$(\Phi + i\Psi)_{comp.} = f(Z^{1/2}) = A_1 Z^{1/2} + A_2 Z + A_3 Z^{3/2} + \dots$$
$$+ B_1 Z^{-1/2} + B_2 Z^{-1} + B_3 Z^{-3/2} + \dots$$

It is clear that the coefficients $A_2 \dots\dots A_n$ must be equal to zero in order that the modification of the wind velocity by the profile will not be felt at infinity. The coefficient A_1 must also be zero because otherwise the modification of the velocity will disappear less quickly with an increase of the distance from the profile than in the Joukowski method—so that the lift would be correspondingly greater. We see from this that the complementary potential referred to the circular profile can contain only negative powers of Z.

It is easily verified that the functions

$$Z^p + \frac{a^{2p}}{Z^p} \quad \text{or} \quad i\left(Z^p - \frac{a^{2p}}{Z^p}\right)$$

where p is a real arbitrary number, treated as complex potentials, give a circular stream-line of radius $r = a$. In order to obtain a complementary function of the form discussed above it is necessary to put $p = 1/4$ so that we have

$$i K u a \frac{Z^{1/4} - \dfrac{a^{1/2}}{Z^{1/4}}}{Z^{1/4} + \dfrac{a^{1/2}}{Z^{1/4}}} = i K u a \frac{Z^{1/2} - a^{1/2}}{Z^{1/2} + a^{1/2}} \tag{1.3}$$

where K is a real number which will be chosen in such a way as to annul the velocity at the point corresponding to the trailing edge of the profile. This function may also be written in the simpler form

$$\frac{2\,i\,K\,u\,a^{3/2}}{Z^{1/2}+a^{1/2}}$$

since a constant term may be added to (1.3) without changing the flow. The complete complex potential for the primitive circle, obtained by adding the complementary term (1.3) to the classic flow represented by (1.2) is then,

$$f(Z) = -u\left(Z\,e^{i\alpha} + \frac{a^2 e^{-i\alpha}}{Z}\right) + i\,K\,u\,a\left(\frac{Z^{1/2}-a^{1/2}}{Z^{1/2}+a^{1/2}}\right)$$

In order to determine the value of the constant K, we shall make use of the condition mentioned above that the velocity at the point corresponding to the trailing edge of the transformed profile must be equal to zero. This is identical with the basic condition of Joukowski's theory. Now in all the work that follows, it will be supposed that the profiles considered are placed so that the trailing edge corresponds to the point $Z = -a$ on the primitive circle. This position is called the "zero position"[1] because the lift is zero when the direction of the air stream at infinity is parallel to the negative x-axis. In other words, we have chosen for the x-axis in the plane of the transformed profile, the axis of zero lift.

The expression for the complex velocity is

$$\frac{df}{dZ} = V_x - i\,V_y = -u\left(e^{i\alpha} - \frac{a^2 e^{-i\alpha}}{Z^2}\right) + \frac{i\,K\,u\,a^{3/2}}{Z^{1/2}(Z^{1/2}+a^{1/2})^2}$$

and, making use of the formula

$$V_c = i\,e^{i\theta}\left[\frac{df}{dZ}\right]_{Z=a\,e^{i\theta}}$$

we get for the velocity on the primitive circle

$$V_c = 2\,u\,\sin(\alpha+\theta) - \frac{K\,u}{2\,(1+\cos\theta/2)}$$

When $Z = -a$, $\theta = \pm(2\,n+1)\,\pi$, where n is any positive integer or zero, $\cos\theta/2 = 0$, and

$$[V_c]_{Z=-a} = -2\,u\,\sin\alpha - \frac{K\,u}{2} = 0$$

so that

$$K = -4\,\sin\alpha$$

The final form for the complex potential is thus:

$$f(Z) = -u\left(Z\,e^{i\alpha} + \frac{a^2 e^{-i\alpha}}{Z}\right) - 4\,i\,u\,a\,\sin\alpha\left(\frac{Z^{1/2}-a^{1/2}}{Z^{1/2}+a^{1/2}}\right) \tag{1.4}$$

while the complex velocity is:

$$\frac{df}{dZ} = -u\left(e^{i\alpha} - \frac{a^2 e^{-i\alpha}}{Z^2}\right) - \frac{4\,i\,u\,a^{3/2}\sin\alpha}{Z^{1/2}(Z^{1/2}+a^{1/2})^2} \tag{1.5}$$

[1] Witoszyński, C., "La mécanique des profils d'aviation", p. 36, E. Chiron, Paris, 1924.

and the expression for the velocity on the primitive circle is:

$$V_c = 2\,u\left[\sin(\alpha + \theta) + \frac{\sin\alpha}{1 + \cos\theta/2}\right] \tag{1.6}$$

It is immediately evident that both the complex potential (1.4) and its derivative (1.5) are multiple-valued functions and we must determine in what manner branch-cuts must be drawn in order that this function may be treated as single-valued and applied to our problem. If we consider a point $Z = r\,e^{i\,\theta}$ situated on any closed curve surrounding the primitive circle, then the term $Z^{1/2}$ in the above expressions has the value $r^{1/2}\,e^{i\,\theta/2}$, but if we pass around this curve in a clockwise direction we must write $Z = r\,e^{i\,(\theta + 2\,\pi)}$ so that $Z^{1/2} = -\,r^{1/2}\,e^{i\,\theta/2}$ and there is a change in the value of the complex potential at this point. It thus becomes necessary to stipulate that we shall never take under consideration a closed curve completely surrounding the primitive circle, or in other words that the polar angle θ must be limited to a range of values not greater than $2\,\pi$. We might, for example, take the interval $-\pi < \theta < \pi$ or $0 < \theta < 2\,\pi$ or in general $-\pi + \theta_1 < \theta < \pi + \theta_1$. If we use the last of these intervals, we find that the velocity on the primitive circle for the lower limit of this range is not in general the same as that obtained for the upper limit. Since there cannot exist a discontinuity of velocity on the profile itself except at the point corresponding to the trailing edge of the transformed profile in the case of single burbling, it is necessary to take $\theta_1 = 0$, in which case the velocity becomes continuous. For $|Z| = a$ we then have $-\pi < \theta < \pi$ and it now remains to determine the limits of θ for the region outside the primitive circle.

Before determining the values of these limits, let us see if there are any points on or outside the primitive circle at which the velocity becomes infinite. It can easily be shown that the only points at which this is possible are those for which the denominator of the complementary term in the complex potential is equal to zero, that is,

$$Z^{1/2} + a^{1/2} = 0$$

from which

$$Z^{1/2} = -\,a^{1/2} = a^{1/2}\,e^{\pm\,i\,\pi}$$

and

$$Z = a\,e^{\pm\,2\,i\,\pi}$$

This value of Z represents a point on the primitive circle but its argument lies outside the range to which we have limited θ, in other words, it is on the second Riemann sheet of the function chosen for the complex potential, and therefore is not to be considered in this problem. The point $Z = 0$ is also a pole of both the complex potential and its derivative, but since it lies inside the primitive circle, it is of no importance here.

In order to determine the limits of the polar angle θ in the region exterior to the primitive circle, we begin with an investigation of the stream-lines of the flow defined by the function (1.4). Putting $Z = r\,e^{i\,\theta}$

in this expression and separating out the imaginary part, we find for the stream function

$$\Psi = -u\,(r - a)\left[\frac{(r+a)}{r}\sin(\alpha+\theta) + \frac{4\,a\sin\alpha}{r + a + 2r^{1/2}a^{1/2}\cos\theta/2}\right] \quad (1.7)$$

By selecting a series of constant values for Ψ, it would be possible to determine any number of stream-lines of the flow, but we, however, are particularly interested in the stream-line $\Psi = 0$ which consists of two parts; the primitive circle and another curve, the nature of which we shall now investigate. The equation of this system of lines is

$$(r - a)\left[\frac{(r+a)}{r}\sin(\alpha+\theta) + \frac{4\,a\sin\alpha}{r + a + 2r^{1/2}a^{1/2}\cos\theta/2}\right] = 0 \quad (1.8)$$

and after extracting the factor $(r - a)$ corresponding to the primitive circle, we get

$$\left[\frac{(r+a)}{r}\sin(\alpha+\theta) + \frac{4\,a\sin\alpha}{r + a + 2r^{1/2}a^{1/2}\cos\theta/2}\right] = 0 \quad (1.9)$$

This stream-line intersects the primitive circle in two points, the arguments of which may be found by putting $r = a$ in the above equation, after which there results

$$\sin(\alpha+\theta) + \frac{\sin\alpha}{1+\cos\theta/2} = 0$$

One root of this expression is $\theta = \pm\pi$ corresponding to the trailing edge of the transformed profile, while the second can be shown to be a small value of θ. Thus, supposing that α is also small, we may write approximately

$$\alpha + \theta + \frac{\alpha}{2} = 0$$

so that

$$\theta \cong -\frac{3}{2}\alpha$$

When r becomes infinitely large, (1.9) reduces to

$$\theta = -\alpha$$

It thus represents a stream-line consisting of two parts, both of which approach asymptotically to the line through the origin parallel to the direction of the air stream at infinity. One branch of this line ends at the point $B\,(\theta = -3/2\,\alpha)$ and the other at $A\,(\theta = \pm\pi)$ as shown in Fig. 4. By means of (1.9) the coordinates of a sufficient number of points on this first line may be obtained, but since on the second branch the values of θ are close to $\pm\pi$, it is more convenient to write

$$\theta = \pm\pi - \theta_1$$

and (1.9) then becomes:

$$-\frac{(r+a)}{r}\sin(\alpha-\theta_1) + \frac{4\,a\sin\alpha}{r + a \pm 2r^{1/2}a^{1/2}\sin\theta_1/2} = 0 \quad (1.10)$$

In this expression the positive sign in the denominator of the second term corresponds to $\theta = +\pi - \theta_1$, while the negative sign corresponds to $\theta = -\pi - \theta_1$. Thus we see that the second branch of the stream-line $\Psi = 0$ which issues from the point $Z = -a$ really consists of two parts, one of which is arrived at by considering the stream-lines above the profile and letting Ψ approach zero, and the other in a similar way by considering the stream-lines underneath the circle. These two lines, which do not intersect except at the point A and at infinity, bound a region in which the flow is not uniquely defined, for if we should try to find any stream-lines in this space we would arrive at two different systems depending on whether we approached this space from above or below the circle. It thus becomes necessary to exclude this region from the plane and this we do by making this double branch of the stream-line $\Psi = 0$, a branch-cut of the function. In this way the complex potential becomes single-valued and the velocity is uniquely defined at every point in the plane outside the circle with the exception of the points lying in the excluded region. These restrictions may be expressed analytically by stipulating that for $|Z| > a$, $-\pi - \varepsilon_1 \leqslant \theta \leqslant \pi - \varepsilon_2$, the angles ε_1 and ε_2 being equal to the values of θ_1 as given by (1.10), in which the plus sign corresponds to ε_2 and the minus sign to ε_1. For $|Z| = a$, ε_1 and ε_2 are equal to zero and we have $-\pi \leqslant \theta \leqslant \pi$ as we have already mentioned before.

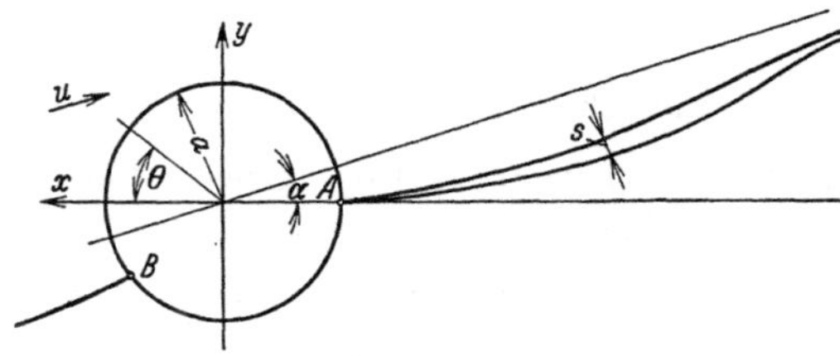

Fig. 4.

In the plane of the transformed profile a similar situation may be shown to exist, for there likewise, a double stream-line issues from the trailing edge, and the flow is completely defined at all points in the region exterior to the profile with the exception of the space between these two lines. This region which is excluded from the plane is known as the "*layer of discontinuity*", since the velocities on its upper and lower boundaries are unequal for the same value of the modulus of Z. Furthermore, it will be found that for the transformed profile, there in general exists a discontinuity of the velocity at the trailing edge. The complex potential function (1.4) which serves to determine this flow is called the "potential of the layer of discontinuity", or more simply, "the discontinuous potential".

Let us now return to (1.10) which represents the stream-lines bounding the layer of discontinuity for the primitive circle and investigate more in detail the geometric characteristics of this region. It is particularly interesting to determine the thickness of the layer and also at what point it is a maximum. We shall calculate the value of this thickness approximately, limiting ourselves to the consideration of small values of the angles θ_1 and α. Only the case of the primitive circular profile will be considered here. Equation (1.10) may then be written in the approximate form:

$$-\frac{(r+a)}{r}(\alpha-\theta_1)+\frac{4\,a\,\alpha}{(r+a)\left(1\pm\frac{r^{1/2}a^{1/2}}{r+a}\theta_1\right)}=0$$

Now the term $\frac{1}{1\pm\frac{r^{1/2}a^{1/2}\theta_1}{r+a}}$ may be expanded in a power series in θ_1 and since the quantity $\frac{r^{1/2}a^{1/2}\theta_1}{r+a}$ has a small value, we may neglect the higher powers in this series. Rejecting the terms which are of a degree higher than the second in θ_1 and α, we finally obtain

$$-\frac{(r+a)}{r}(\alpha-\theta_1)+\frac{4\,a\,\alpha}{r+a}\left[1\mp\frac{r^{1/2}a^{1/2}}{r+a}\theta_1\right]=0$$

Solving this equation for θ_1, we get after considerable simplifying

$$\theta_1=\alpha\left(\frac{r-a}{r+a}\right)^2\left[1\pm\frac{4\,r^{3/2}a^{3/2}\alpha}{(r+a)^3}\right]$$

The larger value of θ_1 corresponds to the upper boundary of the layer of discontinuity and the smaller value to the lower. These values may be written separately as:

$$\theta_{1\,u}=\varepsilon_2=\alpha\left(\frac{r-a}{r+a}\right)^2\left[1+\frac{4\,r^{3/2}a^{3/2}\alpha}{(r+a)^3}\right]$$

$$\theta_{1\,l}=\varepsilon_1=\alpha\left(\frac{r-a}{r+a}\right)^2\left[1-\frac{4\,r^{3/2}a^{3/2}\alpha}{(r+a)^3}\right]$$

and for their difference we get:

$$\varepsilon_2-\varepsilon_1=\frac{8\,r^{3/2}a^{3/2}(r-a)^2\alpha^2}{(r+a)^5}$$

and multiplying by r we obtain for the thickness of the layer of discontinuity:

$$s=r(\varepsilon_2-\varepsilon_1)=\frac{8\,r^{5/2}a^{3/2}(r-a)^2\alpha^2}{(r+a)^5} \qquad (1.11)$$

In order to determine the value of r for which s is a maximum, we set the derivative of s with respect to r equal to zero, first substituting $r^{1/2}/a^{1/2}=t$ so as to avoid having to deal with the cumbersome fractional exponents. The expression thus obtained finally reduces to

$$t^4-14\,t^2+5=0$$

from which

$$t^2=\frac{r}{a}=7\pm\sqrt{44}$$

We are interested only in the value of r greater than a, that is,

$$r=(7+\sqrt{44})\,a\cong 14\,a$$

Substituting this value of r in the expression (1.11) we obtain for the maximum thickness of the layer:

$$s_{max}=\frac{8\,t^5(t^2-1)^2\,a\,\alpha^2}{(t^2+1)^5}\cong 1.31\,a\,\alpha^2$$

For the transformed profile this maximum thickness is practically the same as for the circular profile, since for large values of Z we have approximately $z \simeq Z$. Thus for the rectilinear profile, for example, which has a chord $l = 4a$, we get when $\alpha = 10^0$

$$s_{max} \simeq 0.01\, l$$

Up to the present time we have been concerned principally with the flow around a circular profile as defined by the discontinuous potential, and it is now of interest to examine the flow around other types of profiles. For this purpose we shall investigate the velocity distribution on different transformed profiles making use of the formula

$$v_p = V_c \frac{dS}{ds} \tag{1.12}$$

in which v_p denotes the velocity on the transformed profile, V_c that on the primitive circle, and dS/ds is the ratio of a differential element of arc of this circle to the corresponding element of the transformed profile.

We begin with the determination of the velocity for a simple case, *i. e.*, the rectilinear profile, the transformation function for which may be written in the form: $z = Z + \frac{a^2}{Z}$ when the profile is in the zero position. The derivative of this function is

$$\frac{dz}{dZ} = 1 - \frac{a^2}{Z^2}$$

and after putting $Z = ae^{i\theta}$, $dz = dx + idy$, $dZ = dX + idY$, we get

$$\frac{dx + i\,dy}{dX + i\,dY} = 1 - e^{-2i\theta}$$

The conjugate of this value is

$$\frac{dx - i\,dy}{dX - i\,dY} = 1 - e^{+2i\theta}$$

and the product of these two expressions gives

$$\frac{\overline{dx}^2 + \overline{dy}^2}{\overline{dX}^2 + \overline{dY}^2} = \left(\frac{ds}{dS}\right)^2 = 4\sin^2\theta$$

from which

$$\frac{ds}{dS} = 2\sin\theta$$

Now the velocity on the primitive circle is given by the formula (1.6) which may be transformed into the following form:

$$V_c = 2u\left[\frac{2\sin(\alpha + \theta/2) + \sin(\alpha + \theta)}{1 + \cos\theta/2}\right]\cos\frac{\theta}{2}$$

so that the velocity on the rectilinear profile is

$$v_R = \frac{u\,[2\sin(\alpha + \theta/2) + \sin(\alpha + \theta)]}{2\,(1 + \cos\theta/2)\sin\theta/2} \tag{1.13}$$

This velocity becomes infinite at the point $\theta = 0$, corresponding to the sharp leading edge of the profile, so that strictly speaking the flow around the rectilinear profile cannot exist physically. This profile may, however, be considered as the limiting case of a thin symmetric profile and the results obtained for it will be treated as a close approximation to those of the latter.

We shall now calculate the velocity at the trailing edge, first putting $\theta = -\pi$ and then $\theta = +\pi$. We thus obtain

$$[v_R]_{T.E.} = u\left(\cos\alpha \mp \frac{1}{2}\sin\alpha\right) \tag{1.14}$$

in which the positive sign corresponds to the lower surface of the profile and the negative sign to the upper surface. We thus see that the velocity near the trailing edge is larger on the bottom than on the top, while over the greater part of the surface the contrary is true and the pressures on the bottom are larger than those on top. This phenomenon may be explained in this way, that a profile set at a certain positive angle of attack slows down the flow of the fluid underneath the wing, causing an increase in the pressure, but at the trailing edge the fluid in some way tears itself loose from the surface of the profile, producing the same effect as is obtained by the outflow from a vessel of fluid under pressure.

Let us now determine the expression for the velocity on a Joukowski profile. Taking the transformation function in the form

$$z = \frac{(Z+a)^2}{Z + k\,a\,e^{i\mu}}$$

we get for the ratio of the differential elements of arc

$$\frac{ds}{dS} = \frac{4\cos\theta/2\sqrt{k^2 - k\cos\mu + k\cos(\theta-\mu) + \sin^2\theta/2}}{1 + k^2 + 2k\cos(\theta-\mu)}$$

Substituting this value in the formula (1.12), we get for the velocity on a Joukowski profile:

$$v_J = \frac{\mp u\,[2\sin(\alpha+\theta/2) + \sin(\alpha+\theta)]\,[1 + k^2 + 2k\cos(\theta-\mu)]}{2(1+\cos\theta/2)\sqrt{k^2 - k\cos\mu + k\cos(\theta-\mu) + \sin^2\theta/2}} \tag{1.15}$$

The $\mp$ sign has been introduced in this expression to take care of the fact that on the lower surface of the transformed profile a positive velocity is in the direction of θ decreasing, hence the negative sign must be used, while on the upper surface the contrary is true and we employ the positive sign.

Putting $\theta = \pm\pi$ in (1.15) we obtain for the velocity at the trailing edge

$$[v_J]_{T.E.} = u\left(\cos\alpha \mp \frac{1}{2}\sin\alpha\right)\sqrt{1 + k^2 - 2k\cos\mu} \tag{1.16}$$

the upper sign of $(1/2)\sin\alpha$ corresponding to the upper surface and the lower sign to the bottom. Here again it is seen that the velocity underneath the profile at the trailing edge is greater than that on the upper surface.

As a final example we shall calculate the velocity on the so-called bi-polar profile[1]. The transformation function for such a profile when in the zero position is $z = Z\left(1+\frac{a}{Z}\right)^2\left(1-\frac{k\,a\,e^{i\mu}}{Z}\right)$

and $\frac{d\,s}{d\,S} = 4\cos\frac{\theta}{2}\sqrt{\sin^2\frac{\theta}{2}+k^2+k\cos(2\,\theta-\mu)-k\cos(\theta-\mu)}$

so that

$$v_B = \frac{\pm\,u\,[2\sin(\alpha+\theta/2)+\sin(\alpha+\theta)]}{2\,(1+\cos\theta/2)\sqrt{\sin^2\theta/2+k^2+k\cos(2\,\theta-\mu)-k\cos(\theta-\mu)}} \qquad (1.17)$$

and at the trailing edge

$$[v_B]_{T.E.} = \frac{u\,(\cos\alpha \mp 1/2\sin\alpha)}{\sqrt{1+k^2+2\,k\cos\mu}} \qquad (1.18)$$

where, as in previous cases, the upper sign corresponds to the upper surface and the lower sign to the lower surface of the profile.

From the formulas (1.16) and (1.18) defining the velocity at the trailing edges of the Joukowski and bi-polar profiles, it is seen that these values are smaller than for the rectilinear profile. Furthermore, it may be shown that in general this trailing edge velocity decreases when the thickness increases, although it should be noted that these calculations are valid only for relatively thin profiles set at a small angle of attack. For thick profiles or for high values of the angle of attack, the flow becomes one of double burbling instead of single burbling and the discontinuous potential can no longer be applied.

2. Calculation of the Air Forces for Single Burbling. Having completed the development of the discontinuous potential and discussed some of its general properties, we shall now proceed to the calculation of the air forces acting on several different profiles due to the flow defined by this function. It should be noted that these calculations when made by means of the formulas of Blasius are in general much more difficult than those encountered in the Joukowski theory, because of the fact that in the former case we have to deal with fractional powers of the variable Z. For this reason, it is sometimes more convenient to make use of a graphical method based on the direct application of the fundamental formulas:

$$P_y = -\frac{\varrho}{2}\oint v_p^2\,d\,x; \qquad P_x = -\frac{\varrho}{2}\oint v_p^2\,d\,y \qquad (2.1)$$

in which P_y and P_x denote the components of the air reaction parallel to the coordinate axes, acting vertically upward and horizontally to the right, respectively, when positive. The quantity ϱ is the mass density of the air, v_p is the velocity at a point on the transformed profile, and $d\,x$ and $d\,y$ are the projections on the axes of the differential element

[1] Witoszyński, C., "La mécanique des profils d'aviation", p. 40, E. Chiron, Paris, 1924.

of arc, ds, of the profile contour. The span of the wing is taken as unity so that P_y and P_x represent the forces per unit length of the wing.

This method is illustrated in Fig. 5 in which is drawn a typical Joukowski profile. By means of formula (1.15) we have calculated the values of the square of the velocity for a series of values of θ and have laid them off on a suitable scale below and to the right of the profile. The points on the profile corresponding to the different values of θ are, of course, determined from the transformation function. Each of the two diagrams thus obtained consists of two curves, one corresponding to the upper surface and the other to the lower surface of the profile,

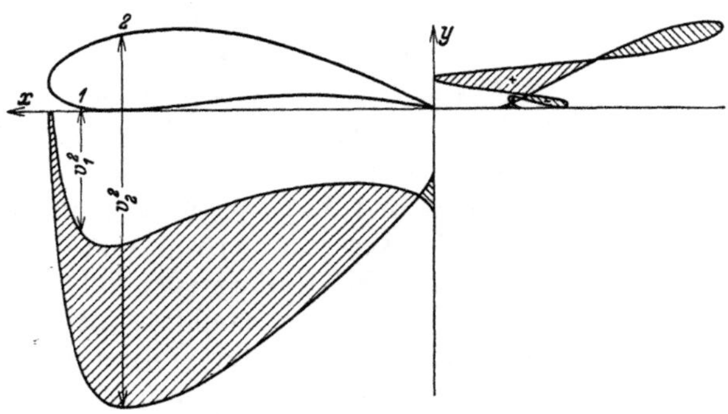

Fig. 5.

and the difference of the areas between them and the axes multiplied by the mass density, ϱ, represents twice the value of the expressions for the force components in (2.1). The diagram below the profile gives the value of P_y while that on the right gives P_x. It is easily seen from this figure that the velocities on the lower surface of the profile are smaller than those on the upper with the exception of the region close to the trailing edge where the contrary is true. The numerical values of the required areas may be obtained with a planimeter or by means of Simpson's rule, but in any case this method is very laborious for not only must the calculations be made separately for each profile, but also for each value of the angle of attack which is to be considered.

We shall now discuss another method for determining the forces per unit length of the profile, making use of the formula of Blasius and calculating the desired values by analytical methods. This formula is usually written in the form

$$P_y - i\,P_x = -\frac{\varrho}{2}\oint\left(\frac{df}{dZ}\right)^2\frac{dZ}{dz}\,dZ$$

in which df/dZ is the complex velocity, that is, the derivative of the complex potential of the flow around the primitive circle, and dZ/dz represents the reciprocal of the derivative of the transformation function.

It is more convenient, however, to calculate the lift and drag directly, that is, the components of the air reaction perpendicular and parallel to the direction of the air stream. Denoting these values by P_{y1} and P_{x1}, respectively, we may write

$$P_{y1} - i\,P_{x1} = -\frac{\varrho\, e^{-i\alpha}}{2} \oint \left(\frac{df}{dZ}\right)^2 \frac{dZ}{dz}\, dZ \tag{2.2}$$

where α is the angle between the velocity vector and the x-axis, and substituting the value (1.5) for the complex velocity, we get:

$$P_{y1} - i\,P_{x1} = -\frac{\varrho\, u^2 e^{-i\alpha}}{2} \oint \left[e^{i\alpha} - \frac{a^2 e^{-i\alpha}}{Z^2} + \frac{4\,i\,a^{3/2} \sin\alpha}{Z^{1/2}(Z^{1/2}+a^{1/2})^2}\right]^2 \frac{dZ}{dz}\, dZ$$

The integration is to be effected along the primitive circle from $Z = a e^{-i\pi}$ to $Z = a e^{+i\pi}$. This integral may be split into two parts,

$$I_1 = -\frac{\varrho\, u^2 e^{-i\alpha}}{2} \oint \left[e^{i\alpha} - \frac{a^2 e^{-i\alpha}}{Z^2}\right]^2 \frac{dZ}{dz}\, dZ$$

and $$I_2 = -4\,\varrho\, u^2 e^{-i\alpha} \sin\alpha \oint \left[\frac{i\,a^{3/2}}{Z^{1/2}(Z^{1/2}+a^{1/2})^2}\left(e^{i\alpha} - \frac{a^2 e^{-i\alpha}}{Z^2}\right) - \right.$$
$$\left. - \frac{2\,a^3 \sin\alpha}{Z\,(Z^{1/2}+a^{1/2})^4}\right] \frac{dZ}{dz}\, dZ$$

The first of these integrals corresponds to the flow obtained with the basic circular potential (1.2) and therefore has a value equal to zero, as may easily be shown by writing dZ/dz in the form $1 + A_2'/Z^2 + A_3'/Z^3 \ldots$ and applying the method of residues. In the second integral we shall first eliminate the fractional powers of the variable by means of the substitution $Z = a\,\zeta^2$, where $\zeta = \tau + i\sigma$, is a new complex variable. We thus obtain:

$$P_{y1} - i\,P_{x1} = -8\,\varrho\, u^2 a \sin\alpha \oint \left\{\left[\frac{i\,(\zeta^2+1)(\zeta-1)}{\zeta^4(\zeta+1)} - \right.\right.$$
$$\left.\left. - \frac{2\,e^{-i\alpha}\sin\alpha}{\zeta\,(\zeta+1)^2}\left\{\frac{1}{\zeta^3} + \frac{1}{(\zeta+1)^2}\right\}\right] \frac{dZ}{dz}\, d\zeta\right\}. \tag{2.3}$$

The same substitution is, of course, to be made in the expression for dZ/dz. This integral is now to be evaluated along a semi-circle of unit radius in the ζ-plane, from $\zeta = -i$ to $\zeta = +i$, that is, along the path ABC in Fig. 6. Since this path is not a closed curve, it is not possible to evaluate the integral by Cauchy's method of residues. The above expression may, however, be transformed into an integral of a real variable in the following way. The integrand in expression (2.3) does not contain any poles outside the region bounded by the semi-circle ABC and the segment of the imaginary axis AC. Any poles of the expression dZ/dz which may exist on the semi-circle itself (for example, the point corresponding to the sharp leading edge in the case of the rectilinear profile) will be considered as being inside it. The above integrand is therefore regular in this region, and consequently the path of integration may be replaced by any ordinary curve joining the end-points A and C of the original path and lying entirely in this region of regularity. We shall suppose that the path ABC is replaced by the path $AD - DEF - FC$ where DEF is a semi-circle of arbitrary radius R, with center at the origin and lying in the left hand half of the plane, and AD and FC are segments of the imaginary axis. By means of this transformation the original integral is changed into three separate ones, along the paths AD, DEF, and FC. In evaluating the second integral along the semi-circle DEF, we may let the radius R approach infinity, in which case this integral vanishes since the integrand becomes an infinitesimal of at least the second order. There remains only the integrals along the lines AD and FC, it being necessary to note, however, that the points D and F now lie at $\mp\infty$ respectively.

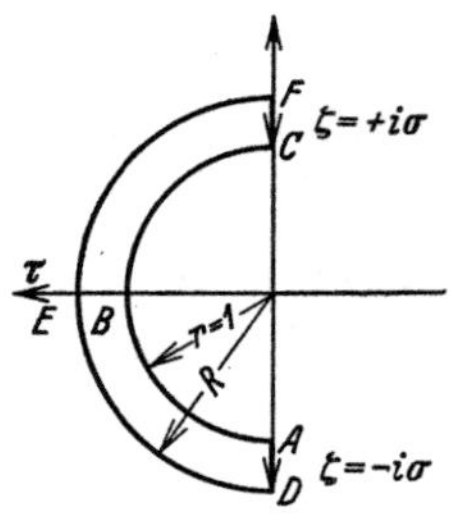

Fig. 6.

In order to evaluate these two integrals, we substitute $\zeta = -i\sigma$ in the integral along AD and $\zeta = +i\sigma$ in that along FC, the limits in both cases being unity and infinity. The expression for dZ/dz in general contains only integral powers of Z and therefore only even powers of ζ, hence in it we need to substitute only $\zeta^2 = -\sigma^2$. After making these substitutions and combining the two integrals thus obtained, we get after simplifying:

$$\left.\begin{aligned} P_{y1} - i P_{x1} = 16\,\varrho\, u^2 a \sin\alpha \int_1^\infty \Bigg[\frac{(\sigma^2-1)^2}{\sigma^4(\sigma^2+1)} - \\ - \frac{2\, i\, e^{-i\alpha}(\sigma^2-1)^2(3\sigma^2+1)\sin\alpha}{\sigma^4(\sigma^2+1)^4}\Bigg]\frac{dZ}{dz}\,d\sigma \end{aligned}\right\} \tag{2.4}$$

3. The Lift Force. Before we can proceed any further with these calculations, it is necessary to know the form of the transformation function in order to determine the value of dZ/dz. We shall now take up the determination of the lift and drag for several different cases. If we divide the expression for dZ/dz into its real and imaginary parts, that is

$$\frac{dZ}{dz} = \mathfrak{R}\left(\frac{dZ}{dz}\right) + i\,\mathfrak{J}\left(\frac{dZ}{dz}\right)$$

we then find for the lift:

$$\begin{aligned} P_{y1} = 16\,\varrho\, u^2 a \sin\alpha \int_1^\infty \Bigg[\left\{\frac{(\sigma^2-1)^2}{\sigma^4(\sigma^2+1)} - \frac{2(\sigma^2-1)^2(3\sigma^2+1)\sin^2\alpha}{\sigma^4(\sigma^2+1)^4}\right\}\mathfrak{R}\left(\frac{dZ}{dz}\right) + \\ + \left\{\frac{2(\sigma^2-1)^2(3\sigma^2+1)\sin\alpha\cos\alpha}{\sigma^4(\sigma^2+1)^4}\right\}\mathfrak{J}\left(\frac{dZ}{dz}\right)\Bigg]\,d\sigma \end{aligned} \tag{3.1}$$

Let us now carry out the computations for the two simplest possible cases, the circular and the rectilinear profiles, which may be considered as the thickest and thinnest profiles respectively. Of course, the calculations for the circle can have no real meaning since the actual flow around such a profile is one of double rather than single burbling. However, the results will give us a basis of comparison on which to estimate the characteristics of more general forms of profiles. For the circle the transformation function is $z = Z$ so that $dZ/dz = 1$, and the expression for the lift becomes:

$$P_{y1} = 16\,\varrho\,u^2\,a\,\sin\alpha \int_1^\infty \left[\frac{(\sigma^2-1)^2}{\sigma^4(\sigma^2+1)} - \frac{2(\sigma^2-1)^2(3\sigma^2+1)\sin^2\alpha}{\sigma^4(\sigma^2+1)^4}\right] d\sigma$$

which after evaluation of the integral gives:

$$P_{y1} = \frac{16}{3}\varrho\,u^2\,a\,\sin\alpha\left[3\pi - 8 + \left(14 - \frac{9}{2}\pi\right)\sin^2\alpha\right]$$

or since α is in practice a small angle, we may write approximately

$$P_{y1} \cong \frac{16}{3}(3\pi - 8)\,\varrho\,u^2\,a\,\sin\alpha\,. \tag{3.2}$$

In the case of the rectilinear profile, we may take the transformation function in the form

$$z = Z + \frac{a^2}{Z}$$

so that

$$\frac{dZ}{dz} = \frac{Z^2}{Z^2 - a^2}$$

and after making the necessary substitution, $Z = -a\sigma^2$, we get

$$\frac{dZ}{dz} = \frac{\sigma^4}{\sigma^4 - 1}$$

Substituting this value in formula (3.1), we obtain for the lift

$$P_{y1} = 16\,\varrho\,u^2\,a\,\sin\alpha \int_1^\infty \left[\frac{\sigma^2-1}{(\sigma^2+1)^2} - \frac{2(\sigma^2-1)(3\sigma^2+1)\sin^2\alpha}{(\sigma^2+1)^5}\right] d\sigma$$

$$= 8\,\varrho\,u^2\,a\,\sin\alpha\left[1 - \left(1 - \frac{9\pi}{32}\right)\sin^2\alpha\right]$$

$$\cong 8\,\varrho\,u^2\,a\,\sin\alpha \tag{3.3}$$

It is thus seen that the difference in the lift for the circular and rectilinear profiles is quite small, and for ordinary profiles having a thickness ratio as high as 25 per cent, we may take the lift as being practically the same as that for the rectilinear profile. This assumption has been confirmed in several cases by actual calculations[1].

[1] Neumark, S., Sur les formes diverses du potentiel servant à calculer les forces agissant sur les profils d'aviation, *Travaux de l'Institut Aérodynamique de Varsovie*, Fascicule I, Varsovie, 1927. Thompson, M. J., The Effect of a Hinged Flap on the Aerodynamic Characteristics of an Airfoil, *Travaux de l'Institut Aérodynamique de Varsovie*, Fascicule IV, Varsovie, 1930.

It is interesting to compare these results with those obtained by the Joukowski theory in which case the lift is given by the formula:

$$P_{y1} = \varrho\, C u$$

where $C = 4\pi\, u\, a\, sin\, \alpha$ is the circulation around the profile. If the expression for the lift as given by the discontinuous potential were written in the same form as that obtained by the Joukowski theory, it is seen that this lift would correspond to a value of $C' = 8u\, a\, sin\, \alpha$, and it may be demonstrated that this is exactly the value of the circulation around the profile as defined by the discontinuous potential. In order to calculate this circulation we must determine the difference of the values of the complex potential at the points $Z = a e^{+i\pi}$ and $Z = a e^{-i\pi}$, it being necessary to consider only the complementary term, since the circulation due to the basic circular potential is zero. We thus obtain

$$C = 8u\, a\, sin\, \alpha \tag{3.4}$$

The circulation around the layer of discontinuity is found to be equal to this same value but with opposite sign. Calculating the circulation along the lower boundary from the point corresponding to the trailing edge to infinity, we find

$$f\,(a e^{-i\pi}) - f\,(\infty)$$

and along the upper boundary

$$f\,(a e^{+i\pi}) - f\,(\infty)$$

The difference of these values gives the circulation around the layer,

$$f\,(a e^{-i\pi}) - f\,(a e^{+i\pi}) = -\,8\, u\, a\, sin\, \alpha$$

It is thus seen that the sum of the circulations around the circle and the layer is equal to zero, which is to be expected since the circulation around a circle of infinitely large radius surrounding the whole plane should be equal to zero. In the case of the transformed profiles, these values will be the same, for it is known that the circulation along corresponding stream-lines is not affected by a conformal transformation. These results are in accordance with our conception of the layer of discontinuity as replacing a system of vortices of negative circulation in the rear of the profile.

4. The Form Drag. We are now ready to determine the form drag acting on a unit length of an infinitely long wing by means of the discontinuous potential. For this purpose it is necessary to separate out the imaginary part of the expression (2.4) which is:

$$P_{x1} = -16\,\varrho\, u^2 a\, sin\,\alpha \int_1^\infty \left[\left\{\frac{(\sigma^2-1)^2}{\sigma^4(\sigma^2+1)} - \frac{2(\sigma^2-1)^2(3\sigma^2+1)\, sin^2\alpha}{\sigma^4(\sigma^2+1)^4}\right\} \mathfrak{J}\left(\frac{dZ}{dz}\right) - \right.$$
$$\left. - \left\{\frac{2(\sigma^2-1)^2(3\sigma^2+1)\, sin\,\alpha\, cos\,\alpha}{\sigma^4(\sigma^2+1)^4}\right\} \mathfrak{R}\left(\frac{dZ}{dz}\right)\right] d\sigma$$

Here the term in $sin^3 \alpha$ may be neglected when α is small and the drag may in general be written in the form:

$$\left.\begin{aligned} P_{x1} \cong -16\,\varrho\, u^2 a \sin\alpha \int_1^\infty \Bigg[\left\{\frac{(\sigma^2-1)^2}{\sigma^4(\sigma^2+1)}\right\} \mathfrak{J}\left(\frac{dZ}{dz}\right) - \\ -\left\{\frac{2(\sigma^2-1)^2(3\sigma^2+1)\sin\alpha\cos\alpha}{\sigma^4(\sigma^2+1)^4}\right\}\mathfrak{R}\left(\frac{dZ}{dz}\right)\Bigg] d\sigma \end{aligned}\right\} \quad (4.1)$$

We again carry out the calculations for the circle and the rectilinear profile. In the first case we have as before $dZ/dz = 1$ and

$$P_{x1} = 32\,\varrho\, u^2 a \sin^2\alpha \cos\alpha \int_1^\infty \left[\frac{(\sigma^2-1)^2(3\sigma^2+1)}{\sigma^4(\sigma^2+1)^4}\right] d\sigma$$

which finally reduces to

$$\begin{aligned} P_{x1} &= \frac{8}{3}(9\pi - 28)\,\varrho\, u^2 a \sin^2\alpha\cos\alpha \\ &\cong 0.732\,\varrho\, u^2 a \sin^2\alpha\cos\alpha \end{aligned} \quad (4.2)$$

For the rectilinear profile we have $dZ/dz = \sigma^4/(\sigma^4 - 1)$ so that

$$\begin{aligned} P_{x1} &= 32\,\varrho\, u^2 a \sin^2\alpha\cos\alpha \int_1^\infty \left[\frac{(\sigma^2-1)(3\sigma^2+1)}{(\sigma^2+1)^5}\right] d\sigma \\ &= 8\left(1 - \frac{9\pi}{32}\right)\varrho\, u^2 a \sin^2\alpha\cos\alpha \cong 0.93\,\varrho\, u^2 a \sin^2\alpha\cos\alpha \end{aligned} \quad (4.3)$$

We thus see that for the circular profile, the drag is considerably smaller than for the straight-line segment, the difference amounting to about 20 per cent. For the usual profiles employed in aeronautics a mean value between these two is obtained, the accurate calculation of which is a long process. However, these profiles resemble the rectilinear profile quite closely and such calculations as have been made[1] show that it is permissible to take the coefficient in the expression for the drag as equal to 0.9.

The results of measurements in the wind tunnel are not in very good agreement with these calculations, due undoubtedly to the fact that we have not taken into consideration the effect of surface friction nor the influence of the finite length of the wing. The case of a wing of finite aspect ratio will be discussed later on. The forces due to surface friction could be found by using the discontinuous potential as a basis for determining the velocity on the profile and then summing the proper components of the elementary frictional forces. The solution of this phase of the problem is a very difficult one and at present we do not possess the results in a general form.

In the calculation of the resistance for the above cases, we had to do only with symmetric profiles for which the reciprocal of the derivative

[1] See p. 15, footnote.

of the transformation function, dZ/dz has real coefficients, so that the term in $\sin\alpha$ in formula (4.1) disappears. It may be shown that this is true of all symmetric profiles while for asymmetric profiles dZ/dz is in general complex. Consequently the general expression for the drag must be written in the form:

$$P_{x1} = \varrho\, u^2\, a\, [D_1 \sin\alpha + D_2 \sin^2\alpha \cos\alpha]$$

but with $D_1 \neq 0$, it will be found[1] that there is always a range of small negative values of α for which the resistance is negative, which, of course, is absurd. In order to avoid this difficulty a more general form of the discontinuous potential was considered, such that the complex velocity takes the form[2],

$$\frac{df}{dZ} = -u\left(e^{i\alpha} - \frac{a^2 e^{-i\alpha}}{Z^2}\right) - 4\, i\, u\, a \sin\alpha \left[\frac{E_1 a^{1/2}}{Z^{1/2}(Z^{1/2}+a^{1/2})^2} + \right.$$
$$\left. + \frac{2 E_2 a}{(Z^{1/2}+a^{1/2})^4} + \frac{4 E_3 Z^{1/2} a^{3/2}}{(Z^{1/2}+a^{1/2})^6} + \frac{8 E_4 Z a^2}{(Z^{1/2}+a^{1/2})^8} + \frac{16 E_5 Z^{3/2} a^{5/2}}{(Z^{1/2}+a^{1/2})^{10}}\right]$$

the numbers E_1, E_2, E_3, E_4 and E_5 being arbitrary constants which were determined by stipulating that: (1) the velocity at the point on the primitive circle corresponding to the trailing edge of the transformed profile must be equal to zero, (2) the term in $\sin\alpha$ in the drag must equal zero, (3) the circulation around the profile must be the same as for the original discontinuous potential and (4) the velocities at the point $\theta = 0$ on the primitive circle as given by the original and modified complementary terms must be equal, this equality being taken as an index of the agreement between the velocity distributions. This gave four conditions for the determination of the five constants and the fifth one was chosen so as to obtain the best possible agreement with the original velocity distribution. It was found that the numbers E_1, E_2, E_3, E_4 and E_5 were practically constant for different profiles, and that the results obtained with this new function were approximately the same as those obtained by means of the earlier discontinuous potential provided that in the latter calculations, the correction was introduced that the coefficient of the term in $\sin\alpha$ in the drag must always be equal to zero. Thus the above expression for the drag must then be modified so as to read:

$$P_{x1} \cong 32\, \varrho\, u^2\, a \sin^2\alpha \cos\alpha \int_1^\infty \left[\frac{(\sigma^2-1)^2 (3\sigma^2+1)}{\sigma^4(\sigma^2+1)^4}\, \mathfrak{R}\left(\frac{dZ}{dz}\right)\right] d\sigma \qquad (4.4)$$

---

[1] THOMPSON, M. J., The Effect of a Hinged Flap on the Aerodynamic Characteristics of an Airfoil, *Travaux de l'Institut Aérodynamique de Varsovie*, Fascicule IV, p. 24, Varsovie, 1930.

[2] THOMPSON, M. J., A Note on the Discontinuous Potential, *Travaux de l'Institut Aérodynamique de Varsovie*, Fascicule VI, p. 9, Varsovie, 1932.

5. Determination of Moment and Center of Pressure. The next important step in this work is the determination of the moment of the air reaction and the position of the center of pressure. For the calculation of the first of these values the second formula of Blasius,

$$M_0 + i N_0 = -\frac{\varrho}{2} \oint \left(\frac{df}{dZ}\right)^2 \frac{dZ}{dz} z\, dZ \tag{5.1}$$

may be used, in which M_0 denotes the moment of the air forces about the origin in the plane of the transformed profile, N_0 has no physical meaning, and z represents the transformation function. A positive value of M_0 indicates that the moment is acting in a clockwise direction. The direct application of this formula leads to a complex integral which cannot be reduced to an integral of a real variable as in the case of the calculations for the lift and drag, so that it is more convenient to make use of a modified form of the above formula. It is evident that for the circular profile the moment about the origin, which coincides with the center, is equal to zero, so that we may write

$$i N_c = -\frac{\varrho}{2} \oint \left(\frac{df}{dZ}\right)^2 Z\, dZ$$

where N_c denotes the value of N_0 for the circle. Now subtracting this value from the original formula (5.1) we get

$$M_0 + i (N_0 - N_c) = -\frac{\varrho}{2} \oint \left(\frac{df}{dZ}\right)^2 \left(\frac{dZ}{dz} z - Z\right) dZ \tag{5.2}$$

The expression $(dZ/dz)\, z - Z$ is of a degree one less than dZ/dz, hence in evaluating the integrals by the method of residues, there is in general an advantage in using this new formula, and in the case of the discontinuous potential, the integral of the terms involving fractional powers of Z may be reduced to an integral of a real variable.

We now proceed to the calculation of the moment by means of the discontinuous potential and, after substituting the expression (1.5) for df/dZ, formula (5.2) becomes:

$$M_0 + i (N_0 - N_c) = -\frac{\varrho u^2}{2} \oint \left[\left(e^{i\alpha} - \frac{a^2 e^{-i\alpha}}{Z^2}\right)^2 + \right.$$

$$\left. + 8 \left(e^{i\alpha} - \frac{a^2 e^{-i\alpha}}{Z^2}\right) \frac{i a^{3/2} \sin\alpha}{Z^{1/2} (Z^{1/2} + a^{1/2})^2} - \frac{16 a^3 \sin^2\alpha}{Z (Z^{1/2} + a^{1/2})^4}\right] \left[\frac{dZ}{dz} z - Z\right] dZ$$

As is well known, the moment due to the basic circular potential, corresponding to the first term within the square brackets, is not equal to zero so that this quantity may not be rejected as in the calculations for the force components. Since the expression $(dZ/dz)\, z - Z$ will in general contain only integral powers of Z, the integral containing this first term may be most conveniently evaluated by the method of residues, while in the remaining part, the same substitutions as in the previous computations on pages 13, 14 are to be applied. The final result after simplification is:

$$M_0 + i\,(N_0 - N_c) = -\frac{\varrho\,u^2}{2}\oint\left[\left(e^{i\alpha} - \frac{a^2 e^{-i\alpha}}{Z^2}\right)^2\left(\frac{dZ}{dz}z - Z\right)\right]dZ +$$

$$+ 16\,\varrho\,u^2\,a\,\sin\alpha\int\limits_1^\infty\left[\frac{(\sigma^2-1)^2 e i\alpha}{\sigma^4(\sigma^2+1)} - \frac{2\,i\,(\sigma^2-1)^2(3\,\sigma^2+1)\sin\alpha}{\sigma^4(\sigma^2+1)^4}\right]\left[\frac{dZ}{dz}z - Z\right]d\sigma$$

Since the transformation function for a profile may always be written in the form

$$z = Z + M_0'\,a + \frac{M_1'\,a^2}{Z} + \frac{M_2'\,a^3}{Z^2} + \dots$$

the constants M_0', M_1', ... being in general complex, the quantity $(dZ/dz)\,z - Z$ may be expressed as:

$$\frac{dZ}{dz}z - Z = N_0' + \frac{N_1'\,a^2}{Z} + \frac{N_2'\,a^3}{Z^2} + \dots$$

hence the value of the first integral above is found by the method of residues to be:

$$J_1 = -\frac{\varrho\,u^2}{2}\oint\left[\left(e^{i\alpha} - \frac{a^2 e^{-i\alpha}}{Z^2}\right)^2\left(\frac{dZ}{dz}z - Z\right)\right]dZ = -\pi\,i\,\varrho\,u^2\,a^2\,e^{2i\alpha}\,N_1'$$

or writing $N_1' = m_1 + i n_1$, we find that the quantity which J_1 contributes to the moment is

$$\mathfrak{R}\,(J_1) = \pi\,\varrho\,u^2\,a^2\,(n_1\cos 2\,\alpha + m_1\sin 2\,\alpha)$$

In the second integral we write

$$\frac{dZ}{dz}z - Z = \mathfrak{R}\left(\frac{dZ}{dz}z - Z\right) + i\,\mathfrak{J}\left(\frac{dZ}{dz}z - Z\right)$$

and we get after separating out the real part:

$$\mathfrak{R}\,(J_2) = 16\,\varrho\,u^2\,a\,\sin\alpha\int\limits_1^\infty\left[\left\{\frac{(\sigma^2-1)^2\cos\alpha}{\sigma^4(\sigma^2+1)}\right\}\mathfrak{R}\left(\frac{dZ}{dz}z - Z\right) - \right.$$

$$\left. - \left\{\frac{(\sigma^2-1)^2\sin\alpha}{\sigma^4(\sigma^2+1)} - \frac{2\,(\sigma^2-1)^2(3\,\sigma^2+1)\sin\alpha}{\sigma^4(\sigma^2+1)^4}\right\}\mathfrak{J}\left(\frac{dZ}{dz}z - Z\right)\right]d\sigma$$

The combination of these two terms finally gives for the moment:

$$\left.\begin{aligned} M_0 = \varrho\,u^2\,a\Bigg[\Bigg\{\pi\,m_1\,a + 8\int\limits_1^\infty\frac{(\sigma^2-1)^2}{\sigma^4(\sigma^2+1)}\,\mathfrak{R}\left(\frac{dZ}{dz}z - Z\right)d\sigma\Bigg\}\sin 2\,\alpha\, - \\ -2\Bigg\{\pi\,n_1\,a + 8\int\limits_1^\infty\Bigg[\frac{(\sigma^2-1)^2}{\sigma^4(\sigma^2+1)} - \\ - \frac{2\,(\sigma^2-1)^2(3\,\sigma^2+1)}{\sigma^4(\sigma^2+1)^4}\Bigg]\mathfrak{J}\left(\frac{dZ}{dz}z - Z\right)d\sigma\Bigg\}\sin^2\alpha + \pi\,n_1\,a\Bigg] \end{aligned}\right\} \quad (5.3)$$

Let us now apply this formula to the rectilinear profile as an example. Taking $z = Z + a^2/Z$ as the transformation function, we find

$$\frac{dZ}{dz}z - Z = \frac{2\,Z\,a^2}{Z^2 - a^2} = 2\,a^2\left(\frac{1}{Z} + \frac{a^2}{Z^3} + \frac{a^4}{Z^5} + \dots\right)$$

or after the substitution $Z = -a\sigma^2$

$$\frac{dZ}{dz}z - Z = \frac{-2\,a\,\sigma^2}{\sigma^4 - 1}$$

Thus $m_1 = 2, \quad n_1 = 0, \quad \mathfrak{R}\left(\frac{dZ}{dz} z - Z\right) = \frac{-2a\sigma^2}{\sigma^4 - 1}, \quad \mathfrak{I}\left(\frac{dZ}{dz} z - Z\right) = 0$

and the above expression becomes:

$$M_0 = 2 \varrho u^2 a^2 \left[\pi - 8 \int_1^\infty \frac{(\sigma^2 - 1)}{\sigma^2 (\sigma^2 + 1)^2} d\sigma \right] \sin 2\alpha$$

After determining the value of the integral and combining terms, we have:

$$M_0 = -2 \varrho u^2 a^2 (3\pi - 12) \sin 2\alpha \tag{5.4}$$

We may now calculate the moment arm of the force with respect to the origin of coordinates which coincides with the mid-point of the profile. A sufficiently close approximation is obtained by dividing the above moment by the lift, in which case we get:

$$l' = \frac{-a(3\pi - 12)\cos\alpha}{2} \simeq 1.29\, a$$

Thus the distance from the leading edge to the center of pressure in per cent of the chord, $l = 4a$, is:

$$c.p. = \frac{2a - 1.29\, a}{4a} \cdot 100 = 18\% \tag{5.5}$$

This result agrees much better with experimental data than the value of 25 per cent obtained in the Joukowski theory, although it is again necessary to note that the calculations are valid only for relatively small values of α.

6. The General Case of Plane Motion[1]. Let us now suppose that the profile under consideration executes any plane motion whatsoever. We conceive of a system of rectangular coordinates x, y connected invariably with the profile as shown in Fig. 7. Let u_x, u_y represent the components of the velocity of the origin in the directions of the moving axes, and also let us designate by ω the angular velocity of rotation of the system described, not forgetting that u_x, u_y and ω are functions of the time, which in this case is the only independent variable. We further represent by w_x, w_y the components of the velocity of a fluid particle relative to the profile and referred to the moving axes x, y, and by v_x and v_y the components of the absolute velocity of the fluid particle also referred to the moving axes. Finally let W and V represent the resultant relative and absolute velocities of the particle, respectively. We note that the relative movement of the fluid with respect to the moving axes x, y, is a rotational motion and that the angular velocity of rotation is equal to ω.

Fig. 7.

[1] See CARAFOLI, E., "Aérodynamique des ailes d'avions", Sect. 2, p. 27, E. Chiron, Paris, 1928.

Now it is known that the components of the absolute acceleration g_x and g_y, parallel to the the moving axes, may be expressed as follows[1]:

$$\left.\begin{aligned} g_x &= \frac{d\,w_x}{d\,t} + \frac{d\,u_x}{d\,t} - \omega^2\,x - y\,\frac{d\,\omega}{d\,t} - \omega\,u_y - 2\,\omega\,w_y \\ g_y &= \frac{d\,w_y}{d\,t} + \frac{d\,u_y}{d\,t} - \omega^2\,y + x\,\frac{d\,\omega}{d\,t} + \omega\,u_x + 2\,\omega\,w_x \end{aligned}\right\} \quad (6.1)$$

By making use of the known hydrodynamical equations[2]:

$$\frac{d\,w_x}{d\,t} = \frac{\partial\,w_x}{\partial\,t} + \frac{1}{2}\,\frac{\partial\,W^2}{\partial\,x} + 2\,\omega\,w_y$$

$$\frac{d\,w_y}{d\,t} = \frac{\partial\,w_y}{\partial\,t} + \frac{1}{2}\,\frac{\partial\,W^2}{\partial\,y} - 2\,\omega\,w_x$$

and observing that in our case the total derivatives du_x/dt, du_y/dt, and $d\omega/dt$, do not differ from the partial derivatives $\partial u_x/dt$, $\partial u_y/dt$, and $\partial\omega/\partial t$, we may write

$$\left.\begin{aligned} g_x &= \frac{\partial\,w_x}{\partial\,t} + \frac{1}{2}\,\frac{\partial\,W^2}{\partial\,x} + \frac{\partial\,u_x}{\partial\,t} - \omega^2\,x - y\,\frac{\partial\,\omega}{\partial\,t} - \omega\,u_y \\ g_y &= \frac{\partial\,w_y}{\partial\,t} + \frac{1}{2}\,\frac{\partial\,W^2}{\partial\,y} + \frac{\partial\,u_y}{\partial\,t} - \omega^2\,y + x\,\frac{\partial\,\omega}{\partial\,t} - \omega\,u_x \end{aligned}\right\} \quad (6.2)$$

Let us introduce the symbols φ_a and ψ_a for the velocity potential and the stream function for the absolute motion of the fluid. The relative motion of the fluid with respect to the profile is not irrotational so that the corresponding velocity potential does not exist, but there does exist a stream function ψ_r. Furthermore, let us designate by ψ_t the stream function corresponding to the translatory movement of the origin of the moving system of coordinates, x, y. We then have:

$$v_x = w_x + u_x - \omega\,y = \frac{\partial\,\varphi_a}{\partial\,x}$$

$$v_y = w_y + u_y + \omega\,x = \frac{\partial\,\varphi_a}{\partial\,y}$$

$$u_x = \frac{\partial\,\psi_t}{\partial\,y}$$

$$u_y = -\frac{\partial\,\psi_t}{\partial\,x}$$

By means of the preceding formulae the components of the acceleration may be written as follows:

$$g_x = \frac{\partial^2\,\varphi_a}{\partial\,t\,\partial\,x} + \frac{1}{2}\,\frac{\partial\,W^2}{\partial\,x} - \omega^2\,x + \omega\,\frac{\partial\,\psi_t}{\partial\,x}$$

$$g_y = \frac{\partial^2\,\varphi_a}{\partial\,t\,\partial\,y} + \frac{1}{2}\,\frac{\partial\,W^2}{\partial\,y} - \omega^2\,y + \omega\,\frac{\partial\,\psi_t}{\partial\,y}$$

---

[1] Appell, P., "Traité de la mécanique rationelle", vol. I, chap. 2, Sect. IV, Gauthier-Villars, Paris, 1902.

[2] Witoszyński, C., "Aerodynamika" pp. 20—21, Warsaw, 1928.

If we suppose that the field of extraneous forces in the fluid considered is non-existent, we obtain the equations of motion of the fluid in the following form, where p is the pressure and ϱ the mass density of the fluid:

$$-\frac{1}{\varrho}\frac{\partial p}{\partial x} = \frac{\partial^2 \varphi_a}{\partial t\, \partial x} + \frac{1}{2}\frac{\partial W^2}{\partial x} - \omega^2 x + \omega \frac{\partial \psi_t}{\partial x}$$
$$-\frac{1}{\varrho}\frac{\partial p}{\partial y} = \frac{\partial^2 \varphi_a}{\partial t\, \partial y} + \frac{1}{2}\frac{\partial W^2}{\partial y} - \omega^2 y + \omega \frac{\partial \psi_t}{\partial y}$$

These equations immediately give the integral defining the pressure p:

$$\frac{p}{\varrho} + \frac{\partial \varphi_a}{\partial t} + \frac{W^2}{2} - \frac{\omega^2 r^2}{2} + \omega\, \psi_t = f(t)$$

where $r^2 = x^2 + y^2$ and $f(t)$ is an arbitrary function of the time. Now observing that $\psi_a = \psi_r + \psi_t - \omega\, r^2/2$, we may also write the equation determining the pressure in the following form:

$$\frac{p}{\varrho} + \frac{\partial \varphi_a}{\partial t} + \frac{W^2}{2} + \omega\, \psi_a - \omega\, \psi_r = f(t) \tag{6.3}$$

If it is a question of calculating the components of the pressure acting on a unit length of the wing in any direction whatsoever, we need consider only the three terms $\partial \varphi_a/\partial t$, $W^2/2$ and $\omega\, \psi_a$ because ψ_r being constant for the entire profile contour, and $f(t)$ being independent of the position, the corresponding integrals vanish:

a) Example of the rectilinear profile. Let us picture first the motion of the fluid set up by the rotation of the profile. For this purpose, we introduce the elliptic coordinates ξ, η, where

$$z = x + i\, y = \frac{l}{2} \cosh(\xi + i\, \eta)$$ [1]

$$x = \frac{l}{2} \cosh \xi \cos \eta, \qquad y = \frac{l}{2} \sinh \xi \sin \eta$$

$$r^2 = x^2 + y^2 = \frac{l^2}{8} (\cosh 2\xi + \cos 2\eta)$$

$$h = \sqrt{\left(\frac{\partial x}{\partial \xi}\right)^2 + \left(\frac{\partial y}{\partial \xi}\right)^2} = \sqrt{\left(\frac{\partial x}{\partial \eta}\right)^2 + \left(\frac{\partial y}{\partial \eta}\right)^2} = \frac{l}{2}\sqrt{\cosh^2 \xi - \cos^2 \eta} =$$
$$= \frac{l}{2}\sqrt{\sinh^2 \xi + \sin^2 \eta}$$

$$v_\xi = \frac{1}{h}\frac{\partial \varphi}{\partial \xi} = \frac{1}{h}\frac{\partial \psi}{\partial \eta}$$
$$v_\eta = \frac{1}{h}\frac{\partial \varphi}{\partial \eta} = -\frac{1}{h}\frac{\partial \psi}{\partial \xi}$$

In these formulae l denotes the length of the profile. The equations $\xi = const.$ and $\eta = const.$ define ellipses and hyperbolas respectively.

The complex potential corresponding to the absolute movement of the fluid is[2]

$$-\frac{i\, \omega\, l^2}{16} e^{-2(\xi + i\eta)} \tag{6.4}$$

---

[1] Lamb, H., "Hydrodynamics", 5th ed., p. 79, Cambridge University Press, 1924.
[2] Same p. 83.

which corresponds to a velocity potential

$$-\frac{\omega l^2}{16} e^{-2\xi} \sin 2\eta \tag{6.5}$$

and to a stream function $$-\frac{\omega l^2}{16} e^{-2\xi} \cos 2\eta \tag{6.6}$$

In the relative motion with respect to the profile we may speak only of the stream function which becomes:

$$-\frac{\omega l^2}{16} e^{-2\xi} \cos 2\eta + \frac{\omega r^2}{2} = \frac{\omega l^2}{16}[(1-e^{-2\xi}) \cos 2\eta + \cosh 2\xi] \tag{6.7}$$

Let us now pass to the translatory motion of the profile and suppose that the discontinuous complement which we developed in **1** for the case of uniform rectilinear motion, may also be applied to the case of a general movement, with only one difference, that $u = \sqrt{u_x^2 + u_y^2}$ and α are now functions of the time. In the section mentioned we have defined the complex potential for the primitive circular profile by the following

formula $$-u\left(Ze^{-i\alpha} + \frac{a^2 e^{-i\alpha}}{Z}\right) + i K a u\left(\frac{Z^{1/2} - a^{1/2}}{Z^{1/2} + a^{1/2}}\right)$$

where the complex variable $Z = X + iY$ refers to the plane of the primitive profile.

Introducing the elliptical coordinates and making use of the transformation function for the rectilinear profile,

$$z = Z + \frac{a^2}{Z}$$

we obtain the complex potential referred to the plane of this profile in the following form: $(l = 4a)$.

$$-\frac{ul}{4}\{e^{\xi + i(\eta+\alpha)} + e^{-\xi - i(\eta+\alpha)}\} + \frac{iKul}{4} \tanh\frac{(\xi + i\eta)}{4} \tag{6.8}$$

or better $$-\frac{ul}{2}\cosh\{\xi + i(\eta+\alpha)\} + \frac{iKul}{4}\tanh\frac{(\xi+i\eta)}{4} \tag{6.9}$$

This complex potential takes account of the *relative* movement produced by the translation of the profile and does not consider its rotation. It may be decomposed into a real part, representing the velocity potential,

$$-\frac{ul}{2}\cosh\xi\cos(\eta+\alpha) - \frac{Kul}{4}\frac{\sin\frac{\eta}{2}}{\cosh\frac{\xi}{2} + \cos\frac{\eta}{2}} \tag{6.10}$$

and an imaginary part giving the stream function.

$$-\frac{ul}{2}\sinh\xi\sin(\eta+\alpha) + \frac{Kul}{4}\frac{\sinh\frac{\xi}{2}}{\cosh\frac{\xi}{2} + \cos\frac{\eta}{2}} \tag{6.11}$$

It is seen that the last expression becomes zero for the points situated on the profile contour for which we have $\xi = 0$.

In order to obtain the functions corresponding to the absolute motion of the fluid set up by the translation of the profile, it is necessary to reject the term $e^{\xi + i(\eta + \alpha)}$ in the expression (6.8). Thus for the absolute motion of the fluid, without considering that caused by the rotation of the profile, we obtain the complex potential:

$$-\frac{ul}{4} e^{-\xi - i(\eta + \alpha)} + \frac{iKul}{4} \tanh \frac{(\xi + i\eta)}{4} \tag{6.12}$$

the velocity potential:

$$-\frac{ul}{4} e^{-\xi} \cos(\eta + \alpha) - \frac{Kul}{4} \frac{\sin \frac{\eta}{2}}{\cosh \frac{\xi}{2} + \cos \frac{\eta}{2}} \tag{6.13}$$

and the stream function:

$$+\frac{ul}{4} e^{-\xi} \sin(\eta + \alpha) + \frac{Kul}{4} \frac{\sinh \frac{\xi}{2}}{\cosh \frac{\xi}{2} + \cos \frac{\eta}{2}} \tag{6.14}$$

In order to define the total movement caused by the translation and rotation of the rectilinear profile, it is necessary to complete the expression (6.12) by the complex potential (6.4). In this way we obtain the total complex potential representing the absolute motion of the fluid arising from the translation and rotation of the profile:

$$\varphi_a + i\psi_a = -\frac{ul}{4} e^{-\xi - i(\eta + \alpha)} + \frac{iKul}{4} \tanh \frac{(\xi + i\eta)}{4} - \frac{i\omega l^2}{16} e^{-2(\xi + i\eta)} \tag{6.15}$$

$$\varphi_a = -\frac{ul}{4} e^{-\xi} \cos(\eta + \alpha) - \frac{Kul}{4} \frac{\sin \frac{\eta}{2}}{\cosh \frac{\xi}{2} + \cos \frac{\eta}{2}} - \frac{\omega l^2}{16} e^{-2\xi} \sin 2\eta \tag{6.16}$$

$$\psi_a = +\frac{ul}{4} e^{-\xi} \sin(\eta + \alpha) + \frac{Kul}{4} \frac{\sinh \frac{\xi}{2}}{\cosh \frac{\xi}{2} + \cos \frac{\eta}{2}} - \frac{\omega l^2}{16} e^{-2\xi} \cos 2\eta \tag{6.17}$$

We have already noted that for the total relative motion neither a complex potential nor a velocity potential exists. There exists only the stream function, which is obtained by adding again to (6.17) the imaginary part of the term $e^{\xi + i(\eta + \alpha)}$, rejected above, and writing instead of the last term, the expression (6.7). This gives:

$$\left.\begin{aligned} \psi_r = -\frac{ul}{2} \sinh \xi \sin(\eta + \alpha) + \frac{Kul}{4} \frac{\sinh \frac{\xi}{2}}{\cosh \frac{\xi}{2} + \cos \frac{\eta}{2}} + \\ + \frac{\omega l^2}{16} \{(1 - e^{-2\xi}) \cos 2\eta + \cosh 2\xi\} \end{aligned}\right\} \tag{6.18}$$

It is now a question of determining the relative velocity W on the profile. It is known that this velocity is equal to the component w_η for $\xi = 0$. Determining w_η by means of the formula $w_\eta = -\frac{1}{h}\frac{\partial \psi_r}{\partial \xi}$ and putting $\xi = 0$, we obtain:

$$\frac{l}{2} \sin\eta \cdot W = \frac{ul}{2} \sin(\eta + \alpha) - \frac{Kul}{8} \frac{1}{1 + \cos\frac{\eta}{2}} - \frac{\omega l^2}{8} \cos 2\eta$$

The trailing edge of the profile is located at the point $\eta = \pi$ and for this value of η, the right-hand side of the preceding equation should become zero in order that the velocity at this point will be finite. In this way we determine the value of the constant K:

$$K = -4 \sin\alpha - \frac{\omega l}{u}$$

Introducing this value in the preceding expression, we obtain the following final value for the velocity on the profile contour:

$$W = \frac{u\left[\sin(\eta + \alpha) + \frac{\sin\alpha}{1 + \cos\eta/2}\right] + \frac{\omega l}{4}\left[\frac{1}{1 + \cos\eta/2} - \cos 2\eta\right]}{\sin\eta} \tag{6.19}$$

The formulas (6.16) and (6.17) give for the points situated on the profile:

$$\varphi_{a,p} = -\frac{ul}{4}\left[\cos(\eta + \alpha) - \frac{4\sin\alpha\sin\eta/2}{1 + \cos\eta/2}\right] - \frac{\omega l^2}{4}\left[\frac{\sin\eta/2}{1 + \cos\eta/2} + \frac{\sin 2\eta}{4}\right] \tag{6.20}$$

and

$$\psi_{a,p} = \frac{ul}{4} \sin(\eta + \alpha) - \frac{\omega l^2}{16} \cos 2\eta \tag{6.21}$$

From the first of these expressions, we get:

$$\left.\begin{aligned} \frac{\partial \varphi_{a,p}}{\partial t} = &-\frac{\partial u}{\partial t} \cdot \frac{l}{4}\left[\cos(\eta + \alpha) - \frac{4\sin\alpha\sin\eta/2}{1 + \cos\eta/2}\right] + \frac{\partial \alpha}{\partial t} \cdot \\ &\cdot \frac{ul}{4}\left[\sin(\eta + \alpha) + \frac{4\cos\alpha\sin\eta/2}{1 + \cos\eta/2}\right] - \frac{\partial\omega}{\partial t} \cdot \frac{l^2}{4}\left[\frac{\sin\eta/2}{1 + \cos\eta/2} + \frac{\sin 2\eta}{4}\right] \end{aligned}\right\} \tag{6.22}$$

The formulas (6.20), (6.21), and (6.22) furnish a means of calculating the pressure with the aid of (6.3). If we make the necessary substitutions in this expression and then calculate the lift (component normal to the relative wind in the translatory motion) for a unit length of the wing in the usual way, we obtain:

$$\left.\begin{aligned} P_{y1} = \varrho\Big[&2u^2 l\alpha + \left(\frac{1}{2} + \frac{\pi}{8}\right) u\omega l^2 + \left(4 - \frac{7\pi}{8}\right)\left(\alpha\frac{\partial u}{\partial t} + u\frac{\partial \alpha}{\partial t}\right) l^2 - \\ &- \left(1 - \frac{\pi}{4}\right) l^3 \frac{\partial\omega}{\partial t}\Big] \end{aligned}\right\} \tag{6.23}$$

The lift coefficient C_y is as follows:

$$\left.\begin{aligned} C_y = 4\alpha + \left(1+\frac{\pi}{4}\right)\frac{\omega l}{u} + \left(8-\frac{7\pi}{4}\right)\left(\alpha\frac{\partial u}{\partial t} + u\frac{\partial \alpha}{\partial t}\right)\frac{l}{u^2} - \\ -2\left(1-\frac{\pi}{4}\right)\frac{l^2}{u^2}\cdot\frac{\partial \omega}{\partial t} \end{aligned}\right\} \qquad (6.24)$$

The calculations for the drag force will not be discussed at this time.

We recall, in conclusion, that the above calculations refer to the case of the motion of an airplane following a curved trajectory, situated in the vertical plane and that there exists in this case the following relation between u, ω, and α:

$$\omega = \frac{u}{R_c} + \frac{\partial \alpha}{\partial t} \qquad (6.25)$$

where R_c is the radius of curvature of the path and is also a function of the time. This equation must be taken into consideration along with (6.24) in the determination of the lift coefficient in any particular case.

7. Influence of the Finite Span of the Wing. Up to the present time we have considered only the case of the two-dimensional or plane motion of an airplane wing, which was assumed to be of infinite span. The results obtained for the lift and moment acting on a unit span of this wing in the case of uniform translatory motion, agree fairly well with experimental data, but the value of the profile drag calculated by the discontinuous potential is much too small. We shall now develop a new theory in which we shall attempt to take into consideration the influence of the finite span of the wing so as to obtain a correspondingly larger value for the drag.

For simplicity we shall in what follows consider only the circular profile, although as we have mentioned before, the actual flow around such a wing cannot be represented by the discontinuous potential. However, as we have already seen, the results obtained for the circular profile are in close approximation to those of ordinary profiles to which the discontinuous potential may be applied. We further simplify this complicated problem by replacing the flow with the layer of discontinuity by a simpler arrangement of vortices so chosen that the lift and drag for the infinitely long wing shall remain the same as in the previous calculations. In the case of the discontinuous potential we showed that the complementary term added to the basic circular potential gave a circulation around the profile equal to

$$C = 8\, u\, a \sin \alpha$$

Furthermore, there existed in the rear of the profile the so-called layer of discontinuity, the circulation around which was equal to the above value with the sign changed. Let us now try to replace this layer by a point vortex of circulation — C, whose axis is a straight line parallel to the span of the circular wing and at a distance R from its center. In any right section of the cylinder, say the plane X, Y, the center of this vortex will for simplicity be supposed to lie on the line through the origin parallel to the direction of the velocity at infinity, so that its coordinates will be $(-R\cos\alpha,\ R\sin\alpha)$.

Now it is known that if the circular profile is to remain a stream-line, we must introduce another vortex inside the primitive circle, whose circulation is $+C$ and whose center is at the image point in the circle of the center of the external one. The coordinates of its center are thus $-(a^2/R)\cos\alpha,\ (a^2/R)\sin\alpha$.

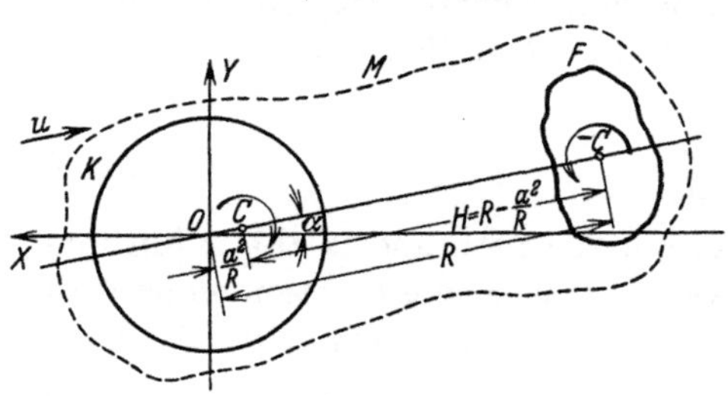

Fig. 8.

By means of this internal vortex, we now obtain a circulation around the profile equal to $+C$, while that around an infinitely large circle enclosing the entire coordinate plane is equal to zero.

The complex potential representing the continuous classical flow around the circle modified by this vortex pair, is

$$f(Z) = -u\left(Ze^{i\alpha} + \frac{a^2 e^{-i\alpha}}{Z}\right) + \frac{iC}{2\pi}\log\frac{(Z + Re^{-i\alpha})}{\left(Z + \frac{a^2}{R}e^{-i\alpha}\right)} \tag{7.1}$$

and the complex velocity is

$$\frac{df}{dZ} = -u\left(e^{i\alpha} - \frac{a^2}{Z^2}e^{-i\alpha}\right) + \frac{iC}{2\pi(Z + Re^{-i\alpha})} - \frac{iC}{2\pi\left(Z + \frac{a^2}{R}e^{-i\alpha}\right)}$$

Substituting this last expression in the first formula of Blasius and noting that $dZ/dz = 1$ for the circular profile, we obtain:

$$P_{y1} - iP_{x1} = \frac{-\varrho e^{-i\alpha}}{2}\oint\left[u\left(e^{i\alpha} - \frac{a^2 e^{-i\alpha}}{Z^2}\right) - \frac{iC}{2\pi(Z + Re^{-i\alpha})} + \right.$$
$$\left. + \frac{iC}{2\pi\left(Z + \frac{a^2}{R}e^{-i\alpha}\right)}\right]^2 dZ$$

As in the previous calculations the integration is to be effected around the circle $Z = a e^{i\theta}$, which leads to the determination of the residues of Cauchy corresponding to the poles of the integrand lying inside the circle; *i. e.*, the points $Z = 0$ and $Z = -(a^2/R)e^{-i\alpha}$.

It is more convenient, however, to evaluate this integral by replacing the circle K, by any arbitrary closed curve F (see Fig. 8) enclosing the external pole $Z = -Re^{-i\alpha}$ but no others. The integral in the above expression around this curve F is equal to that around the primitive circle with the sign changed. It may be shown that the integral around another auxiliary curve M surrounding both K and F is equal to zero, and since F and M may be chosen in such a way that the combination of F with K gives the equivalent of the contour M, then it is immaterial whether the residues are calculated for the poles inside the circle or inside the contour F. We may write the expression for the lift and drag in the form:

$$P_{y1} - i\,P_{x1} = \frac{+\varrho e^{-i\alpha}}{2} \oint_F \left[\left\{ u\left(e^{i\alpha} - \frac{a^2 e^{-i\alpha}}{Z^2}\right) + \frac{i\,C}{2\pi\left(Z + \frac{a^2}{R} e^{-i\alpha}\right)} \right\} - \right.$$

$$\left. - \frac{i\,C}{2\pi\,(Z + R\,e^{-i\alpha})} \right]^2 dZ$$

in which we have grouped the terms so that the expression within the curved brackets does not possess a pole inside the contour F. The square of this term therefore has a residue equal to zero, and while the square of the second term possesses a pole of the second order at the point $Z = -R\,e^{-i\alpha}$, its residue is also zero. There remains, then, only the double product

$$P_{y1} - iP_{x1} = -\frac{C\,i\,\varrho\,e^{-i\alpha}}{2\pi} \oint \left[u\left(e^{i\alpha} - \frac{a^2 e^{-i\alpha}}{Z^2}\right) + \frac{i\,C}{2\pi\left(Z + \frac{a^2}{R} e^{-i\alpha}\right)} \right] \frac{dZ}{(Z + R e^{-i\alpha})}$$

$$= \varrho\,C\,u\left(1 - \frac{a^2}{R^2}\right) - \frac{i\,\varrho\,C^2}{2\pi R\left(1 - \frac{a^2}{R^2}\right)}$$

so that

$$P_{y1} = \varrho\,C\,u\left(1 - \frac{a^2}{R^2}\right) \tag{7.2}$$

$$P_{x1} = \frac{\varrho\,C^2}{2\pi R\left(1 - \frac{a^2}{R^2}\right)} \tag{7.3}$$

If R is taken sufficiently large, (7.2) gives for the lift a value practically the same as was obtained by the original discontinuous potential, and we shall see shortly that this is the case. In order to obtain a similar agreement for the profile drag, it is necessary to equate the value (7.3) to that obtained before for the circle in (4.2). After substituting the proper value of C in (7.3) this relation gives:

$$\frac{32\,\varrho\,u^2\,a^2\,\sin^2\alpha}{\pi R\left(1 - \frac{a^2}{R^2}\right)} = 0.732\,\varrho\,u^2\,a\,\sin^2\alpha$$

which reduces to a quadratic equation in R

$$0.023\,\pi\,R^2 - a\,R - 0.023\,\pi\,a^2 = 0$$

the solution of which is $\quad R \simeq 14a$

With this value of R, the lift and drag for a unit length of the circular wing are practically the same as calculated by the discontinuous potential.

It is interesting to note that the layer of discontinuity was found to have its maximum thickness at approximately this same distance from the origin.

If in (7.3) we put $R\,(1 - a^2/R^2) = H$, which is the distance between the centers of the two vortices, we have:

$$P_{x\,1} = \varrho\, C \frac{C}{2\,\pi\, H}$$

Now the coefficient $C/2\,\pi\, H$ represents the velocity due to a vortex of circulation C at a distance H from its center so that we may say that this is the value of the velocity of the center of the internal vortex due to the presence of the external one. If we put $C/2\,\pi\, H = \Omega$, we have,

$$[P_{x\,1}]_\infty = \varrho\, C\, \Omega \tag{7.4}$$

which is similar in form to the Joukowski formula for the lift. We have introduced the symbol ∞ to indicate that this formula applies to a wing of infinite length.

In the case of a wing of finite length, we shall suppose that the same internal and external vortices exist as before, with the exception that they are now of finite length L equal to the span of the wing. But since vortices cannot be terminated in finite space, we shall further assume that the internal vortex is bent at the wing tips, and then flows laterally along a straight line in the rear of the wing to where it joins the ends of the external one. In this way a closed rectangular vortex is obtained, which is shown in plan view in Fig. 9. For simplicity the angles at the points where the vortex is bent are taken as equal to 90^0.

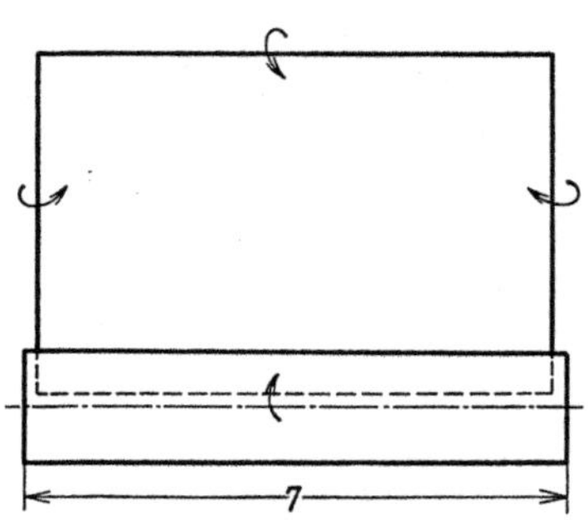

Fig. 9.

With such a simplified representation of this phenomenon, the next step is to determine the influence of the change in form of the external vortex on the velocity Ω. It can be foreseen that this velocity increases and by taking this increase into consideration in formula (7.4), it is possible to estimate roughly the increase in drag due to the finite length of the wing. The problem thus becomes analogous to a well-known one in electrodynamics concerning the magnetic action of a current flowing through a conductor of arbitrary form, for the solution of which the law of Biot-Savart may be employed.

Now in the theory of vortices there exists a formula analogous to that of Biot-Savart by means of which it is possible to determine the velocity at a given point P due to the existence of a vortex filament.

The elementary velocity dv at a given point due to an infinitesimal element ds of a vortex filament of circulation C is

$$dv = \frac{C}{4\pi} \frac{\sin\beta \cdot ds}{\delta^2}$$

where δ is the distance from the element ds to the point under consideration and β is the angle between this line and the tangent to the vortex filament as shown in Fig. 10. If a perpendicular r is dropped from P to the tangent at ds, and a second one from the end of ds nearest P on the line from P to the vortex element, then it is possible to show that

$$dv = \frac{C}{4\pi r} \sin\beta \, d\beta \qquad (7.5)$$

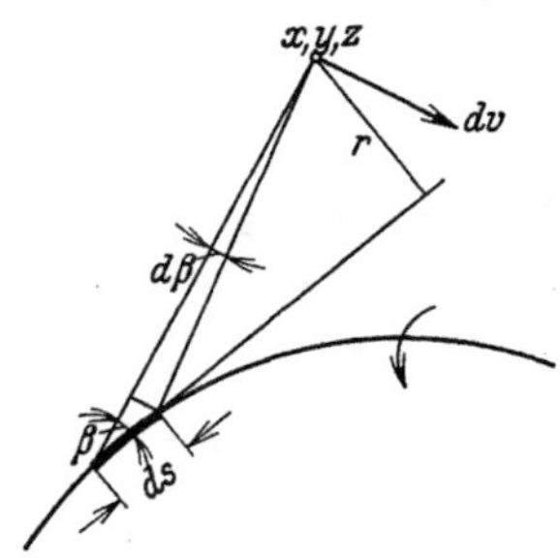

Fig. 10.

This formula is completely analogous to the law of Biot-Savart expressing the magnetic action of an element of a conductor carrying an electric current. In order to determine the velocity at P due to the whole filament, it is necessary to integrate the above expression along the vortex line.

In the case of a finite segment of a linear vortex filament, the formula (7.5) gives after integration,

$$v_{AB} = \frac{C}{4\pi r} (\cos\beta_A + \cos\beta_B) \qquad (7.6)$$

where β_A and β_B are the interior angles of the triangle PAB as shown in Fig. 11[1].

We now return to our problem of the wing of finite length which may be solved by application of the formula (7.6). In the plane of the vortices, our configuration for the circular wing is as shown in Fig. 12. Here it is necessary to consider only the influence of the external

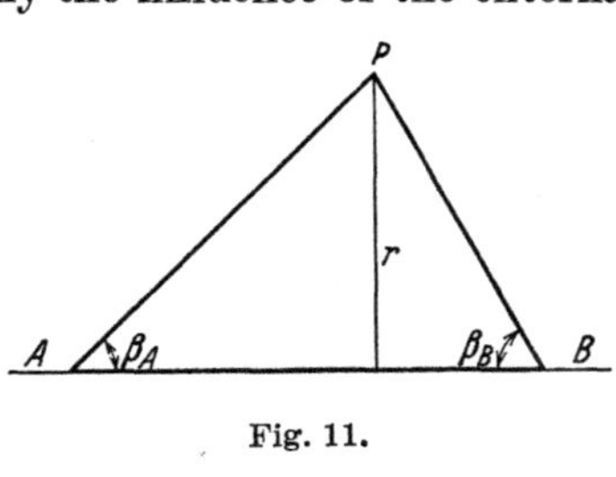

Fig. 11.

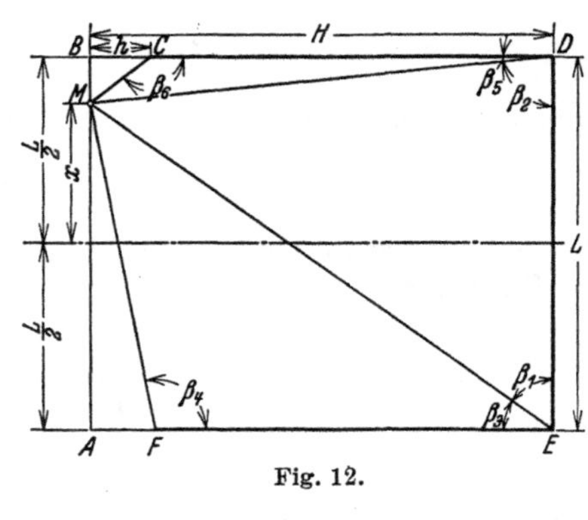

Fig. 12.

part of the vortex $CDEF$, on the internal segment AB, since by symmetry the forces which may arise due to the existence of the segments AF and BC will cancel out. Thus by applying formula (7.6)

[1] See also Division B III 2.

to the segments DE, EF, and CD in turn, we obtain for the velocity at a point M of the internal vortex, situated at a distance x from the plane of symmetry of the wing:

$$\Omega = \frac{C}{4\pi H}(\cos\beta_1 + \cos\beta_2) + \frac{C}{4\pi\left(\frac{L}{2}+x\right)}(\cos\beta_3 + \cos\beta_4) + \\ + \frac{C}{4\pi\left(\frac{L}{2}-x\right)}(\cos\beta_5 + \cos\beta_6)$$

Now replacing the cosines of the different angles by their values in terms of the various lengths, we find:

$$\begin{aligned} \Omega = {} & \frac{C}{4\pi H}\left[\frac{\frac{L}{2}+x}{\sqrt{\left(\frac{L}{2}+x\right)^2+H^2}} + \frac{\frac{L}{2}-x}{\sqrt{\left(\frac{L}{2}-x\right)^2+H^2}}\right] \\ & + \frac{C}{4\pi\left(\frac{L}{2}+x\right)}\left[\frac{H}{\sqrt{\left(\frac{L}{2}+x\right)^2+H^2}} - \frac{h}{\sqrt{\left(\frac{L}{2}+x\right)^2+h^2}}\right] \\ & + \frac{C}{4\pi\left(\frac{L}{2}-x\right)}\left[\frac{H}{\sqrt{\left(\frac{L}{2}-x\right)^2+H^2}} - \frac{h}{\sqrt{\left(\frac{L}{2}-x\right)^2+h^2}}\right] \end{aligned} \tag{7.7}$$

It is now possible to determine the value of the drag. For an element of the wing of width dx, the profile drag, is expressed, by analogy with formula (7.4), by

$$\varrho\, C\, \Omega\, d\, x$$

so that for the whole wing, the drag is

$$[P_{x1}]_L \cdot L = \varrho\, C \int_{-L/2}^{L/2} \Omega\, d\, x$$

For a section of length L of an infinitely long wing, the drag is

$$[P_{x1}]_\infty \cdot L = \varrho\, C \frac{C}{2\pi H} L$$

so that we get for the ratio between the drag for unit lengths of the wing of finite length and one of infinite length:

$$\lambda = \frac{[P_{x1}]_L}{[P_{x1}]_\infty} = \frac{2\pi H}{C L}\int_{-L/2}^{L/2} \Omega\, d\, x$$

After substituting the value (7.7) for Ω and carrying out the necessary integration we find:

$$\lambda = \sqrt{1+\frac{H^2}{L^2}} - \frac{H}{L} + \frac{H}{L}\log\left[\frac{\sqrt{1+\frac{H^2}{L^2}}-\frac{H}{L}}{\sqrt{1+\frac{h^2}{L^2}}-\frac{h}{L}}\cdot\frac{H}{h}\right] \tag{7.8}$$

The quantities H/L and h/L may be expressed in the forms:

$$\frac{H}{L} = \frac{H}{a} \cdot \frac{a}{l} \cdot \frac{l}{L}; \qquad \frac{h}{L} = \frac{h}{a} \cdot \frac{a}{l} \cdot \frac{l}{L}$$

h being the distance from the center of the internal vortex to the trailing edge, and l the chord of the wing. Since $H/a \cong 14$, $h/a \cong 1$, and a/l is determined from the transformation function, we have λ expressed as a function of the aspect ratio L/l. Taking $l/a = 3.8$ as an average value of this ratio, a series of values of λ were calculated for different aspect ratios, the results being shown graphically in Fig. 13. It is thus seen that as the aspect ratio increases, the value of λ decreases although at a slower and slower rate. Furthermore, because of difficulties in construction, it is not possible to select on overly large value of L/l, even though aerodynamically it may be advantageous to do so.

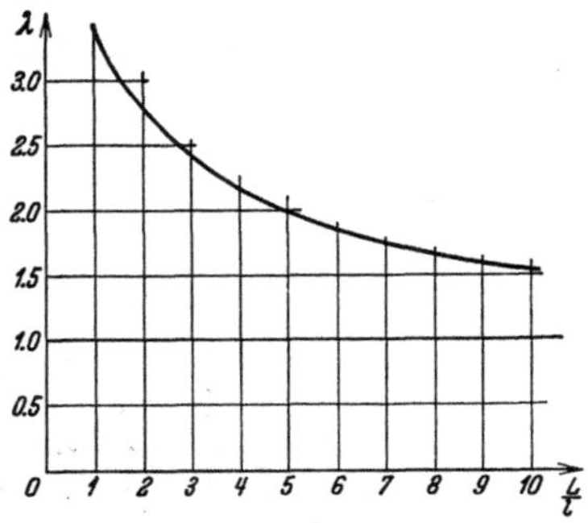

Fig. 13.

Bibliography.

APPELL, P., "Traité de la mécanique rationelle", (Gauthier-Villars, Paris, 1902) Vol. I.

BONDER, J. and SZYMAŃSKI, P., "Sur le multiplan en tandem", Travaux de l'Institut Aérodynamique de Varsovie, Fascicule II, Warsaw, 1928.

— "Contribution à la théorie du biplan", Travaux de l'Institut Aérodynamique de Varsovie, Fascicule II, Warsaw, 1928.

CARAFOLI, E., "Aérodynamique des ailes d'avion" (E. Chiron, Paris, 1928).

JOUKOWSKI, N., "Aérodynamique", (Gauthier-Villars, Paris, 1916).

LAMB, H., "Hydrodynamics" 5th ed., (Cambridge University Press, 1924).

MÜLLER, W., "Mathematische Strömungslehre" (Julius Springer, Berlin, 1928), sect. 37.

NEUMARK, S., "Sur les formes diverses du potentiel servant à calculer les forces qui agissent sur un profil d'aviation", Travaux de l'Institut Aérodynamique de Varsovie, Fascicule I, Warsaw, 1927.

SZYMAŃSKI, P., "Écoulement plan du fluide à travers une pallisade de segments rectilignes", Travaux de l'Institut Aérodynamique de Varsovie, Fascicule II, Warsaw, 1928.

THOMPSON, M. J., "The effect of a hinged flap on the aerodynamic characteristics of an airfoil", Travaux de l'Institut Aérodynamique de Varsovie, Fascicule IV, Warsaw, 1930.

— "A Note on the Discontinuous Potential", Travaux de L'Institut Aérodynamique de Varsovie, Fascicule VI, Warsaw, 1932.

WITOSZYŃSKI, C., "La mécanique des profils d'aviation", (E. Chiron, Paris, 1924).

— "Modification du principe de circulation", Proceeding of the 1st International Congress for Applied Mechanics, Delft, 1924.

— "Aerodynamika", Warszawa, 1928.

DIVISION G
THE MECHANICS OF VISCOUS FLUIDS

By

L. Prandtl,
Göttingen

EDITOR'S PREFACE

The development of aerodynamic theory as based on the ideal or perfect fluid, and as exemplified especially in Volume II (Division E) of this series, has proven of the very highest value in the application of aerodynamics to the problems of aviation, representing, as it does, a limit toward which actual fluids approach as the viscosity is indefinitely decreased. The very extended mathematical theory which has been developed by the use of this ideal form of fluid medium is made possible by the omission of viscosity and compressibility from representation in the mathematical methods employed.

Furthermore, as shown in the present Division, the use of the results thus developed is justifiable in high degree for the treatment of the conditions in a flow far from the boundaries of solid bodies, even with a viscous fluid.

However, when the question concerns the behavior of actual fluids near the boundaries of solid bodies, between which and the fluids there is relative motion, the classical theory offers no help and recourse must be had to other methods of approach.

As shown in the present Division, the domain in which the classical theory thus breaks down, comprises the so-called "boundary layer" close to the solid boundary, together with the "wake"—the eddying and turbulent body of fluid which remains after the body is passed, and which represents the domain of degeneration of the kinetic energy communicated to the fluid by the passage of the body, into the thermal energy of molecular agitation.

For this domain there exists no comprehensive body of theory for the conditions of the fluid as regards the distribution of pressure or the movement of its parts, nor does there seem to be any avenue through which we might hope to reach such a body of theory.

In consequence, the approach must be by other methods as described and discussed by Dr. Prandtl in the following chapters, and comprising clever applications of mathematical methods to physical hypotheses (sometimes intuitive) and checked back by observation and experiment.

Notwithstanding the difficulties inherent in an approach of this character, it is shown that very gratifying results have been reached and that by these clever combinations, a very substantial body of working theory has been developed, and which serves to answer to a reasonable and even in cases to a high degree of approximation, many of the important practical questions which arise in the more refined treatment of the problems of aerodynamics.

Special attention may be directed to the later sections of the Division in which are presented the more recent developments and views on these subjects, in which the author of the present Division has been for many years a pioneer. Thus in Sections 19—26 will be found the more recent developments with reference to turbulent flow, its generation and growth, distribution of velocity and the relation between the latter and the shearing stresses developed in turbulent flow along smooth and rough walls; and again in later sections, discussions of turbulent friction layers in accelerated and retarded flows, resistance of bodies in turbulent flows and experimental methods for the determination of the same.

The work of translating this Division from the German has been carried out chiefly by Dr. Louis Rosenhead, Professor of Applied Mathematics at the University of Liverpool and Fellow of St. John's College, Cambridge, to whom special acknowledgments are here made for this valuable assistance.

W. F. Durand.

---

1. Viscosity. In contrast with the ideal frictionless fluid which is usually taken as the basis of theoretical investigation, natural fluids always exhibit a certain resistance to alterations of form. This resistance is comparatively small for those fluids which are important for technical purposes, such as water, and also for gases, but, as will be shown, it is not negligible. In the case of other fluids such as oil, glycerin, honey, etc., this resistance to alterations of form is quite considerable. In order to appreciate correctly the nature of such resistances we shall first consider the behavior of "viscous" fluids.

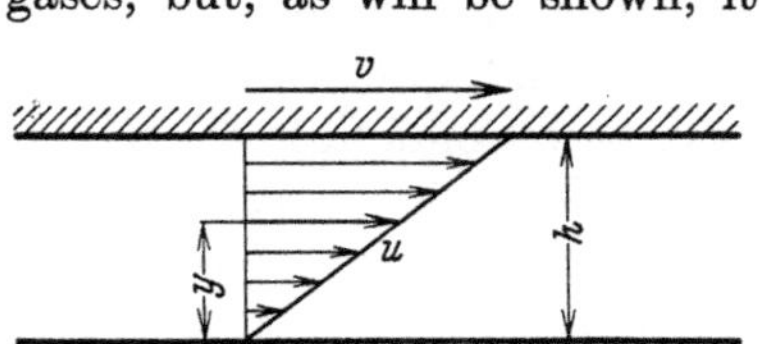

Fig. 1.

The following experiment is perhaps the simplest for our purpose: Suppose a suitable quantity of viscous fluid is put between two parallel plates which are at a distance h from each other (see Fig. 1). Should one of these plates be displaced without altering their mutual distance, it is well-known that a resistance, due to the viscous fluid, is experienced, and that to a first approximation it is proportional to the relative velocity of the two plates.

If the distance h between the plates is varied, the resistance for any particular relative velocity is found to be inversely proportional to the distance between the plates. Closer investigation shows that the viscous fluid adheres to both plates; that is, the layers of fluid immediately in contact with the plates have no velocity relative to them. The fluid is displaced in such a way that the various layers of fluid slide uniformly over one another. If y denotes the distance of a layer of fluid from the lower plate, which is supposed to be at rest, and v the velocity of the upper plate relative to the lower one, the velocity u of the layer at distance y is given by

$$u = v \frac{y}{h} \tag{1.1}$$

(for $y = 0$ this equation gives $u = 0$, for $y = h$, $u = v$, in accordance with the previous considerations). The results dealing with the resistance produced by the displacements in the above experiments can be explained by supposing that throughout the whole of the moving viscous fluid, there are tangential stresses (shearing forces) opposing alterations in shape. The result of the experiment is to show that the shear (τ) is proportional to the velocity of change of the initially right angles of the particle of fluid; that is, in our case τ is proportional to du/dy. In the case considered, this by (1.1) is equal to v/h, so that the above statement is in agreement with the previous assertions and gives a resistance proportional to the velocity and inversely proportional to the thickness of the layer of fluid. The factor of proportion attached to du/dy naturally becomes greater as the fluid becomes more viscous, and therfore furnishes a measure of the viscosity. Let this be denoted by μ[1]. We therefore establish the relation satisfied by the tangential stress (force per unit area) of a fluid in a condition of sliding motion as described, and as given by the equation

$$\tau = \mu \frac{du}{dy} \tag{1.2}$$

It must be emphasized that the motion considered represents a very simple and special case. The extensions necessary for the general case will be introduced in 7. Nevertheless, the results obtained above enable us to deal with a series of simple cases. We shall first consider one of the most familiar of these, the flow through a tube of circular cross-section.

2. The Poiseuille Flow. The following statements regarding the flow through a long straight tube of circular cross-section may be made: All particles which lie on a circular cylinder of radius y concentric with the tube have the same velocity and this velocity is of such a magnitude that the innermost particles flow most swiftly and the particles next to the wall of the tube are at rest. The movement is such that the

[1] In the literature of Physics the symbol η is sometimes used to denote the same quantity.

respective cylinders of particles of fluid slide like a number of rolls of paper which have been placed inside one another. Each such cylinder flows forward with a velocity greater than that of the cylinder which is around it. Movements of this kind, in which neighboring layers glide over one another as if they were solid bodies, are called "laminar movements" (from lamina, a thin foil).

In order to maintain such a motion, a difference of pressure between the ends of the tube must be operative. Corresponding to the relative velocity between two neighboring layers there is a shear which is connected with the rate of decrease of velocity du/dy, as given by (1.2). The shear on the other hand is directly connected with the fall of pressure in the tube. A cylinder of radius y and length l (see Fig. 2) is in equilibrium under the action of the following forces. There are pressures p_1 from the left, p_2 from the right, on the end surfaces of area πy^2, giving a resultant $(p_1 - p_2)\pi y^2$. The pressures on the curved surface are normal to it and obviously have no component in the direction of x. The tangential stresses however, supply a force $\tau \cdot 2\pi y \cdot l$. Equating these

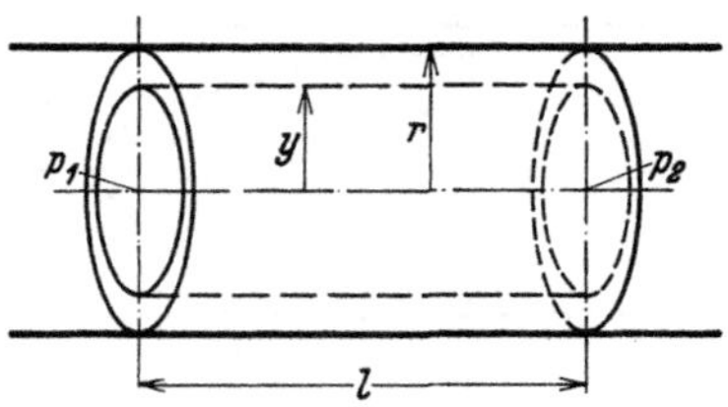

Fig. 2.

two expressions we have
$$\tau = \frac{p_1 - p_2}{l} \cdot \frac{y}{2} \tag{2.1}$$

Since the velocity in this case decreases from the interior to the exterior of the fluid, (1.2) must be taken with the minus sign, and integration gives
$$u = \frac{p_1 - p_2}{\mu l} \cdot \left(const. - \frac{y^2}{4}\right)$$
This constant must be so determined that $u = 0$ at the boundary, and as the radius of the tube is r, we have
$$u = \frac{p_1 - p_2}{4\mu l}(r^2 - y^2) \tag{2.2}$$

It can be seen that the distribution of velocity follows a parabolic law. The velocity at the center $(y = 0)$ is therefore
$$u_1 = \frac{p_1 - p_2}{l} \cdot \frac{r^2}{4\mu}$$

It is now quite easy to determine the total amount of flow through the tube. This, expressed as the volume flowing per unit time, is obtained from the equation $Q = \int u\, dA$, or geometrically, as the volume of the paraboloid (Fig. 3). It is known that the volume of a paraboloid equals half the area of the base $\times$ height. Hence
$$Q = \frac{1}{2} u_1 \pi r^2 = \frac{\pi}{8} \frac{p_1 - p_2}{l} \frac{r^4}{\mu} \tag{2.3}$$

This relation, that is, that the quantity of outflow is proportional to the rate of decrease of pressure $(p_1 - p_2)/l$, and to the fourth power of the radius (or diameter) of the tube, was obtained experimentally by Hagen[1] in 1839, and independently by Poiseuille[2] in 1840. The work of the latter became more widely known than that of the former, so that formula (2.3) is usually called "Poiseuille's Law". The calculation as given here was first obtained by G. Wiedemann[3] in 1856.

This formula allows us to test the frictional relation of (1.2) by experiment. It is only necessary to observe the flow through a thin tube such as shown in Fig. 4. If the radius and length of the tube are determined, the difference of pressure between the ends of the tube is given by $p_1 - p_2 = g\varrho h$, where g is the acceleration due to gravity, ϱ the density of the fluid, and h the pressure head. If the flow per unit time is measured by collecting the outflowing fluid in a vessel, the coefficient of viscosity μ can be determined. If the experiment be repeated with different heights of head, h, the relation between the velocity and the difference of pressure, and hence with the shear, can be shown. Repetition of the experiment with tubes of other diameters proves that the liquid does in fact adhere to the wall of the tube and hence that the constant of integration has been correctly determined.

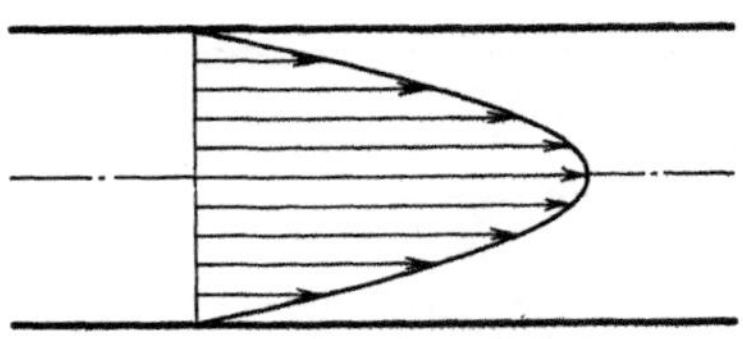

Fig. 3.

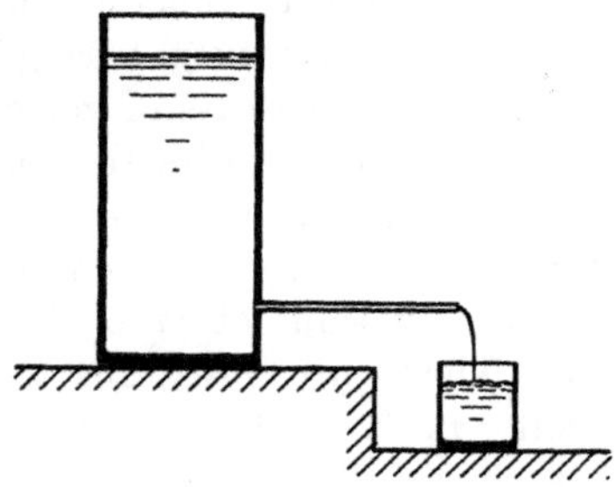

Fig. 4.

When carrying out the experiment in practice, a correction is necessary due to the fact that the fluid is accelerated on entering the tube, and, when it is inside it only gradually assumes the distribution of velocities calculated above. In the case of a well-rounded entrance, these circumstances require an additional fall in pressure of about $2.2\,(\varrho/2)\,(Q/\pi r^2)$ to be introduced[4]. This correction is very slight for small tubes, but for larger tubes or fluids of small viscosity, its effect is quite considerable. If the tubes are not sufficiently long it may happen that the distribution of

[1] Pogg. Ann., vol. 46, p. 423, 1839.

[2] Compt. Rend., vol. XI, p. 961, 1041, 1840; vol. XII, p. 112, 1841; in greater detail Mem. Sav. Etrang, vol. IX, 1846.

[3] Pogg. Ann., vol. 99, p. 218, 1856.

[4] The investigation of the flow near the entrance of the tube has been carried out in great detail by L. Schiller. See his account in the Zeitschr. f. angew. Math. u. Mech., vol. II, p. 96, 1922.

velocity required has not been formed even at the end of the tube, and it must then be calculated on the basis of a more exact theory.

Allowing for these possible sources of error it has been found that the law expressed by (2.3) is valid for most fluids with a high degree of accuracy.

There are exceptions however, for the most part when dealing with the so-called colloidal fluids. For these, instead of (1.2) a relation is found of such a nature that for increasing rates of change of shape du/dy, the shear τ does not increase as rapidly as it would with simple proportion, or, to express the same fact in another way, the coefficient μ slowly decreases for increasing rates of change of shape, or increasing shear. Such fluids are called "abnormal fluids". These deviations belong to the chemistry of colloids rather than to hydrodynamics, and will not be here discussed.

3. General Theory of Stress. As mentioned previously, we have only considered a very special case of the change of shape of a fluid. In order to prepare the way for the laws which will be obtained in the general case, we shall now discuss the theory of a general state of stress. Stresses are internal forces. The laws of statics can be applied to them only by taking a section, or considering a separating surface through the body in question, and by considering the resultant force, or the force per unit area, which the particles of the body to the right of the separating surface (for example) exert on the particles to the left. (The force which the particles to the left of the section exert on the particles to the right is, of course, according to the principle of the equality of action and reaction, simply equal in magnitude, and opposed in direction, to the above force). A force of this character is called a stress.

In order to characterize the state of stress at a point of the body it must be possible to take sections in all directions through this point. Our problem will therefore be solved when we can give the stress for every conceivable direction of the section. The information necessary to enable us to do so emerges from the following simple consideration. We assume that we already know the stresses corresponding to three surfaces forming a solid angle. Any fourth position of a surface can obviously be so taken as to cut out a small tetrahedron from the body. It follows from the equilibrium of this tetrahedron that the only unknown force is that on the triangular surface which belongs to the fourth separating surface. The other three forces are by hypothesis known. The force which is in equilibrium with these can however be determined by the laws of statics, that is, by the assumptions and by the condition that the tetrahedron must be in equilibrium. Division of the force by the corresponding area will then give the stress. Hence the state of stress at a point of the body is given when the stresses for three surfaces forming a solid angle are given.

These relations will now be illustrated by a formal example. We choose the three surfaces forming a solid angle at right angles to each other, and take the lines in which they intersect as the x, y, z axes. The tetrahedron is cut out by some oblique section, whose normal makes angles α, β, γ, with the x, y, z axes (see Fig. 5). For this specially chosen form, the areas of the three faces which lie on the planes of reference are $A \cos\alpha$, $A \cos\beta$, $A \cos\gamma$ where A is the area of the oblique section. $A \cos\alpha$ is the surface which is normal to the x axis. The angle of inclination between any surface and the surface A is in fact the same as the angle between their normals, from which the above relation follows. If $\boldsymbol{p}_x$, $\boldsymbol{p}_y$, $\boldsymbol{p}_z$ denote the vectorial stresses per unit area on the three faces formed by the planes of reference, and $\boldsymbol{p}$ denotes the vectorial stress on the oblique surface, the forces on the surfaces are obtained by multiplying the stresses by the areas. Equilibrium requires that the sum of the four force-vectors should vanish, hence

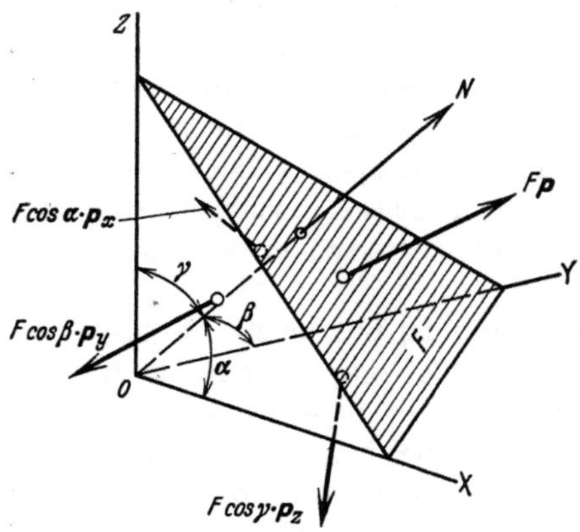

Fig. 5.

$$\boldsymbol{p} \cdot A + \boldsymbol{p}_x A \cos\alpha + \boldsymbol{p}_y A \cos\beta + \boldsymbol{p}_z A \cos\gamma = 0$$

A vanishes from the formula and $\boldsymbol{p}$ is given directly in terms of $\boldsymbol{p}_x$, $\boldsymbol{p}_y$, $\boldsymbol{p}_z$ when the three angles α, β, γ are given. Each of the three vectors $\boldsymbol{p}_x$, $\boldsymbol{p}_y$, $\boldsymbol{p}_z$ can be given numerically by its components which for $\boldsymbol{p}_x$ can be denoted by p_{xx}, p_{xy}, p_{xz} according to a scheme which is very easily understood. Thus p_{xy} is the y component of the stress on the surface perpendicular to the x axis. Altogether nine quantities are obtained, which characterize the state of stress. Three of these have both suffixes identical; they are perpendicular to the surfaces on which they act and are therefore called normal stresses. The other six components have directions along the surfaces and are therefore called tangential stresses. Notations are often used in which normal and tangential stresses are distinguished by different letters. In the theory of elasticity normal stresses are frequently denoted by σ (where $p_{xx} = \sigma_x$) and tangential stresses by τ (where $p_{yx} = \tau_{yx}$). The tangential stress τ considered in the first and second sections is identical with what is here called τ_{yx}.

Further, it can be shown that of the six tangential stresses those which have the same suffixes, but in reversed order, are equal. This can be seen very simply by considering the equilibrium of a small parallelepiped with respect to rotations, as shown in Fig. 6. Let the sides of the parallelepiped be dx, dy, dz. The surface on which τ_{xy} acts is perpendicular to the x axis, and therefore equal to $dy\,dz$. The force $\tau_{xy}\,dy \cdot dz$ appears twice, but with opposite sign, on opposing sides of the parallelepiped. The two forces form a couple of moment $\tau_{xy} \cdot dx \cdot dy \cdot dz$. Another couple about the same axis is given by two equal and opposite forces τ_{yx} which act on the surfaces at right angles to the y axis. Similar considerations show that the moment of this couple equals $\tau_{yx} \cdot dx \cdot dy \cdot dz$. If the sign of the shear is so arranged that the positive direction on a surface coincides with that of the axis given by the second suffix, we find that the two couples act in opposing senses. Equilibrium demands that the two moments should be equal and opposite, and therefore that $\tau_{yx} = \tau_{xy}$. There remain thus, six different stress magnitudes: σ_x, σ_y, σ_z, τ_{xy}, τ_{xz}, τ_{yz}.

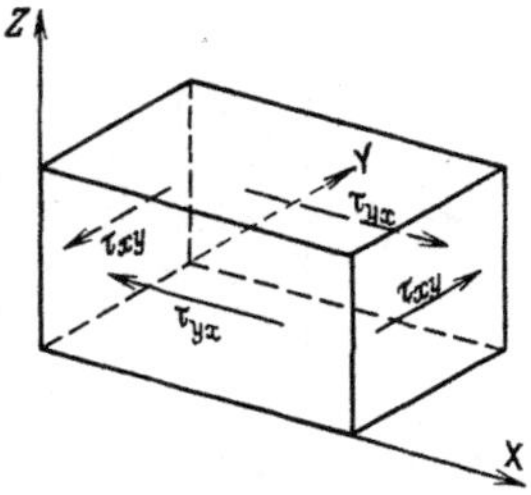

Fig. 6.

The state of stress of the ideal fluid is characterized by the fact that the stress acts perpendicularly to the separating surface everywhere (no friction). If to the considerations made in connection with Fig. 5 we add the condition that the stress $\boldsymbol{p}$ should always be perpendicular to the surface A, it is easy to show that the stresses $\boldsymbol{p}_x$, $\boldsymbol{p}_y$, $\boldsymbol{p}_z$, which are of course perpendicular to their surfaces, are each equal to $\boldsymbol{p}$. Since in most cases it is a case of pressure stress, we have

$$\sigma_x = \sigma_y = \sigma_z = -p,$$

and

$$\tau_{xy} = \tau_{xz} = \tau_{yz} = 0.$$

In order to be able to pass from fluids of small viscosity to non-viscous fluids, the state of stress is split up into a hydrostatic pressure in all directions $p = -1/3\ (\sigma_x + \sigma_y + \sigma_z)$ and an additional state of stress arising out of the viscosity and having components σ'_x, σ'_y, σ'_z, τ_{xy}, τ_{xz}, τ_{yz}, where

$$\sigma'_x = \sigma_x - 1/3\ (\sigma_x + \sigma_y + \sigma_z), \text{ etc.}$$

4. Equilibrium of Non-homogeneous States of Stress. If in some finite region the components of stress $\sigma_x \ldots \tau_{yz}$, are the same everywhere we say that we are dealing with a homogeneous state of stress. A non-homogeneous state is given when some or all of these six magnitudes vary from place to place, that is, are functions of the co-ordinates. In this case it is important to know the forces which result from the rate of change of stresses with the position. Assume again a parallelepiped with sides of length dx, dy, dz (see Fig. 7). We must now take account of the change which any stress, *e.g.* σ_x, undergoes, as we proceed from the left surface towards the right. Since the rate of change of the stress in the x direction is $\partial\sigma_x/\partial x$, the consequent change along dx is $(\partial\sigma_x/\partial x)\,dx$.

To obtain the resultant we must observe that the surface perpendicular to the x axis has an area $dy \cdot dz$. On the left hand surface we have $\sigma_x \cdot dy \cdot dz$ acting to the left; on the right hand surface acting to the right we have $\left(\sigma_x + \frac{\partial \sigma_x}{\partial x} dx\right) \cdot dy \cdot dz$. The resultant is $(\partial \sigma_x / \partial x)\, dx \cdot dy \cdot dz$. There are also other x components to be considered. We have τ_{yx} acting on the surface perpendicular to the y axis and τ_{zx} on the surface perpendicular to the z axis. These stresses also vary as we proceed from one side to the other to the extent of $(\partial \tau_{yx}/\partial y)\, dy$ and $(\partial \tau_{zx}/\partial z)\, dz$ respectively. Multiplication by the corresponding surfaces $dx \cdot dz$ and $dx \cdot dy$ gives resulting forces in the direction of the x axis of amounts $(\partial \tau_{yx}/\partial y)\, dy \cdot dx \cdot dz$ and $(\partial \tau_{zx}/\partial z)\, dz \cdot dx \cdot dy$. These three terms contain the parallelepiped volume $dx\, dy\, dz$. Dividing by this quantity we obtain the force per unit volume in the direction of

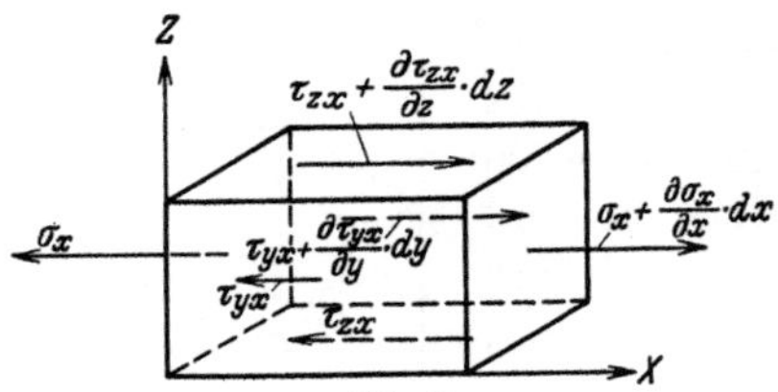

Fig. 7.

the x axis as
$$\frac{\partial \sigma_x}{\partial x} + \frac{\partial \tau_{yx}}{\partial y} + \frac{\partial \tau_{zx}}{\partial z}$$

There are corresponding expressions for the y and z components, *viz.*
$$\frac{\partial \tau_{xy}}{\partial x} + \frac{\partial \sigma_y}{\partial y} + \frac{\partial \tau_{zy}}{\partial z}$$
and
$$\frac{\partial \tau_{xz}}{\partial x} + \frac{\partial \tau_{yz}}{\partial y} + \frac{\partial \sigma_z}{\partial z}$$

When there is equilibrium, these three expressions are to be equated to the components of the external force per unit volume in the negative directions of the x, y, and z axes respectively. If there is no such force (gravity, centrifugal force, or the like) each of the three expressions is to be put equal to zero. In the case of movement the resultants of the external forces and the stress forces as expressed above are to be equated to the density times the x, y, and z components of the acceleration.

Before these considerations can be applied to the equations of motion of a viscous fluid, the general form of the relation between stress and the rate of change of shape must be found. This has already been found for a simple case, in (1.2). It will be advisable first to consider the relation between the system of changes, or the rates of change of shape, and the displacement paths and velocities of the individual points of our assumed continuous medium. This is done in the following section.

5. Theory of Deformation. In the general case the character of the motion of a fluid or gas is given by a velocity vector which varies from point to point. The components u, v, w of this vector are therefore functions of the co-ordinates x, y, z and of the time t. When it is a question, as in the following discussion, of the spatial arrangement only at some definite point in time, the temporal dependence can be neglected. In order to obtain information concerning the changes of shape which are taking place in the neighborhood of some point A, we develop the velocity components u, v, w in a Taylor series. This gives for example, for the component u

$$\begin{aligned} u = u_A &+ \left(\frac{\partial u}{\partial x}\right)_A (x - x_A) + \frac{1}{2}\left(\frac{\partial^2 u}{\partial x^2}\right)_A (x - x_A)^2 + \\ &+ \left(\frac{\partial^2 u}{\partial x \partial y}\right)_A (x - x_A)(y - y_A) + \dots \\ &+ \left(\frac{\partial u}{\partial y}\right)_A (y - y_A) + \frac{1}{2}\left(\frac{\partial^2 u}{\partial y^2}\right)_A (y - y_A)^2 + \\ &+ \left(\frac{\partial^2 u}{\partial y \partial z}\right)_A (y - y_A)(z - z_A) + \dots \\ &+ \left(\frac{\partial u}{\partial z}\right)_A (z - z_A) + \frac{1}{2}\left(\frac{\partial^2 u}{\partial z^2}\right)_A (z - z_A)^2 + \\ &+ \left(\frac{\partial^2 u}{\partial z \partial x}\right)_A (z - z_A)(x - x_A) + \dots \end{aligned}$$

For the immediate neighborhood of A the linear terms only need be kept. Hence

$$\begin{aligned} u - u_A &= (x - x_A)\left(\frac{\partial u}{\partial x}\right)_A + (y - y_A)\left(\frac{\partial u}{\partial y}\right)_A + (z - z_A)\left(\frac{\partial u}{\partial z}\right)_A \\ v - v_A &= (x - x_A)\left(\frac{\partial v}{\partial x}\right)_A + (y - y_A)\left(\frac{\partial v}{\partial y}\right)_A + (z - z_A)\left(\frac{\partial v}{\partial z}\right)_A \\ w - w_A &= (x - x_A)\left(\frac{\partial w}{\partial x}\right)_A + (y - y_A)\left(\frac{\partial w}{\partial y}\right)_A + (z - z_A)\left(\frac{\partial w}{\partial z}\right)_A \end{aligned}$$

It thus appears that the behavior in the immediate neighborhood of A can be given by the nine differential coefficients

$$\left\{\begin{matrix} \dfrac{\partial u}{\partial x} & \dfrac{\partial v}{\partial x} & \dfrac{\partial w}{\partial x} \\ \dfrac{\partial u}{\partial y} & \dfrac{\partial v}{\partial y} & \dfrac{\partial w}{\partial y} \\ \dfrac{\partial u}{\partial z} & \dfrac{\partial v}{\partial z} & \dfrac{\partial w}{\partial z} \end{matrix}\right\}$$

We shall always write this system, giving the change of shape, in the above "matrix" form.

In order to make clear the significance of the individual members of this system, some especially simple changes of shape will first be studied. For example, what does the following mean?

$$u - u_A = a\,(x - x_A); \qquad v - v_A = 0; \qquad w - w_A = 0$$

We see that of the nine differential coefficients only $du/dx = a$ differs from zero. The deformation matrix is therefore

$$\begin{Bmatrix} a & 0 & 0 \\ 0 & 0 & 0 \\ 0 & 0 & 0 \end{Bmatrix}$$

The equations show that the points of a plane perpendicular to the x axis, which all have the same value of x, are shifted by the same amount, $a\,(x - x_A)$. The plane through the point A remains undisplaced, and all parallel planes are displaced without distortion by amounts proportional to their distance from the point A. If a is positive, the planes separate from each other, those to the right of A move to the right, and those to the left of A move to the left (see Fig. 8). The whole state of affairs is called an extension, the measure of which is the elongation undergone by an element of unit length along the x axis. In the example, the extension in the x direction is accordingly $du/dx = a$. If we had $v - v_A = by$ and $u - u_A = w - w_A = 0$, we should have a corresponding extension of extent b in the direction of the y axis. Both extensions can of course occur together, in which case

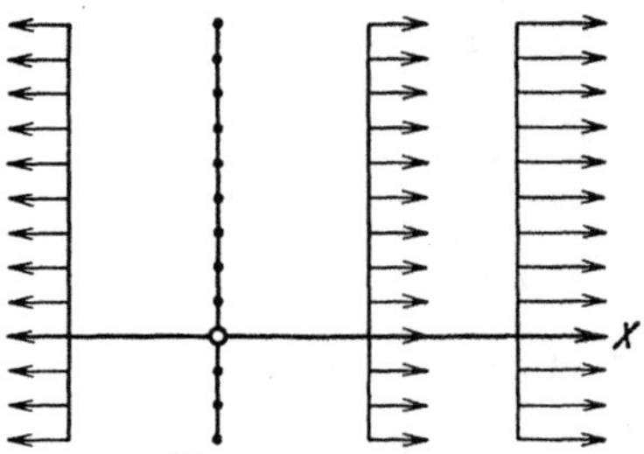

Fig. 8. Extension.

$$u - u_A = a x, \qquad v - v_A = b y, \qquad w - w_A = 0$$

which again would be shown by the deformation matrix

$$\begin{Bmatrix} a & 0 & 0 \\ 0 & b & 0 \\ 0 & 0 & 0 \end{Bmatrix}$$

Here the lines parallel to the x axis and the lines parallel to the y axis are stretched, the first by the coefficient a, the second by the coefficient b, and only those parallel to the z axis remain unstretched. If we add a third extension parallel to the z axis, and of measure c, there will in general be no unstretched line. These extensions however are a type of change of shape corresponding to the diagonal members of the matrix. What do the other members represent? If we write

$$u - u_A = w - w_A = 0, \qquad v - v_A = e \cdot x$$

only dv/dx of the nine differential coefficients is not zero, and the matrix becomes

$$\begin{Bmatrix} 0 & e & 0 \\ 0 & 0 & 0 \\ 0 & 0 & 0 \end{Bmatrix}$$

The interpretation is obviously that all the planes perpendicular to the x axis are again shifted, this time not in the direction of the x axis, but in that of the y axis. Once again the plane through A remains where it was and the shifts of the other planes are proportional to their distances from A. As Fig. 9 shows, all lines which were originally parallel to the x axis are now at an angle ε given by

$$\tan \varepsilon = e$$

There are six different displacements of this type as shown by the form of the matrix. They can be collected into three pairs, both members of which give rotations about the same axis. Thus, as we have just explained, $\partial v/\partial x$ measures the rotation about the z axis of the lines which were originally parallel to the x axis, and correspondingly $\partial u/\partial y$ measures the rotation about the z axis of lines originally parallel to the y axis. If we consider the alteration in the right angle between these two original directions, we obtain (when the deformation is so small that an angle and its tangent are interchangeable) the sum of these two magnitudes, $\partial v/\partial x + \partial u/\partial y$. This expression is called an angular deformation. There are obviously three magnitudes of this kind, the angular deformations about the:

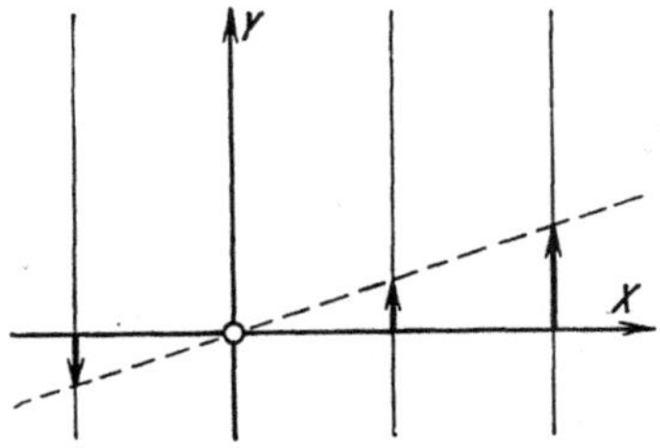

Fig. 9. Angular deformation.

$$x \text{ axis as axis of rotation, } \gamma_{yz} = \frac{\partial w}{\partial y} + \frac{\partial v}{\partial z},$$

$$y \text{ axis as axis of rotation, } \gamma_{zx} = \frac{\partial u}{\partial z} + \frac{\partial w}{\partial x},$$

$$z \text{ axis as axis of rotation, } \gamma_{xy} = \frac{\partial v}{\partial x} + \frac{\partial u}{\partial y}$$

The extensions in the three directions, given by

$$\varepsilon_x = \frac{\partial u}{\partial x}, \text{ (along } x \text{ axis)}$$

$$\varepsilon_y = \frac{\partial v}{\partial y}, \text{ (along } y \text{ axis)}$$

$$\varepsilon_z = \frac{\partial w}{\partial z}, \text{ (along } z \text{ axis)}$$

and the three angular deformations γ_{yz}, γ_{xz}, γ_{xy} are called the components of change of shape (or components of strain). When, for some definite case the change of shape is zero, *i. e.* all the six components of change of shape vanish, it does not necessarily follow that every shift relative to the point A vanishes, for the body can still perform rotations, *viz.* three independent rotations about three coordinate axes, according to well-known theory. For example, consider the matrix

$$\begin{Bmatrix} 0 & \omega & 0 \\ -\omega & 0 & 0 \\ 0 & 0 & 0 \end{Bmatrix}$$

for which all the components of change of shape do in fact vanish. Written out in the form of equations we have

$$\begin{aligned} u - u_A &= -\omega(y - y_A), \\ v - v_A &= +\omega(x - x_A), \\ w - w_A &= 0 \end{aligned}$$

The lines parallel to the x axis experience a rotation ω toward the y axis, the lines parallel to the y axis an equal rotation toward the negative x axis. The angle between two such lines therefore remains the same (see Fig. 10). Any point distant r from the z axis undergoes, as an easy calculation shows, a translation compounded of u and v, such that the resulting translation is perpendicular to r and of amount $r\omega$, so that the system of deformations in question does in fact reduce to a pure rotation about the z axis (provided the displacements are sufficiently small) as shown in Fig. 10.

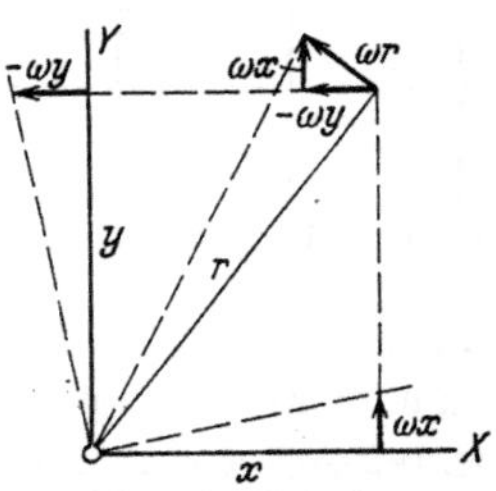

Fig. 10. Rotation.

It is customary also in the general case, where changes of angles may enter, to speak of rotations, meaning the mean values of the rotations of the two axes concerned about the third axis. The three rotation velocity components are therefore (see Division A VI 9)

$$\omega_x = \frac{1}{2}\left(\frac{\partial w}{\partial y} - \frac{\partial v}{\partial z}\right),$$

$$\omega_y = \frac{1}{2}\left(\frac{\partial u}{\partial z} - \frac{\partial w}{\partial x}\right),$$

$$\omega_z = \frac{1}{2}\left(\frac{\partial v}{\partial x} - \frac{\partial u}{\partial y}\right)$$

In the preceding paragraphs we have been speaking of displacements and changes of shape. Exactly corresponding relations hold for the velocities of the various point-masses and the "rates of change of shape" which belong to them. The formulae of these paragraphs need no alteration, but the magnitudes u, v, w, which have been considered hitherto as small displacements, now denote velocities, and $\partial u/\partial x$ etc. are now the corresponding rates of displacement, of rotation etc.

6. Stokes' Theorem for Stresses. We have now to find a generalization of (1.2) corresponding to the general state of deformation. This can be done by connecting the matrix of the components of stress, *viz.*

$$\begin{Bmatrix} \sigma_x & \tau_{xy} & \tau_{xz} \\ \tau_{xy} & \sigma_y & \tau_{yz} \\ \tau_{xz} & \tau_{yz} & \sigma_z \end{Bmatrix}$$

with the matrix of the rates of deformation, *viz.*

$$\begin{Bmatrix} \frac{\partial u}{\partial x} & \frac{\partial v}{\partial x} & \frac{\partial w}{\partial x} \\ \frac{\partial u}{\partial y} & \frac{\partial v}{\partial y} & \frac{\partial w}{\partial y} \\ \frac{\partial u}{\partial z} & \frac{\partial v}{\partial z} & \frac{\partial w}{\partial z} \end{Bmatrix}.$$

A relation of this kind will satisfy the condition of independence of change of coordinate direction. The stress-matrix is symmetric (with respect to the diagonal) but the matrix of the rates of deformation is not. We can however construct a symmetric matrix out of it by adding to it the transposed matrix. By adding the matrix of the pressure p, the correct generalization of (1.2) is obtained as follows:

$$\begin{Bmatrix} \sigma_x & \tau_{xy} & \tau_{xz} \\ \tau_{xy} & \sigma_y & \tau_{yz} \\ \tau_{xz} & \tau_{yz} & \sigma_z \end{Bmatrix} = -\begin{Bmatrix} p & 0 & 0 \\ 0 & p & 0 \\ 0 & 0 & p \end{Bmatrix} + \mu \begin{Bmatrix} \frac{\partial u}{\partial x} & \frac{\partial v}{\partial x} & \frac{\partial w}{\partial x} \\ \frac{\partial u}{\partial y} & \frac{\partial v}{\partial y} & \frac{\partial w}{\partial y} \\ \frac{\partial u}{\partial z} & \frac{\partial v}{\partial z} & \frac{\partial w}{\partial z} \end{Bmatrix} + \mu \begin{Bmatrix} \frac{\partial u}{\partial x} & \frac{\partial u}{\partial y} & \frac{\partial u}{\partial z} \\ \frac{\partial v}{\partial x} & \frac{\partial v}{\partial y} & \frac{\partial v}{\partial z} \\ \frac{\partial w}{\partial x} & \frac{\partial w}{\partial y} & \frac{\partial w}{\partial z} \end{Bmatrix} \tag{6.1}$$

The meaning of such a matrix equation is that the corresponding equations in the ordinary sense hold between the quantities which occupy similar places in the various matrices. Thus the middle place in the first line, or equally well, the first place in the second line of the matrices gives the relation

$$\tau_{xy} = \mu \left(\frac{\partial v}{\partial x} + \frac{\partial u}{\partial y} \right) \tag{6.2}$$

which we recognize as identical with (1.2). For, if v is not, as assumed in that paragraph, zero, the change in angle is equal to $(\partial v/\partial x + \partial u/\partial y)$, which is the expression that appears in the above formula for τ_{xy}. The first place in the first line gives

$$\sigma_x = -p + 2\mu \frac{\partial u}{\partial x} \tag{6.3}$$

The considerations of **3** show that for frictionless fluids we must put $\sigma_x = -p$, and the present calculation shows that the extension du/dx contributes an extra term $2\mu\, \partial u/\partial x$.

The other components are constructed in the same way and it would be superfluous to write them out in detail. These formulae for the stresses in the form given are correct for incompressible fluids. If we calculate the value of (1/3) $(\sigma_x + \sigma_y + \sigma_z)$ we get

$$-p + \frac{2}{3}\mu\left(\frac{\partial u}{\partial x} + \frac{\partial v}{\partial y} + \frac{\partial w}{\partial z}\right)$$

The expression in parenthesis is, however, the change in volume, and therefore by hypothesis zero, so that the mean value of the three normal stresses becomes $-p$. If it is desired to have the pressure p equal to the mean value of the three normal stresses also in the case of flow with change of volume, we can subtract the expression (2/3) μ $(\partial u/\partial x + \partial v/\partial y + \partial w/\partial z)$ from the right hand side of (6.3), and from the similar equations for σ_y and σ_z. This is equivalent to giving a different definition for p. This is however a nicety which will play no part in the present development, so that we shall retain the simple equations (6.2) and (6.3).

It may be mentioned incidentally, that in accordance with Gibbs'[1] procedure, there is a simple way of writing equations similar to (6.1) by replacing each matrix by one symbol. The quantities which are represented by the nine-termed matrices have various names; their general form will here be denoted by affinor. The symmetric forms are also called tensors. Let Σ be the affinor of the state of stress, $\boldsymbol{V}$ the velocity vector and $\nabla \cdot \boldsymbol{V}$ the affinor of the deformation velocities, and let the transposed (according to Gibbs' "conjugate") affinor be $(\nabla \boldsymbol{V})_c$. Equation (6.1) can now be written

$$\Sigma = -pI + \mu \nabla \boldsymbol{V} + \mu (\nabla \boldsymbol{V})_c \tag{6.1a}$$

where I denotes the unit tensor

$$\begin{Bmatrix} 1 & 0 & 0 \\ 0 & 1 & 0 \\ 0 & 0 & 1 \end{Bmatrix}$$

We still need now to calculate the components of force per unit volume which were obtained in 4. Since we have only to perform a simple differentiation, the result can be written down immediately. We will however, write out the terms arising from the two matrices in (6.1) and thus obtain the force in the x direction as

$$-\frac{\partial p}{\partial x} + \mu\left(\frac{\partial^2 u}{\partial x^2} + \frac{\partial^2 u}{\partial y^2} + \frac{\partial^2 u}{\partial z^2}\right) + \mu\left(\frac{\partial^2 u}{\partial x^2} + \frac{\partial^2 v}{\partial x \partial y} + \frac{\partial^2 w}{\partial x \partial z}\right)$$

which can also be written as

$$-\frac{\partial p}{\partial x} + \mu \nabla^2 u + \mu \frac{\partial}{\partial x} \operatorname{div} \boldsymbol{V} \tag{6.4}$$

---

[1] GIBBS-WILSON, Vector Analysis. New York and London, 1902.

in which the symbol ∇^2 is used to denote $\partial^2/\partial x^2 + \partial^2/\partial y^2 + \partial^2/\partial z^2$. When no changes of volume are to be taken into account the last term disappears since $div\, \boldsymbol{v} = 0$. Even for motion with changes of volume this term plays only a subordinate part so that it need not be further considered. Thus in addition to the fall in pressure $-dp/dx$ which we have already met in frictionless fluids, we have a force due to the viscosity whose components in the directions of the x, y, z axes are,

$$\mu \nabla^2 u, \qquad \mu \nabla^2 v, \qquad \mu \nabla^2 w$$

The equations of motion for viscous fluids without change of volume, can now be obtained quite simply from the familiar Euler equations by adding the frictional terms above noted. These equations, found independently by Navier and Stokes, and therefore usually known by the names of these two authors, become, when written in Gibbs' notation,

$$\varrho \left(\frac{\partial v}{\partial t} + \boldsymbol{v} \cdot \nabla \overline{\boldsymbol{v}}\right) = \varrho \boldsymbol{F} - grad\, p + \mu \nabla^2 \boldsymbol{v} \tag{6.5}$$

In addition we have the equation of continuity, which is of the same form as that for ideal fluids:

$$div\, \boldsymbol{v} = 0 \tag{6.6}$$

Written in components these equations are:

$$\left.\begin{aligned}
\varrho \left(\frac{\partial u}{\partial t} + u \frac{\partial u}{\partial x} + v \frac{\partial u}{\partial y} + w \frac{\partial u}{\partial z}\right) &= \varrho X - \frac{\partial p}{\partial x} + \mu \left(\frac{\partial^2 u}{\partial x^2} + \frac{\partial^2 u}{\partial y^2} + \frac{\partial^2 u}{\partial z^2}\right) \\
\varrho \left(\frac{\partial v}{\partial t} + u \frac{\partial v}{\partial x} + v \frac{\partial v}{\partial y} + w \frac{\partial v}{\partial z}\right) &= \varrho Y - \frac{\partial p}{\partial y} + \mu \left(\frac{\partial^2 v}{\partial x^2} + \frac{\partial^2 v}{\partial y^2} + \frac{\partial^2 v}{\partial z^2}\right) \\
\varrho \left(\frac{\partial w}{\partial t} + u \frac{\partial w}{\partial x} + v \frac{\partial w}{\partial y} + w \frac{\partial w}{\partial z}\right) &= \varrho Z - \frac{\partial p}{\partial z} + \mu \left(\frac{\partial^2 w}{\partial x^2} + \frac{\partial^2 w}{\partial y^2} + \frac{\partial^2 w}{\partial z^2}\right)
\end{aligned}\right\} \tag{6.5a}$$

$$\frac{\partial u}{\partial x} + \frac{\partial v}{\partial y} + \frac{\partial w}{\partial z} = 0 \tag{6.6a}$$

7. Reynolds' Laws of Similitude. As yet no general methods are known for solving the Navier-Stokes' equations, and solutions which hold without restrictions for all values of the viscosity have been found only for a small number of special cases. Moreover, the problem of flow with finite viscosity can be attacked only "from the limits" by dealing, on the one hand, with fluids of very large viscosity, and on the other hand, with fluids of almost vanishing viscosity. The methods which have been developed for these special cases are however by no means sufficient for finding by interpolation the state of affairs in the case of medium viscosities. Moreover, the mathematical methods are very difficult and inconvenient, even in the limiting cases. This explains why most attempts to obtain information on the behavior of viscous fluids are from the experimental side. Very important deductions can however be drawn from the Navier-Stokes' equations, in consequence of which the extent of needed research is much reduced. This is brought about through the use of the so-called laws of similitude.

When considering cases of similitude, in mechanics, the following question usually arises: If in two systems the external circumstances exhibit geometrical similarity, what conditions must be satisfied in order that the resulting phenomena should also exhibit geometrical similarity? For two states of flow this similarity will be said to be present when an alteration of the units of time and length makes the results for both states coincide. This implies that the paths of the individual particles, the stream-lines at any instant, the fields of acceleration etc., in the two cases, become geometrically similar. This implies also a similarity between the distributions of pressure (geometric similarity between surfaces of constant pressure, and proportionality of the differences of pressure between corresponding points). We shall denote this complex of similarities and proportionalities by the term dynamic similitude.

What information now can the above differential equations supply as to the conditions for this dynamic similitude? The answer is: We shall have dynamic similitude when alterations of the units of length, time, and if necessary, force, transform the differential equations and boundary conditions in one case into those of the other case, so that the equations completely coincide. Assuming that we have a unique solution of the system of equations, it will furnish a direct description of the circumstances for one case, and a corresponding description for the other case, by a suitable alteration of the units.

The two systems of equations will coincide in this way only if alteration of the units of time and length (and if necessary, of force) converts each member of one equation into the corresponding member of the other equation, multiplied by a constant factor (dividing by this factor makes both equations identical).

For the case of fluids of variable density, the conditions for similitude cannot in general be satisfied. We therefore restrict ourselves to the case of constant density, which is the most important one for technical applications. Here the pressure can be divided into two parts of which the first is in equilibrium with the gravitational forces according to the usual laws of hydrostatics, and the other balances the forces produced by movement. We can therefore neglect the forces of gravity when using the Navier-Stokes' equations for fluids of constant density, providing we understand by "pressure" the second of these components.

We must now investigate how the Navier-Stokes' equations change when new units of time and length are chosen in order to express two similar states of affairs by the same equation. If l_1 be any length in the first system, and l_2 the corresponding length in the second, we may put

$$l_2 = \lambda\, l_1$$

This same relation naturally holds between all other pairs of corresponding points and hence for two corresponding points $x_2 = \lambda\, x_1$, etc.

For two corresponding times t_2 and t_1, we put $t_2 = \tau t_1$, and for two corresponding velocities, $u_2 = \eta u_1$. Since this relation must also hold for the velocity required to traverse the length l in time t we must have

$$\eta = \frac{\lambda}{\tau}$$

(to obtain this the equation $u_2 = l_2/t_2$ is divided by the equation $u_1 = l_1/t_1$). A differential coefficient is a ratio of infinitesimals of two quantities. The infinitesimals of the quantities are expressed in the same units as the quantities themselves and therefore the ratio for converting a differential coefficient in one case to the corresponding differential coefficient in the other is equal to the quotient of the conversion-ratios for denominator and numerator. Hence the conversion-ratio for

$\partial u/\partial t$ is
$$\frac{\eta}{\tau} = \frac{\lambda}{\tau^2}$$

and the ratio for $\partial u/\partial x = \eta/\lambda = 1/\tau$. A second differential coefficient is the differential coefficient of the first differential coefficient. Hence, for example, the dimension factor for $\partial^2 u/\partial x^2 = (1/\lambda)(\eta/\lambda) = 1/\lambda\tau$. By the conditions of the problem, the pressures p_1 and p_2 are determined subject to an additive constant. If this be suitably chosen the relation between the two pressures can be made $p_2/p_1 = \pi$[1] (actually we have only to put the ratio of the pressure gradients equal to π/λ). It may also be noted that the densities ϱ_1 and ϱ_2 of the two fluids, and their viscosities μ_1 and μ_2, may be different, and we may write $\varrho_2/\varrho_1 = \varrho'$ and $\mu_2/\mu_1 = \mu'$.

We wish now to express the magnitudes of the second system in terms of those of the first system. As the preceding discussion shows, expressions in the Navier-Stokes' equations will thereby receive factors involving λ, τ, etc. which always have the same value for a group of terms built up in the same way. Thus the term $\varrho_1 (\partial u_1/\partial t)$ and corresponding terms take the factor $\varrho' (\lambda/\tau^2)$, terms like $\varrho_1 u_1 (\partial u_1/\partial x)$ take the factor $\varrho' (\lambda/\tau)(1/\tau) = \varrho'(\lambda/\tau^2)$; that is, the same factor as before. (It was to be expected that the two kinds of expressions, both representing accelerations multiplied by density, should take the same factor.) The term $\mu_1 \partial^2 u_1/\partial x^2)$ takes the factor $\mu' (1/\lambda\tau)$, and $\partial p_1/\partial x$ the factor π/λ. If similarity is present, the Navier-Stokes' equations for the second system, when transformed in this way, must be converted into the corresponding equations for the first case. This is obviously possible only when the factors which have been attached to the various terms all have the same numerical value. If this is so, these factors can be removed by division, leaving the Navier-Stokes' equations for the first case. The condition for similarity can therefore be formulated as

$$\frac{\varrho' \lambda}{\tau^2} = \frac{\pi}{\lambda} = \frac{\mu'}{\tau \lambda} \tag{7.1}$$

[1] Note that π here signifies a general ratio.

In this pair of equations the condition for the middle term can always be satisfied, if the other two terms are equal, by a suitable choice of the ratio π. The equality of the first and third terms however, gives an important condition. It can be written in the form

$$\frac{\varrho' \lambda^2}{\mu' \tau} = 1 \tag{7.2}$$

The relation between the pressures now becomes

$$\pi = \frac{\varrho' \lambda^2}{\tau^2} = \frac{\mu'}{\tau} \tag{7.3}$$

provided of course that (7.2) is satisfied. If η, the ratio of corresponding velocities be introduced, instead of the ratio of corresponding times, that is by putting $\tau = \lambda/\eta$, (7.2) can also be written as

$$\frac{\varrho' \lambda \eta}{\mu'} = 1 \tag{7.2a}$$

A similar transformation of (7.3) gives

$$\pi = \varrho' \eta^2 = \frac{\mu' \eta}{\lambda} \tag{7.3a}$$

If the states of flow around two bodies, geometrically similar and of length measurements l_1 and l_2, immersed respectively in fluids of densities ϱ_1, ϱ_2 and viscosities μ_1, μ_2 are to be dynamically similar, it follows from (7.2a) that the proportion between the velocities of flow at corresponding points must be such that the relation

$$\frac{\varrho_1 l_1 u_1}{\mu_1} = \frac{\varrho_2 l_2 u_2}{\mu_2} \tag{7.4}$$

is satisfied. We can therefore state as the condition of similitude, that the expression $\varrho l u/\mu$ must have the same value for dynamically similar flows. This law was discovered by Osborne Reynolds[1] and the preceding expression has in his honor been called the Reynolds Number. Obviously we obtain small Reynolds numbers for very viscous fluids (very large μ), and large Reynolds numbers for very slightly viscous fluids (very small μ). The constitution of the number shows however that its value does not depend on the ratio of ϱ to μ alone, but also on the linear dimensions of the body and on the velocity. When we deal with only one fluid, the value of the product of the length and the velocity of the body is the determining factor. For a body twice as large but with half the velocity, there is in this case a state of flow similar to that in the original case.

Equation (7.3a) shows that in the case of dynamic similitude, any differences of pressure are in the ratio of the product of the density and the square of the velocity, or again in the ratio

$$\frac{[\text{viscosity} \times \text{velocity}]}{[\text{length}]}$$

[1] REYNOLDS, O., Phil. Trans. Roy. Soc., 1883, or Collected Papers, vol. II, p. 51.

Of these two expressions the first is connected with the inertia property of fluids and agrees with the well-known relation that the increase of pressure at a stagnation point is equal to $(\varrho/2)\,(u^2 + v^2 + w^2)$, the second is connected with the fundamental property of viscous fluids. It is sufficient to mention that the shearing stresses which were considered in **1** are quantities of the same dimensions as pressures (*viz.* forces per unit area) and that for dynamic similarity the two shearing stresses at corresponding points in the systems compared, are in the same ratio as the corresponding differences of pressure at the two points. After these preliminary remarks it is sufficient to indicate that (1.2) for the shearing stress shows precisely this construction. When a difference of pressure which is derived in the above way from the inertial effect, and is therefore proportional to ϱu^2, is divided by a difference of pressure, or a shear which is caused by viscosity, and is therefore proportional to $\mu u/l$, we again obtain the Reynolds number multiplied however by a numerical factor arising out of the conditions of the particular problem. This shows that the Reynolds number can, with certain reservations, be regarded as the ratio of the inertial effects to the viscous effects. The Reynolds number is the decisive factor in determining in any given flow of fluid whether the inertial effect or the viscous effect predominates. A small Reynolds number denotes predominating viscosity and a large Reynolds number preponderating inertia.

Information concerning the construction of the formula for the resistance can be derived from the expressions for the differences of pressure just discussed. The resistance of a body in a fluid can be expressed as the product of an area and a difference of pressure. Since the area varies as the square of a length, we obtain, considering dynamically similar flows for the present, formulae of the form

$$D = n_1\, l^2\, \varrho u^2 \tag{7.5}$$

and

$$D = n_2\, l \mu\, u \tag{7.6}$$

where n_1, n_2 are numbers, according as we take the inertial effect or the viscosity effect as starting points. The two numbers occurring in these formulae are of course different, but each has a constant value for each class of dynamically similar flows, that is for one Reynolds number. If on the other hand, in considering two flows, we go from one Reynolds number to another, so that dynamic similarity no longer holds, the numbers in the two formulae will in general change. Hence these numerical values are constant for constant Reynolds numbers and change with the Reynolds number; or in other words, they are functions of the Reynolds number. As can be easily seen the following relation holds

$$n_2 = n_1 \frac{\varrho\, l u}{\mu} \tag{7.7}$$

The method naturally tells us nothing about the nature of this function for any particular case (*e. g.* for circular cylinders, rectangular plates or the like). Nevertheless it is a great advantage to have a relation such as (7.7). It is now sufficient to measure the resistance of a body in the same fluid for different velocities, in order to discover how the resistance changes when the magnitude of the body changes, or when the density and viscosity change, *e. g.* through changes in temperature. The following can be said concerning the form of the functions which occur.

For very large Reynolds numbers the magnitude of the viscosity becomes unimportant for the resistance and we must expect a formula which no longer contains μ, as in (7.5). In this case it may therefore be assumed that

$$n_1 = \text{constant}$$

For very small Reynolds numbers, on the other hand, the inertial effect becomes unimportant and we must expect a formula in which the density no longer occurs. This is the case in (7.6) if we assume that here

$$n_2 = \text{constant}$$

Both these laws have in fact been observed experimentally. The first is "Newton's Law of Resistance" and the second is "Stokes' Law of Resistance". Newton's law holds for very large bodies and velocities, Stokes' law for very small bodies and very small velocities, or large viscosity.

Since the ratio of the viscosity μ to the density ϱ always occurs in the Reynolds number $\varrho l u/\mu$, it is convenient to denote this ratio by a special symbol and to give it a name. The ratio μ/ϱ is called the kinematic viscosity and is denoted by ν. The units for ν are particularly simple. It has the dimensions [(length)^2/time] which follows at once from the fact that the Reynolds number, which is the ratio of inertial force to viscous force, must be a pure number. Hence ν must have the same dimensions as the product $l\,u$. We shall use this abbreviation regularly in the future and write the Reynolds number as ul/ν, vd/ν, etc., and also use u, v as characteristic velocities in two cases, $l =$ length in the one case and $d =$ diameter in the other. Where no ambiguity can arise we shall also use the letter R for the Reynolds number.

According to the preceding remarks the two formulae for the law of resistance expressed by (7.5) and (7.6) are of equal value. For most practical applications, the Reynolds numbers are large; Newton's law holds, at least approximately, and the first formula is to be preferred. Nevertheless, in accordance with formula (7.6), states of affairs which correspond to Stockes' law can be expressed by the same formula. We must then put

$$n_1 = \frac{\text{constant}}{R}$$

so that, in accordance with (7.7), n_2 becomes a constant.

8. Dimensional Analysis Applied to the Problem. Instead of starting from dynamic similitude, we could apply the so-called dimensional analysis which is based upon the fact that the physical content of any theory must not depend upon the units which are chosen for the calculations. Considerations of this kind are in their very nature closely related to those of dynamic similitude. In the latter case we look for the conditions under which the equations for two similar states of affairs can be transformed, the one into the other. In the case of dimensional analysis on the other hand, we look for the conditions under which the different equations obtained for one and the same state of affairs with different magnitudes of the units, have the same physical effect. To measure geometrically similar bodies of different sizes with the same units of measurement, and the same body with various units of measurement, are however, absolutely related processes.

When considering the question of dimensions it is usual to start from the fundamental principle that the physically essential relations between the various magnitudes can always be expressed in a form independent of the units of measurement. Such dimensionless magnitudes can then always be treated as numerical ratios between magnitudes of the same kind. In order that such a dimensionless representation may be possible it is necessary that magnitudes in the formulae which are added or equated to one another should always have the same "dimensions", that is, the units of their numerical values must be derived in the same way from the fundamental units. This is only another way of stating the condition for the existence of a dimensionless representation.

In a rational system there are only three fundamental units occurring in Mechanics; these can be taken arbitrarily. We may choose, *e. g.* the units of length, time, and mass, or those of length, time and force; an example of the first kind is the physical C.G.S. system, and, of the second, the technical system of measurements preferred by engineers.

Considerations involving dimensions have the advantage over those of "similarity" in that the former can be applied in circumstances where the differential equations of the problem in question are unknown, but where the well-defined physical magnitudes, on which depend the phenomena whose laws we are investigating, are known. The task of discovering the physical magnitudes connected with the phenomena which are decisive and of eliminating those magnitudes which are of subordinate importance, is the very core of the problem.

The answer supplied by the condition for the possibility of a dimensionless representation varies according to the number of independent magnitudes which are important for the system in question. If there are only three such independent magnitudes present and if all three of the fundamental units are involved in them, the condition will only give one way of compounding powers of the three magnitudes to give a fourth magnitude of prescribed dimensions. If the three magnitudes be called x_1, x_2, x_3, any magnitude y dependent on them will give a formula

$$y = const.\ x_1^{\alpha} \cdot x_2^{\beta} \cdot x_3^{\gamma},$$

and since no further possibilities for constructing the formula with the correct dimension for y can be found, the conclusion is that the actual physical law must be of this form. If there are four independent magnitudes x_1, x_2, x_3, x_4 however, it is always possible to build a dimensionless combination of the form[1]

$$X = x_1 \cdot x_2^{\delta} \cdot x_3^{\varepsilon} \cdot x_4^{\zeta}$$

In this case, a possible form for the law for which we are searching is

$$y = f(X)\, x_1^{\alpha}\, x_2^{\beta}\, x_3^{\gamma}$$

where $f(X)$ is some function of the dimensionless magnitude X, and therefore, from the standpoint of dimensions, may be considered arbitrary. The choice of the three variables x_1, x_2, x_3, out of the four, x_1, x_2, x_3, x_4, is obviously arbitrary. It is equally possible to take a different selection of three of these four magnitudes. This would simply mean that a suitable power of X is divided out of the function $f(X)$ whose multiplication by $x_1^{\alpha}\, x_2^{\beta}\, x_3^{\gamma}$ causes x_1 say, to vanish, and a power of x_4 to appear. Such an alteration would therefore be merely of a formal nature. Information as to the nature of this function $f(X)$ can naturally be obtained only through experiment or an actual theory of the phenomena in question.

We can at once extend the above to more than four independent magnitudes. If, for example, five independent magnitudes are present, two independent dimensionless magnitudes X_1, X_2 can be constructed, and a function of these two independent magnitudes will have to be introduced, *e. g.*

$$y = F(X_1, X_2)\, x_1^{\alpha}\, x_2^{\beta}\, x_3^{\gamma}$$

If we subsequently ask for the conditions for dynamic similitude, we can easily establish the fact that in the case of three independent quantities, geometrically similar systems are also dynamically similar. With four independent magnitudes, dynamic similitude is present in the corresponding case only when the dimensionless combination X has the same value in the two geometrically similar systems. With five independent variables it will in general no longer be possible to produce dynamic similitude unless two of the five magnitudes can be so chosen that the two dimensionless magnitudes X_1, X_2 simultaneously attain the same numerical value.

These general remarks will now be applied to the problem of hydrodynamic resistance. If we are considering a uniform progressive motion

---

[1] See also Division A IV.

in a heavy viscous fluid, the following factors will exert influence, in addition to the shape of the body, which is naturally an essential factor for the results: above all, the size of the body, represented by a linear measurement l, further the velocity v of the movement relative to the medium, the density ϱ of the fluid as measuring its inertia, and the coefficient of viscosity μ. We assume that the liquid is homogeneous so that the effect of gravity is balanced by the statical component of the pressure and hence has no effect on the motion. We further assume that the velocity of motion of the body is small enough to permit neglect of the effect of the compressibility of the fluid. (It is known that the effect of the compressibility becomes appreciable when the velocity of the body approaches the velocity of sound in the medium.) If these subordinate effects can be neglected we have four mutually independent magnitudes l, v, ϱ, μ to deal with. Let the fundamental dimensions be those of length L, time T, and mass M. Then first of all, the dimension of l is simply L, that of the velocity v is L/T and of the density ϱ (mass per unit volume) is M/L^3. It is rather more difficult to determine the unit of viscosity. Equation (1.2) shows that $\mu\,(du/dy)$ is a shear, that is, force per unit area. Now a force is equal to mass times acceleration; hence with the unit of acceleration L/T^2, the dimensions of force are ML/T^2. Hence

$$[\mu]\,\frac{L/T}{L} = \frac{ML/T^2}{L^2}$$

or

$$[\mu] = \frac{M}{LT}$$

(We denote the "dimension" of a magnitude by enclosing this magnitude in square brackets.) We can now readily check the earlier statement as to the dimensions of kinematic viscosity, $\nu = \mu/\varrho$: $[\nu] = L^2/T$.

We will now set ourselves the following problem. It is conjectured, that in a definite case, the resistance D, *i. e.* the force with which the fluid resists the movement of a body, can be represented by a product of powers of the form

$$D = (\text{number})\, l^\alpha\, v^\beta\, \varrho^\gamma\, \mu^\delta$$

The question is—which powers α, β, γ, δ are compatible with D having the correct dimensions. We simply write the formula

$$[D] = [l]^\alpha\,[v]^\beta\,[\varrho]^\gamma\,[\mu]^\delta$$

or

$$\frac{ML}{T^2} = (L)^\alpha \left(\frac{L}{T}\right)^\beta \left(\frac{M}{L^3}\right)^\gamma \left(\frac{M}{LT}\right)^\delta$$

The powers of indices of M, L, T on both sides of this equation must agree. Therefore

$$\begin{aligned} 1 &= \gamma + \delta \\ 1 &= \alpha + \beta - 3\gamma - \delta \\ -2 &= -\beta - \delta \end{aligned}$$

We see, as was to be expected from the previous considerations, that there are three equations and four unknowns. One of these unknowns remains free; we choose δ and, solving, obtain

$$\gamma = 1 - \delta$$
$$\alpha = 2 - \delta$$
$$\beta = 2 - \delta$$

We then have the formula

$$D = (\text{number})\, \varrho\, l^2 v^2 \left(\frac{\mu}{\varrho v l}\right)^\delta$$
$$= (\text{number})\, \varrho\, l^2 v^2 R^{-\delta}$$

We therefore obtain the product (density times area times square of velocity), which is already known to us from Newton's resistance formula, multiplied by an arbitrary power of the Reynolds number. This is a special case of the resistance formula of the preceding paragraph in which a function of the Reynolds number occurred, which the given facts could not further determine. That this function occurs here as (number) $R^{-\delta}$ is merely a consequence of the assumption that a law in the form of powers of l, v, ϱ and μ was expected. The general result of the theory of dimensions is the same as that of similitude, *viz.*

$$D = \varrho\, l^2 v^2 f(R).$$

The resistance laws which occur in practice can often be written as power series of the Reynolds numbers, at least for certain ranges of values. The above form shows that the viscosity is eliminated if δ is chosen equal to zero. This gives Newton's law of resistance. A value $\delta = 1$ on the other hand, makes the density vanish and furnishes the Stokes' law in which the inertia is neglected. At a later stage we shall find formulae for the resistance of a plate moving parallel to its own plane and we shall find, as long as the flow is laminar, that it corresponds to the value $\delta = 1/2$. In the case of the turbulent movement of such a plate we find—only as an approximate formula however—$\delta = 0.15$ to 0.20 for Reynolds numbers which are not too large. It is of great importance to know that only one of the four exponents may be freely chosen and that the others, as shown above, are thereby determined.

Other laws are found for the resistance of ships on the surface of the water, for in this case the acceleration due to gravity plays an important part. The relations become very obscure if the acceleration due to the earth is added as a fifth magnitude to the four so far considered. Most writers therefore avoid an exact study of the effect of viscosity and restrict themselves to very large Reynolds numbers where this, at least approximately, is permissible. We have again four magnitudes, *viz.*, length, velocity, density and acceleration due to the earth. Their dimensions L, L/T, M/L^3, L/T^2 show at once that the density which is the only one of the four in which the unit of mass occurs, is not involved in the dimensionless magnitude. This magnitude can be written as v^2/gl, as is easily verified, and is called the "Froude number" after the discoverer of this law of similarity. The resistance formula in this case then becomes $D = \varrho\, l^2 v^2 f\left(\frac{v^2}{g\,l}\right)$

This formula is in fact in good agreement with experiments on models at shipbuilding research laboratories. The system of waves caused by the ship proves to be very important for the resistance. Equal values of v^2/gl in geometrically similar body forms give geometrically similar systems of waves. The experiments must not however be made with models which are too small, for the differences in the Reynolds numbers become too great and produce large variations in that part of the resistance dependent on viscosity.

A few remarks may be added on the influence of compressibility. The compressibility can be always characterized by the velocity of sound in the medium. The dimensionless magnitude in question is then simply the ratio of the velocity to the velocity of sound. For the rest, things are less simple here, since a true similarity can only be expected when the laws connecting pressure and density are the same in the states of affairs which are compared, *e. g.*, in the case of the formula for adiabatic compression, $p\,V^{\gamma} = \text{constant}$,
if the value of γ is the same in both cases.

9. Properties of the Navier-Stokes' Differential Equations. The Navier-Stokes' differential equations (see **6**) have, in the case of an incompressible homogeneous fluid, which will be the only one here discussed, the following noteworthy properties.

If the velocity vector be derived from a potential φ, by putting

$$\boldsymbol{v} = grad\,\varphi\,,$$

which, in components, is written as

$$u = \frac{\partial \varphi}{\partial x}\,, \qquad v = \frac{\partial \varphi}{\partial y}\,, \qquad w = \frac{\partial \varphi}{\partial z}\,,$$

the equation of continuity

$$div\,\boldsymbol{v} = \frac{\partial u}{\partial x} + \frac{\partial v}{\partial y} + \frac{\partial w}{\partial z} = 0$$

becomes

$$\frac{\partial^2 \varphi}{\partial x^2} + \frac{\partial^2 \varphi}{\partial y^2} + \frac{\partial^2 \varphi}{\partial z^2} = 0$$

or, with the abbreviation already used

$$\nabla^2 \varphi = 0$$

This gives, however, $\nabla^2\, grad\,\varphi = grad\,\nabla^2 \varphi = 0$,

i. e. $\nabla^2 \boldsymbol{v} = 0$, or $\nabla^2 u = \nabla^2 v = \nabla^2 w = 0$.

We therefore see that in motions of this nature the terms which are multiplied by the viscosity μ vanish identically, which is as much as to say that the frictional forces which individually are certainly present are collectively in equilibrium as are, for example, the stresses inside an arbitrarily constrained elastic body. We see also that in the theory of ideal frictionless fluids, the well-known flows for which there are potentials, these being characterized by the absence of rotation of the individual elements of the fluid, will at the same time supply solutions of the Navier-Stokes' differential equations. Unfortunately, owing to another circumstance, the solutions so obtained are of little practical use in most cases. The difficulties arise from the boundary conditions. For the ideal frictionless fluid it is necessary to require only that the relative velocity perpendicular to the surface of a rigid body (or to a rigid wall) should vanish, while sliding along the surface is allowed. In the case of viscous fluids it is necessary to require, in order to be in accordance with experimental results, that also the tangential componens of the velocity at the surface of the body vanish; *i. e.* the outermost layer of fluid adheres completely to the body or to the wall. Those motions of the ideal frictionless fluid for which there is a potential are, however, already completely determined by the condition that the normal component of velocity vanishes, and there is no possibility of making the tangential components vanish in addition to the normal components unless the fluid as a whole is assumed to be at rest relative to the body. We thus see that we cannot in this way reach a solution for a case in which the fluid in any manner moves past the body. The reason for this peculiar behavior is that the order of the differential equations is decreased by the vanishing of the frictional terms, and the possibility of satisfying the boundary conditions is consequently lost. We must therefore conclude that the solutions of the Navier-Stokes' equations represent in general, fields of motion different from those where a potential is present; or in other words, that as a general rule we must reckon with rotations of the elements of the fluid. The knowledge that movements with a potential field are possible solutions remains valuable nevertheless, since we often obtain solutions which exhibit rotations only in the immediate neighborhood of the surface of the body, but are free from rotation at a distance from it. At these greater distances from the body we shall have therefore the same types of flow as are known for the case of the ideal fluid.

For two-dimensional motions, where

$$w = 0; \qquad u = u\,(x,\ y,\ t); \qquad v = v\,(x,\ y,\ t);$$

$$\frac{\partial u}{\partial x} + \frac{\partial v}{\partial y} = 0$$

an intuitive indication of the behavior of the rotation of an element can be given. This is the well-known

$$\omega = \frac{1}{2}\left(\frac{\partial v}{\partial x} - \frac{\partial u}{\partial y}\right)$$

We obtain a result containing this by differentiating the first Navier-Stokes' equation with respect to y, the second with respect to x, and then subtracting the first from the second. With due consideration to the equation of continuity this gives

$$\varrho\left(\frac{\partial\omega}{\partial t}+u\frac{\partial\omega}{\partial x}+v\frac{\partial\omega}{\partial y}\right)=\mu\nabla^2\omega,$$

where ∇^2 stands for $\partial^2/\partial x^2+\partial^2/\partial y^2$. The pressure p has been eliminated by this process. The form of this equation is exactly that of the equation for the distribution of temperature in a heat conducting medium, flowing with velocity components u and v. In order to convert one equation into the other we must replace the viscosity μ by the thermal conductivity, and the density ϱ by the heat capacity per unit volume (density times specific heat). Our idea of the solutions to be expected is obviously improved if we know that the rotation ω spreads in the fluid in exactly the same way as the temperature from some point that has been heated. We obtain a good picture of the flow if we imagine that we are considering the behavior of a hot body in a cold fluid flowing past it. If the flow is very slow the temperature will be able to spread in all directions; with a swifter flow however, only a relatively thin layer of fluid will be warmed before the fluid has passed the body. In all cases a band of higher temperature will be found in the stream beyond the body consisting of those fluid particles which have passed close to the body. Qualitatively, exactly the same can be said of the rotations. If the movement is very slow, rotation will be found everywhere in the fluid. For swifter movements however, the rotation is principally concentrated in the immediate neighborhood of the surface of the body and the particles of fluid at a greater distance remain free from rotation. This point of view is very important and will show us the way to an approximate treatment of rapid motions. But even for relatively slow motions (small Reynolds numbers) more exact investigation will show that at a distance from the body, which in most cases is not very great, the fluid moves without rotation, with the exception of those particles which have previously come close to the body and been set into rotation by it.

10. Some Examples of Exact Solutions. The problem of discovering exact solutions of the Navier-Stokes' differential equations presents in general insuperable difficulties, a condition principally due to the fact that in one part of the terms, the unknown quantities occur multiplied by themselves or by one another, so that the equation is no longer linear. It is therefore impossible to find further solutions by the superposition of special solutions. Nevertheless, in particularly simple cases it is possible to give exact solutions, either because the "quadratic terms" disappear by themselves as a result of the nature of the problem or else because the problem can be simplified by some suitable assumption.

(a) Solutions of the first kind include, for example, those for "laminar flows", for which only one velocity component, say u, is different from zero and depends only on the coordinates perpendicular to the direction of flow, *i. e.* $u = f(y, z, t), \quad v = w = 0$

The name laminar flow is inspired by the circumstance that for steady flows of this kind, all the particles of fluid which have the same value of u can be considered joined together as in a rigid surface. In accordance with the hypotheses these surfaces are general cylinders with axes parallel to the x axis, and the whole movement consists in the individual cylinders sliding over one another each with its own velocity. The word "lamina", as already noted, means approximately "thin sheet or foil". The picture, accordingly, is that of a number of such sheets spread out, or bent into concentric tubes, as the case may be, sliding over one another, whereby each individual sheet completely preserves its form. Continuity is automatically satisfied for such motions since u does not depend on x, and v and w vanish. The Navier-Stokes' equations, after removing the terms which vanish, appear simply as

$$\varrho \frac{\partial u}{\partial t} = \varrho X - \frac{\partial p}{\partial x} + \mu \left(\frac{\partial^2 u}{\partial y^2} + \frac{\partial^2 u}{\partial z^2} \right),$$

$$0 = \varrho Y - \frac{\partial p}{\partial y}$$

$$0 = \varrho Z - \frac{\partial p}{\partial z}$$

We shall assume that X, Y, Z are simply components of gravitation, so that X (component in the direction of the axes of the tubes) as well as the components perpendicular to it, Y and Z, are constant in space. If the first equation is differentiated with respect to x, all the terms except $-d^2p/dx^2$ become zero of themselves, so that this term must also vanish. Similarly by differentiating the second equation with respect to y, and the third with respect to z, we have

$$\frac{\partial^2 p}{\partial y^2} = \frac{\partial^2 p}{\partial z^2} = 0$$

It follows that p must be a linear function of the coordinates and that dp/dx is a constant which we can combine with ϱX to form a single constant. Especially simple is the steady flow for which $du/dt = 0$. Here we have simply,

$$\frac{\partial^2 u}{\partial y^2} + \frac{\partial^2 u}{\partial z^2} = \text{constant} \tag{10.1}$$

a differential equation which also occurs in the theory of the torsion of elastic bars of constant cross section. A membrane-analogy can also be applied here. If ξ be the elevation of a membrane stretched uniformly in all directions by a tension S, and held in equilibrium by an excess of pressure p on one side of the membrane, we obtain the following equation, if we assume the bending to be small:

$$\frac{\partial^2 \xi}{\partial y^2} + \frac{\partial^2 \xi}{\partial z^2} = \frac{p}{S} \tag{10.1a}$$

If we add the boundary conditions of the hydrodynamical problem, *viz*, $u = 0$ at the boundary, this means that the membrane is attached to a rigid boundary lying in the plane $\xi = 0$. The elevations ξ of the membrane above the plane of the boundary are then proportional to the required velocities. The solution for a circular boundary is no other than the "Hagen-Poiseuille flow" already discussed in **2**. The solution is very simple for the case of the equilateral triangle. Here ξ is proportional to the product of the distances of the point in question from the three sides of the triangle. The complicated calculations, which for the most part have been performed for the torsion problem, give the solutions in the case of the rectangle, semicircle, and some other cross sections.

The flow which develops in a circular cylinder at the commencement of the motion when a given difference of pressure is applied to the ends of the tube has only recently become known through a paper given by Herr Szymanski at the International Congress of Mechanics in Stockholm[1]. The diagram representing the results of his calculations is reproduced in Fig. 11. We recognize that under the influence of the difference of pressure the middle parts of the fluid exhibit a velocity increasing with the time but constant across the section, while the friction makes itself felt in the boundary regions by holding back the fluid. Later the middle part too becomes subject to the friction and as time passes, the profile approaches more and more to the parabola.

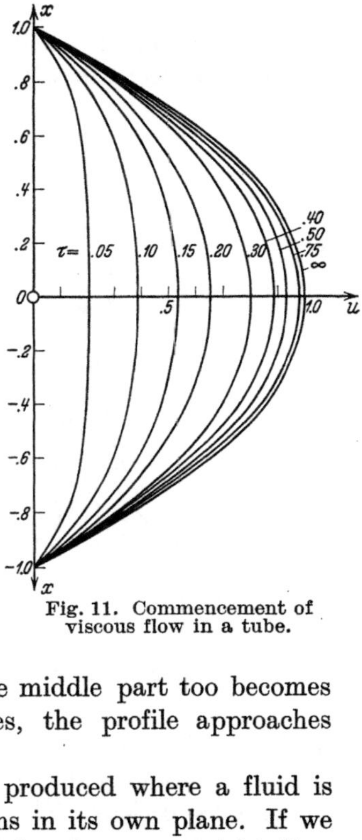

Fig. 11. Commencement of viscous flow in a tube.

(b) A very simple non-steady motion is produced where a fluid is bounded by a wall which executes oscillations in its own plane. If we assume that these oscillations take place only along a straight line, velocities will be induced only in the direction of this line. Taking this direction as the x axis, and a perpendicular to the plane as y axis, we obtain

$$u = u(y, t)$$

If we neglect gravity, no differences of pressure occur, and the differential equation is simply

$$\frac{\partial u}{\partial t} = \nu \frac{\partial^2 u}{\partial y^2} \tag{10.2}$$

(note the use of ν for μ/ϱ)

[1] Proceedings of the 3rd International Congress for Applied Mechanics, Stockholm, vol. I, p. 249, 1930.

A solution of this equation known for about the last hundred years in connection with the theory of heat conduction, is

$$u = A\, e^{-ky} \cos (nt - ky) \tag{10.3}$$

where in the present case k must be put equal to $\sqrt{n/2\nu}$. It can be easily verified, by differentiating the above equation, that the differential equation is in fact satisfied by this solution. It corresponds, as putting $y = 0$ shows, to a motion of the wall of amount

$$u_0 = A \cos nt$$

that is, a harmonic oscillation of the wall $y = 0$. The amplitude of the oscillations decrease with increasing distance from the wall, combined

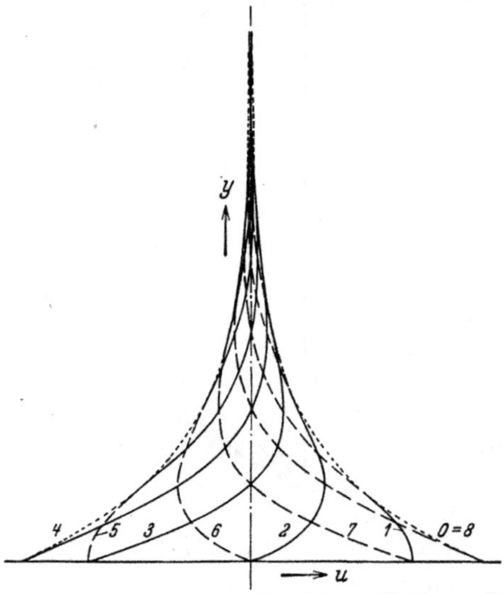

Fig. 12. Flow in the vicinity of an oscillating wall.

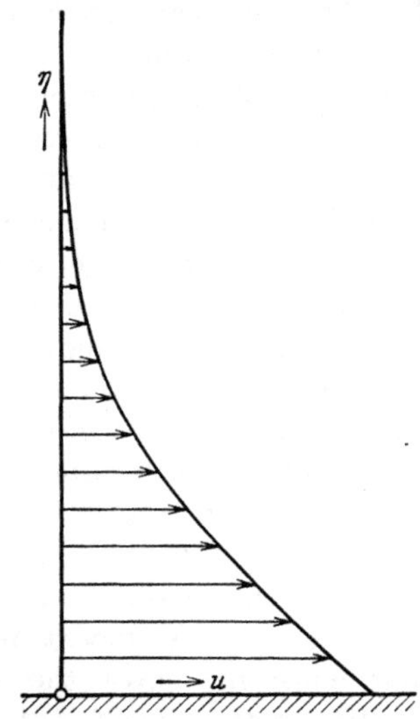

Fig. 13. Flow in the vicinity of a wall suddenly set in motion.

with a shift of phase such that the layers which are farther away lag in their motion behind the nearer ones. Fig. 12 represents the motion at various periods. The distance after which the same phase recurs (a kind of "wave-length") is $2\pi/k = 2\pi \sqrt{2\nu/n}$. The smaller the viscosity and the higher the frequency, the smaller is this distance.

Since the differential equation is linear it is at once permitted to superpose several oscillations with various frequencies and amplitudes, so that these formulae have also supplied the solution for arbitrary periodic motion of the wall parallel to its own plane.

Another solution of (10.2), also taken from the theory of heat conduction, is of interest. If an unbounded plane wall suddenly begins to move in its own plane with a constant velocity, the distribution of velocities is represented by the following function:

$$u = \frac{2 u_0}{\sqrt{\pi}} \int_{\eta}^{\infty} e^{-\eta^2}\, d\eta \tag{10.4}$$

where $\eta = y/2\sqrt{\nu t}$. The properties of this function are shown in Fig. 13. The thickness of the layer set in motion is therefore of the order $2\sqrt{\nu t}$. This result can also be found from dimensional analysis. The only combination of a kinematic viscosity and a time that gives a length is $\sqrt{\nu t}$, so that this expression occurs in all cases where a certain time t determines the state of motion. In the preceding example we might have introduced the period of oscillation T equal to $2\pi/n$ instead of the frequency n, thus giving the semi-wave-length

$$= \pi \sqrt{2\nu \cdot \frac{T}{2\pi}} = \sqrt{\pi \nu T}$$

It may be observed that both the above solutions can be easily rewritten for the case in which the wall is at rest and the fluid performs oscillations parallel to the wall, or where at the instant $t = 0$, it commences to move with velocity u_0. A velocity must then be added to the whole of the mass of fluid so that the points at $y = 0$ are at rest.

(c) A different type of solution is given by putting $u = -f(x)$, $v = y f'(x)$.

The condition of continuity is thus automatically satisfied. Further light on the character of this motion is gained by considering a potential flow which satisfies this scheme as given by putting $f(x) = a x$. We then have

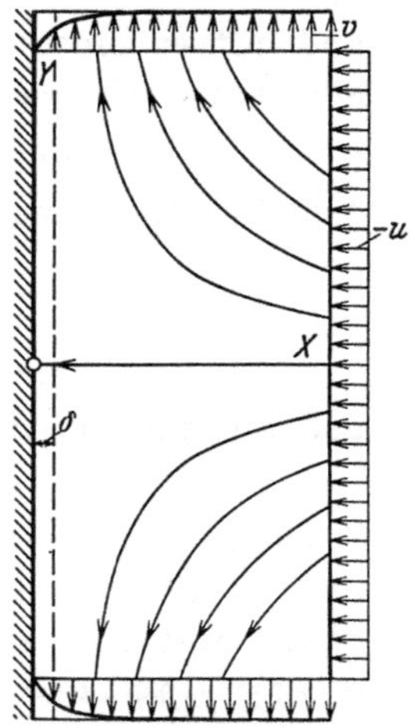

Fig. 14. Flow in the vicinity of a stagnation point.

$$u = -a x, \qquad v = a y,$$

which is the well-known flow at a stagnation point. The corresponding pressure is

$$p = p_1 - \varrho \frac{a^2}{2}(x^2 + y^2)$$

as given immediately by the Bernoulli equation. We will now consider the problem of finding the flow in the vicinity of a stagnation point for a viscous fluid having the plane $x = 0$ as a fixed wall with $u = v = 0$ at the wall, but such that at large distances from it the motion is essentially the same as that given by the previous potential flow. Fig. 14 gives a picture of this motion. The boundary conditions are obviously as follows: For $x = 0$ we must have $f(x) = 0$, and $f'(x) = 0$; and for $x = \infty$ let $f'(x)$ be equal to a. The pressure will be written as

$$p = p_0 - \frac{1}{2}\varrho a^2 [F(x) + y^2]$$

With these values the Navier-Stokes' equations give

$$f f' + 0 = \frac{a^2}{2} F'(x) - \nu f'' \tag{10.5}$$

$$y(-f f'' + f'^2) = a^2 y + \nu y f''' \tag{10.6}$$

From the first of these equations we can find the unknown function $F(x)$, which was introduced into the expression for the pressure p. The equation which is of importance for the problem is the second, in which it can at once be seen that y can be cancelled throughout. Let the differential equation which remains be transformed so that all the numerical factors of the various terms become equal to unity. This is accomplished by taking as independent variable $\xi = \alpha x$, where α is to be determined. We write $f(x) = A\,\varphi(\xi)$

and further denote differentiation with respect to ξ by a prime. Substitution in the second equation gives

$$A^2 \alpha^2 (-\varphi\varphi'' + \varphi'^2) = a^2 + \nu A \alpha^3 \varphi'''$$

Equating the three factors

$$A^2 \alpha^2 = a^2 = \nu A \alpha^3,$$

we have

$$A = \sqrt{\nu a}, \qquad \alpha = \sqrt{\frac{a}{\nu}}$$

and the differential equation becomes

$$\varphi''' + \varphi\varphi'' - \varphi'^2 + 1 = 0 \tag{10.7}$$

For large values of ξ, for which the frictional effects have disappeared, φ'' and φ''' vanish, so that φ' becomes equal to ± 1. By hypothesis $\partial f/\partial x$ must here be equal to a. But according to our assumptions $\partial f/\partial x = A\,\alpha\varphi'$, and $A\,\alpha = a$, so that we see that the boundary condition for $\xi = \infty$ is correct if we take the positive sign for φ'. But with $\varphi = 0$, $\varphi' = 0$ for $\xi = 0$ and $\varphi' = 1$ for $\xi = \infty$, the function $f(\xi)$ is completely determined. We shall consider the mathematical treatment of this problem later in **15**. The function φ' which is of especial interest to us here, since it gives the sliding velocity at the wall (v) is shown in Fig. 23 by the heavier of the two curves (in this diagram f_1' stands for φ' and η for ξ. We see that for ξ approximately equal to 2.6, φ' becomes practically constant[1]. This means a distance of

$$x = \frac{2.6}{\alpha} = 2.6\sqrt{\frac{\nu}{a}}$$

This solution is valid, and this point must be emphasized, only for the case here considered, in which the flow is directed on to the plate and in which it splits up into two streams in opposite directions. If the velocities are reversed in direction so that the flow comes from both sides and then leaves the plate, the above solution does *not* hold. With this reversal the convection and pressure terms of (10.6) preserve their sign, and only the sign of the term with ν changes. We therefore obtain a differential equation for φ' in which the φ''' term has the opposite sign, but the other signs remain the same. The differential equation

[1] The error is about 1/2 per cent.

so modified has however no solution compatible with the given boundary conditions. Hence the motion in one direction has a solution capable of physical interpretation, but the one in the opposite direction has not.

From the last three examples we obtain the uniform rule that the thickness of the layer in which the frictional effects take place is proportional to $\sqrt{\nu}$, and this layer therefore becomes very thin for small kinematic viscosities. All rotational motions take place in this layer. Outside it the motion is free from rotation.

It may be further noted that the case of a uniformly rotating plane disk with axis of rotation perpendicular to the disk is also capable of an exact solution similar to those preceding. A calculation of this kind has been carried through by Th. von Kármán[1].

(d) The following group of solutions is also worthy of note: Let a plane flow be assumed for which all the stream-lines are radii intersecting in a point. Let the velocities on the various radii be different, and therefore a function of the polar angle. A radius along which the velocity is zero then represents a radial wall. Continuity is satisfied if the velocity along one and the same radius is made inversely proportional to the distance from the center. If u be the radial velocity, we must then have

$$u = \frac{1}{r} f(\theta)$$

A transformation of the Navier-Stokes' equations to polar coordinates and introducing the previous assumption gives, as we may briefly observe, the differential equation

$$f^2 + k + \nu (4f + f'') = 0 \tag{10.8}$$

which may be solved by elliptic integrals (k is a parameter connected with the pressure gradient). The solutions have been given by G. Hamel[2]. Hamel incidentally shows that it is possible, without considerable additional difficulty, to calculate flows in which the stream-lines are logarithmic spirals. We shall not enter into a discussion of the details here. It may be observed however, that for flows directed toward the center, that is, accelerated flows, a reasonable solution can always be obtained. These motions are of such a nature that for small viscosity the flow in the interior of the field coincides with that for the simple case of flow towards a sink ($u = const./r$), but in the neighborhood of a radial wall becomes zero. On the other hand, for outward flow, which is of the retarded type, solutions capable of a practical interpretation can be given only for relatively narrow wedge-shaped channels with an angle of wedge of the order of magnitude $\sqrt{\nu/f_{max}}$. Here again the

[1] Zeitschr. f. angew. Math. u. Mech., vol. I, p. 244, 1921.

[2] Hamel, G., Jahresbericht der Deutsch. Mathematiker-Vereinigung, 25, 1916, also Tollmien, Handbuch d. Exp.-Phys. IV, 1, p. 257.

phenomenon of the previous example reappears whereby practical solutions without restriction exist for one direction but not for the other.

(e) As a last example we may discuss the dissolution of a vortex filament through fluid friction. The stream-lines are here concentric circles. This problem can be handled best by the aid of the heat-conduction analogy. A vortex-filament concentrated along a line corresponds to a warm point in a conducting flat plane; the constant circulation of the vortex corresponds therefore to a constant quantity of heat. The analogous problem can be stated as follows: At time $t = 0$ a definite quantity of heat is brought to a point of the plane—*e. g.* by bringing a hot body into contact with it for a moment—and this quantity of heat now spreads through the sheet by conduction. The solution is known. Let ω be the element of rotation, r the distance from the axis of the vortex and t the time from the commencement of the rotation; the distribution of ω is then given by

$$\omega = \frac{A}{t} e^{-r^2/4\nu t} \tag{10.9}$$ [1]

The circulation around a circle of radius r, according to Stokes' theorem, is equal to double the surface integral of ω inside the circle. Putting the surface element equal to $2\pi r\,dr$, the circulation becomes

$$\Gamma = 4\pi \int_0^r \omega\, r\, dr$$
$$= 8\pi\nu A\,(1 - e^{-r^2/4\nu t})$$

The factor A can be expressed in terms of $\Gamma_1 = 8\pi\nu A$, the circulation around an infinitely large circle. The final result is

$$\Gamma = \Gamma_1 (1 - e^{-r^2/4\nu t}) \tag{10.10}$$

The velocity is however obtained from the circulation by dividing by $2\pi r$, so that

$$u = \frac{\Gamma_1}{2\pi r}(1 - e^{-r^2/4\nu t}) \tag{10.11}$$

[1] A verification of this solution can be made as follows: First

$$\nabla^2 \omega = \frac{\partial^2 \omega}{\partial x^2} + \frac{\partial^2 \omega}{\partial y^2},$$

is to be calculated on the assumption that ω depends only on $r = \sqrt{x^2 + y^2}$. This gives

$$\nabla^2 \omega = \frac{\partial^2 \omega}{\partial r^2} + \frac{1}{r}\frac{\partial \omega}{\partial r}$$

We must now have $$\frac{\partial \omega}{\partial t} = \nu \nabla^2 \omega.$$

But we find $$\frac{\partial \omega}{\partial t} = \left(-\frac{A}{t^2} + \frac{A r^2}{4\nu t^3}\right) e^{-r^2/4\nu t}$$

and further $$\frac{\partial \omega}{\partial r} = -\frac{A r}{2\nu t^2} e^{-r^2/4\nu t}$$

Also $$\frac{\partial^2 \omega}{\partial r^2} = \left(-\frac{A}{2\nu t^2} + \frac{A r^2}{4\nu^2 t^3}\right) e^{-r^2/4\nu t}$$

so that the differential equation is really satisfied.

The function $e^{-r^2/4\nu t}$ tends to zero very rapidly for large arguments so that constant circulation, *i. e.* the usual circulatory motion with a potential is obtained. But in the region where $e^{-r^2/4\nu t}$ is different from zero there are rotational motions and the velocities are smaller than in the potential flow. At different times the radial distance at which the velocity is some definite fraction of the velocity in the potential case, *e. g.* 80 per cent of the velocity, is therefore again proportional to $\sqrt{\nu t}$, in accordance with the earlier discussion of the dimensions of the quantities involved. For small radial distances the exponential function can be expanded in a series

$$e^{-r^2/4\nu t} = 1 - \frac{r^2}{4\nu t} + \frac{1}{2}\frac{r^4}{(4\nu t)^2} - \dots$$

and so we obtain
$$u = \frac{\Gamma_1}{2\pi}\left(\frac{r}{4\nu t} - \frac{r^3}{2(4\nu t)^2} + \dots\right) \tag{10.12}$$

The velocity therefore tends to zero on approaching the axis, as must be the case. On the axis, (10.9) gives

$$\omega = \omega_0 = \frac{A}{t} = \frac{\Gamma_1}{8\pi\nu t}$$

and naturally for very small radii we must have $u = \omega_0 r$. We see from (10.12) that this is the case. Altogether we see from (10.11) that the circulatory motion at any radial distance which is not too small remains unaltered until the diffusion of the vortex core reaches the point in question, after which the velocity gradually becomes less. Fig. 15 gives the velocity as a function of the radial distance for various consecutive intervals of time.

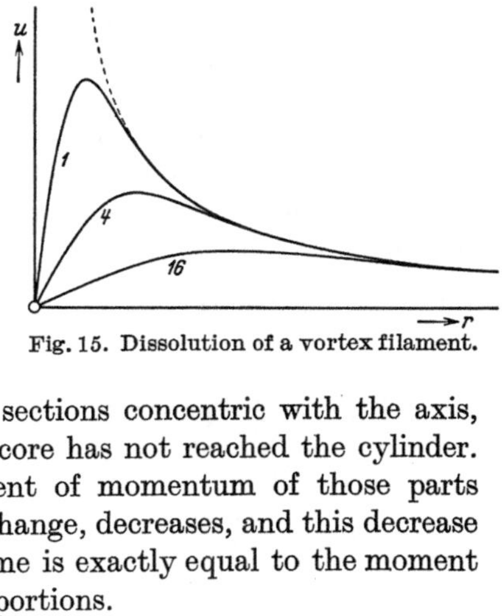

Fig. 15. Dissolution of a vortex filament.

If the frictional forces are investigated, the same constant moment of rotation is found in all cylindrical sections concentric with the axis, so long as the diffusion of the vortex core has not reached the cylinder. As the vortex core grows, the moment of momentum of those parts of the fluid which are affected by the change, decreases, and this decrease in the angular momentum per unit time is exactly equal to the moment of the frictional forces in the outer portions.

11. Theory of Very Slow Motions. The exact solutions given in the preceding paragraph cover only a set of rather special cases; most of them deal with flow along a plane wall, or otherwise, the form of the stream-lines is given as some very simple curve. The majority of the problems which arise in practice cannot, at present, be solved in this way, and must be attacked by approximate methods. Such approximations are possible:

1) In the case of a predominantly large viscosity where the effects of inertia can be entirely neglected.

2) In the case of very small viscosities if the deviations from the motion of frictionless fluids are restricted to regions in the neighborhood of the wall or fixed body respectively, in which case the friction need be considered only in these limited regions, thus leading to certain simplifications in the calculations.

In both cases it is fundamentally possible to obtain at a later stage approximate values for the quantities neglected in the first calculations, and hence to improve the approximation. This, however, makes the discussion in general so complicated that such refinements are commonly omitted. This method therefore leaves the whole of the middle region of the Reynolds numbers without solutions. When the flow drags particles of rotating fluid from the neighborhood of the wall or solid body into the interior of the fluid, as very often happens, we have no useful solutions, even for large Reynolds numbers, except for a few cases where this occurs in very regular fashion.

For the very smallest Reynolds numbers (*i. e.* very small velocities and dimensions of the body or very great viscosities, respectively) the frictional terms outweigh the inertial terms in such degree that the latter can be completely neglected. In the case of steady flow this can be seen very simply from the Navier-Stokes' equations. The inertial terms contain the velocities to the second order, the frictional terms to the first, so that as the velocity diminishes, the inertial terms decrease more rapidly than the frictional terms. Otherwise it is obvious that the frictional terms predominate if the viscosity increases sufficiently. If the terms in which the density ϱ occurs are therefore removed, we obtain the following equations

$$\mathit{grad}\ p = \mu \nabla^2 \boldsymbol{v}, \tag{11.1}$$

$$\mathit{div}\ \boldsymbol{v} = 0 \tag{11.2}$$

or, written in components

$$\left.\begin{aligned} \frac{\partial p}{\partial x} &= \mu \nabla^2 u \\ \frac{\partial p}{\partial y} &= \mu \nabla^2 v \\ \frac{\partial p}{\partial z} &= \mu \nabla^2 w \end{aligned}\right\} \tag{11.1a}$$

$$\frac{\partial u}{\partial x} + \frac{\partial v}{\partial y} + \frac{\partial w}{\partial z} = 0 \tag{11.2a}$$

where, as before

$$\nabla^2 = \frac{\partial^2}{\partial x^2} + \frac{\partial^2}{\partial y^2} + \frac{\partial^2}{\partial z^2}$$

Taking the divergence of (11.1) or, in other words, differentiating the first of (11.1a) with respect to x, the second with respect to y, and the third with respect to z, and adding the three equations, it follows from (11.2a), since the operators ∇^2 and *div* can be interchanged, that

$$div\ grad\ p = \nabla^2 p = 0 \tag{11.3}$$

Functions of space coordinates whose ∇^2 derivative vanishes are studied in great detail in the theory of electrostatic and gravitational potential. There is a great number of such functions and appropriate choices can be taken as bases for calculating many types of flow.

We shall here deal with the simplest of such cases—the passage of a sphere through a viscous fluid. The solution was given by Stokes[1]. It is to be expected that the pressure on the side where the fluid flows toward the sphere (negative x) will be greater, and on the side where it flows away from it, less, than the undisturbed pressure. The pressure disturbances must die away at infinity. The simplest function satisfying (11.3) and having the required properties is

$$p = -\frac{A\,x}{r^3} \tag{11.4}$$

where [2]

$$r = \sqrt{x^2 + y^2 + z^2}$$

We will now look for a flow of viscous fluid for which the pressure satisfies (11.4) thus furnishing certain information regarding $\nabla^2 u$, $\nabla^2 v$ and $\nabla^2 w$ through the use of (11.1a).

We have,

$$\left.\begin{aligned} \nabla^2 u &= \frac{A}{\mu}\left(\frac{3\,x^2}{r^5} - \frac{1}{r^3}\right) \\ \nabla^2 v &= \frac{A}{\mu}\left(\frac{3\,x\,y}{r^5}\right) \\ \nabla^2 w &= \frac{A}{\mu}\left(\frac{3\,x\,z}{r^5}\right) \end{aligned}\right\} \tag{11.5}$$

We will first verify the symmetry of the three velocity components which involves the symmetry of their ∇^2 derivatives. It can be seen from Fig. 16 that u, starting as the undisturbed velocity tends to zero as the sphere is reached, and then on the other side, starts at zero and increases to the undisturbed velocity. Hence u must be an even function of x, and also of y and z, since the flow must be symmetrical about the plane of yz and also about the axis of x. The velocity component v must be even in z, but odd in x and y (for positive y it is positive on the

[1] Stokes, Camb. Trans., vol. IX, 1851, or Collected Papers, vol. III, p. 55.

[2] The simplest solution of (11.3) corresponds to the electric potential of a point charge and is $1/r$. All partial differential coefficients of $1/r$ with respect to the coordinates are further solutions. If it is observed when differentiating that

$$\frac{\partial r}{\partial x} = \frac{x}{r}, \text{ etc.}$$

it is found that

$$\frac{\partial}{\partial x}\left(\frac{1}{r}\right) = -\frac{x}{r^3}$$

which corresponds to the above expression.

inflow side, negative on the outflow side, and exactly reversed for negative y). The same holds for w, except that y and z must be interchanged. It is seen that these requirements of symmetry are exactly satisfied by (11.5).

We have now to integrate these equations. The most practical method is to experiment with the ∇^2 derivatives of several expressions having the required properties of symmetry. The number of suitable expressions is not at all extensive, if it be observed that the ∇^2 derivation changes the dimension of the expressions the same as division by r^2. The expressions in (11.5) are all of dimension $1/r^3$. We must therefore consider expressions of dimension $1/r$. The expression $1/r$ cannot be used, since its ∇^2 derivative vanishes. The next simplest expression having the required symmetry for the component u is x^2/r^3. This gives

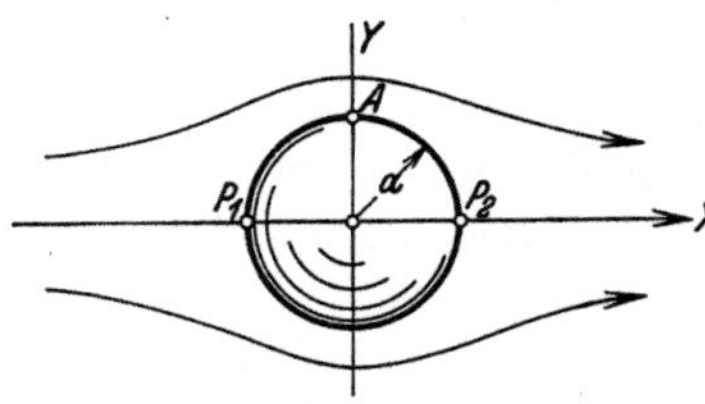

Fig. 16.

$$\nabla^2\left(\frac{x^2}{r^3}\right)=\frac{2}{r^3}-\frac{6\,x^2}{r^5},$$

and we at once find that this expression, when multiplied by $-A/2\mu$ supplies exactly the $\nabla^2 u$ of (11.5). An expression with the symmetries required for v is $x\,y/r^3$. Calculation gives

$$\nabla^2\left(\frac{x\,y}{r^3}\right)=-\frac{6\,x\,y}{r^5}$$

Here too we must multiply by $-A/2\mu$ to obtain the value of $\nabla^2 v$. A corresponding result holds for $\nabla^2 w$. We must now further demand of this solution, which with the expressions chosen satisfies (11.1), that it satisfy also (11.2), and further, that on the surface of a definite sphere all these velocity components vanish. The terms so far chosen will not do this and we must therefore add more terms in such a way that (11.1) will still be satisfied. This will be the case if we add terms whose ∇^2 derivatives vanish identically. Bearing in mind the symmetry of u we may add a constant for this component, and the following functions multiplied by certain constants:

$$\frac{1}{r}, \quad \text{and} \quad \frac{\partial^2}{\partial x^2}\left(\frac{1}{r}\right)=\frac{3\,x^2}{r^5}-\frac{1}{r^3}$$

For v we may add the following function multiplied by a constant:

$$\frac{\partial^2}{\partial x\,\partial y}\left(\frac{1}{r}\right)=\frac{3\,x\,y}{r^5}$$

and for w similarly

$$\frac{\partial^2}{\partial x\,\partial z}\left(\frac{1}{r}\right)=\frac{3\,x\,z}{r^5}$$

It happens that these functions are the same as those occurring in (11.5) and we therefore deduce that $\nabla^2\cdot\nabla^2 u=0$, and similarly $\nabla^2\cdot\nabla^2 v=\nabla^2\cdot\nabla^2 w=0$. This is a relation which can be seen to hold in general by writing out the ∇^2 derivatives of (11.1) or (11.1a), respectively. After these preliminaries we can write

$$\left.\begin{aligned} u &= -\frac{A}{2\mu}\frac{x^2}{r^3} + u_0 + \frac{B}{r} + C\left(\frac{3x^2}{r^5} - \frac{1}{r^3}\right) \\ v &= -\frac{A}{2\mu}\frac{xy}{r^3} + 3D\frac{xy}{r^5} \\ w &= -\frac{A}{2\mu}\frac{xz}{r^3} + 3D\frac{xz}{r^5} \end{aligned}\right\} \quad (11.6)$$

The first term is in each case that which satisfies (11.5), the rest are the additions previously mentioned.

We first evaluate $\partial u/\partial x, \partial v/\partial y, \partial w/\partial z$, and add them together. This gives

$$-\frac{A}{2\mu}\frac{x}{r^3} - \frac{Bx}{r^3} + (9C + 6D)\frac{x}{r^5} - [15Cx^2 + 15D(y^2+z^2)]\frac{x}{r^7} \equiv 0.$$

Putting $D = C$ the last term simplifies to $-15C\,x/r^5$ and cancels with the term previous to it. Further if B is put equal to $-A/2\mu$ the condition of continuity is satisfied for all values of x and r, and we have surely obtained a solution of the simplified Navier-Stokes' equations. We must now make the velocities vanish for a given radius $r = a$. It is more convenient to consider the components v or w before considering u. We see at once that v and w both vanish for $r = a$ if we put $D = A a^2/6\mu$ and from above we get the same value for C.

Inserting the values of B and C in the first of (11.6) and putting r equal to a, it is seen that the terms containing x^2 vanish identically. The terms not containing the factor x^2 supply the connection between the constant A, hitherto undetermined, and the velocity at infinity u_0.

For A we obtain $$A = \frac{3}{2}\mu a u_0.$$

We have in fact succeeded in this case in finding a solution for the sphere in a viscous fluid. The velocity components and the pressure can now be written down explicitly, involving both the radius of the sphere and the velocity at infinity u_0:

$$\left.\begin{aligned} u &= u_0\left[\frac{3}{4}\frac{ax^2}{r^3}\left(\frac{a^2}{r^2} - 1\right) + 1 - \frac{1}{4}\frac{a}{r}\left(3 + \frac{a^2}{r^2}\right)\right] \\ v &= u_0\frac{3}{4}\frac{axy}{r^3}\left(\frac{a^2}{r^2} - 1\right) \\ w &= u_0\frac{3}{4}\frac{axz}{r^3}\left(\frac{a^2}{r^2} - 1\right) \end{aligned}\right\} \quad (11.7)$$

$$p = -\frac{3}{2}\frac{\mu u_0 a x}{r^3} \quad (11.8)$$

It is now possible to calculate the resistance. This is a compound of the difference of pressure and the frictional forces. The pressure on the surface of the sphere is $$p = -\frac{3}{2}\frac{\mu u_0 x}{a^2}$$

The pressure is respectively greatest and least at the points P_1 and P_2 in Fig. 16 where $x = \pm a$. At these points which we will call the "poles"

$$p = \pm \frac{3}{2} \frac{\mu u_0}{a}$$

There are no frictional forces at these two points since here, as can easily be verified

$$\frac{\partial u}{\partial x} = \frac{\partial v}{\partial y} = \frac{\partial w}{\partial z} = 0$$

(the other differential coefficients of the first order are already zero on the axis on account of symmetry). We do however obtain frictional forces at other points on the sphere. These are greatest at the "equator", *i. e.* at $x = 0$. We will calculate $\partial u/\partial y$ for the point $x = 0$, $y = a$, $z = 0$ (A in Fig. 16). We write $x = 0$ and $r = y$ before differentiating and obtain

$$u = u_0 \left(1 - \frac{3}{4} \frac{a}{y} - \frac{1}{4} \frac{a^3}{y^3}\right)$$

so that

$$\frac{\partial u}{\partial y} = u_0 \left(\frac{3}{4} \frac{a}{y^2} + \frac{3}{4} \frac{a^3}{y^4}\right)$$

Putting $y = a$ this gives $\partial u/\partial y = 3/2\, u_0/a$ and the shear at the equator is therefore

$$\tau = \frac{3}{2} \frac{\mu u_0}{a}$$

This value is exactly equal to the surplus or deficit in pressure, respectively, at the poles. An easy calculation shows that the resultant of the pressure differences relative to the pressure in the undisturbed fluid, and of the frictional forces, has the same magnitude at all other points on the surface and is always in the direction of the x axis. Hence the whole resistance is obtained by multiplying the given force per unit area of the surface of the sphere by $4 \pi a^2$. This gives

$$D = 6 \pi \mu u_0 a \tag{11.9}$$

This formula is well-known as "Stokes' Formula". It may be noted that one third of this resistance arises from the differences in pressure and two thirds from frictional forces.

Corresponding calculations can be effected also for the ellipsoid. Instead of commencing with the potential function $1/r$ (potential of a sphere) we start with the gravitation potential of an ellipsoid of uniform mass. For the calculations, which are rather complicated, the reader may be referred to Lamb's Hydrodynamics § 339. The case of a plane circular disc which is a special case of the ellipsoid is of great interest. The resistance in that case is, according to Oberbeck[1],

$$D = 16 \mu u_0 a \tag{11.10}$$

where a is the radius of the disc.

[1] Oberbeck, Über stationäre Flüssigkeitsbewegungen mit Berücksichtigung der inneren Reibung. Crelles' Journal, vol. 81, p. 62, 1876.

By going to the limit, the case of the infinitely long cylinder with axis perpendicular to the direction of flow can be derived from the ellipsoid. The flow is two-dimensional and can also be directly calculated as in the case of the sphere. It is necessary to start with the two-dimensional potential $log\ r$ instead of $1/r$, where $r = \sqrt{x^2 + y^2}$. The result for the pressure is here $p = -\frac{A\,x}{r^2}$

This gives the following formulae for the velocity when the cylinder has a radius a,

$$\left.\begin{aligned} u &= \frac{A}{4\,\mu}\left[2\,log\,\frac{r}{a} + \left(1 - \frac{a^2}{r^2}\right)\left(1 - \frac{x^2}{r^2}\right)\right] \\ v &= -\frac{A}{2\,\mu}\left(1 - \frac{a^2}{r^2}\right)\frac{x\,y}{r^2} \end{aligned}\right\} \qquad (11.11)$$

The difficulty arises however that the velocity at a great distance becomes

$$u_0 \sim \frac{A}{2\,\mu}\,log\,\frac{r}{a} \qquad (11.11\,a)$$

i. e., it is not constant but becomes logarithmically infinite for large distances, so that we are not in a position to connect the flow with a definite velocity of the undisturbed fluid. This is due to the fact that the frictional effects due to the infinitely long cylinder also exert an influence infinitely far at right angles to the axis. The problem only becomes determinate when the fluid is bounded by a fixed wall or the like.

The resistance can be calculated as for the sphere. Once again the shear τ at the point A of Fig. 16 is as great as the excess or deficit respectively, of pressure at the points P_1 and P_2, *viz.* A/a. For a part of the cylinder of length l, the surface is equal to $2\,\pi\,a\,l$, and therefore the resistance, which is equal to the surface area times the force per unit area, can be written as

$$D = 2\,\pi\,A\,l \qquad (11.12)$$

12. Oseen's Improvement of the Preceding Theory. In order to improve the solution of the preceding section, the method of successive approximation obviously suggests itself. Using the values of the velocities obtained, the neglected inertial terms can be calculated, thus giving the modifications of the velocities which are necessary in order to bring these inertial terms (small by hypothesis) into agreement with the pressures and viscosity forces. Such a refinement would be specially desirable in the case of the infinitely long cylinder where the results are not very satisfactory. It is found however, that the attempt to calculate these corrections fails because at some distance from the sphere or cylinder respectively the corrections become larger than the first approximations with which the process starts—in other words the process is divergent. This shows that something must be wrong with the theory so far developed. The cause of the trouble was discovered by Oseen who also gave a method for improving the theory. Briefly the situation is as follows:

If the orders of magnitude of the viscosity terms and the inertial terms at large distances from the disturbing object are considered, the results are, for example for the sphere, of an order of magnitude $\mu \frac{u_0 a}{r^3}$ for the viscosity terms (of the type $\mu \nabla^2 u$) and $\varrho \frac{u_0^2 a}{r^2}$ for the inertial terms $\left(\text{of type } \varrho u \frac{\partial u}{\partial x}\right)$. The ratio of the inertial terms to the viscosity terms, which in Stokes' calculations was assumed to be negligibly small, appears as

$$\frac{\varrho u_0 r}{\mu} = \frac{u_0 r}{\nu},$$

that is, it equals a Reynolds number containing the radius r. It is clear that for increasing r this ratio can become arbitrarily large, even though the Reynolds number of the sphere $u_0 a/\nu$ is very small, so that in every case there is a distance at which the postulate of the theory, *viz.* that the inertial terms can be neglected in comparison with the viscosity terms, is no longer correct. The theory must therefore be completed by a development of the inverse kind, *viz.* a development which starts based on the preponderating effect of the inertial terms. This expansion then furnishes a different state of flow at large distances from the body and it will thus be possible to clear up completely the paradoxical behavior of the infinitely long cylinder.

We must first establish the following facts: at a very great distance from the body, we have the undisturbed velocity u_0; at finite but large distances from the body, small disturbing velocities u', v', w' are added to u_0. The inertial terms $u\,(\partial u/\partial x)$, $v\,(\partial u/\partial y)$, $w\,(\partial u/\partial z)$ after writing $u = u_0 + u'$, $v = v'$, $w = w'$ can be split into two groups of different orders of magnitude. One group with the factor u_0, *e. g.* $u_0\,(\partial u'/\partial x)$ contains small velocities of the first order, the second group, *e. g.* $u'\,(\partial u'/\partial x)$ or $v'\,(\partial v'/\partial y)$ etc., contains small velocities in both factors, and is therefore of the second order. The idea of the approximation due to C. W. Oseen[1] is to consider only terms of the first order, whereby a linear problem is obtained for which mathematical solutions are ready. The solutions obtained in this way are obviously approximations for a region far from the body where in fact the disturbing velocities due to the body are small compared with u_0. In the special case under consideration, where in the neighborhood of the body the viscosity terms predominate, and, at distances where this is not the case the velocities added to u_0 are already sufficiently small, the result can be allowed to hold for the whole region. This is so since in the interior region the neglected inertial terms are of the second order relative to the viscosity terms, and in the exterior region they are small relative to the inertial

[1] Über die Stokes'sche Formel und über eine verwandte Aufgabe in der Hydrodynamik. Arkiv för matematik, astronomi och fysik, Bd. 6, No. 29, 1910.

terms of the first order. The results so obtained can naturally be used only for Reynolds numbers $u a/\nu$ which are small relative to unity. The additions we have mentioned change (11.1a) as follows:

$$\left.\begin{aligned} \varrho u_0 \frac{\partial u'}{\partial x} + \frac{\partial p}{\partial x} &= \mu \nabla^2 u' \\ \varrho u_0 \frac{\partial v'}{\partial x} + \frac{\partial p}{\partial y} &= \mu \nabla^2 v' \\ \varrho u_0 \frac{\partial w'}{\partial x} + \frac{\partial p}{\partial z} &= \mu \nabla^2 w' \end{aligned}\right\} \tag{12.1}$$

the equation of continuity is as before

$$\frac{\partial u'}{\partial x} + \frac{\partial v'}{\partial y} + \frac{\partial w'}{\partial z} = 0 \tag{12.2}$$

Solutions of these equations can be obtained by the following method due to H. Lamb[1].

The solution is composed of two parts

$$u' = u_1 + u_2,$$
$$v' = v_1 + v_2,$$
$$w' = w_1 + w_2,$$

where u_1, v_1, w_1 are derived from a potential

$$u_1 = \frac{\partial \varphi}{\partial x}, \qquad v_1 = \frac{\partial \varphi}{\partial y}, \qquad w_1 = \frac{\partial \varphi}{\partial z} \tag{12.3}$$

and are chosen so that they exactly balance the differences of pressure. By constructing the divergence of (12.1) and by using (12.2), we can show that here also, $\nabla^2 p = 0$. If we put

$$\nabla^2 \varphi = 0, \tag{12.4}$$

the frictional terms $\mu \nabla^2 u_1$ etc., vanish, and on the other hand

$$p = \varrho u_0 \frac{\partial \varphi}{d x} \tag{12.5}$$

so that $\nabla^2 p = 0$, and equations (12.1) are therefore also satisfied.

The portions u_2, v_2, w_2 must now serve to balance the frictional terms $\mu \nabla^2 u_2$ etc. with the inertial forces $\varrho u_0 (\partial u_2/\partial x)$. A solution is obtained by using the relations,

$$\left.\begin{aligned} u_2 &= \frac{\partial \chi}{\partial x} - \frac{u_0}{\nu} \chi, \\ v_2 &= \frac{\partial \chi}{\partial y} \\ w_2 &= \frac{\partial \chi}{\partial z} \end{aligned}\right\} \tag{12.6}$$

where χ satisfies the differential equation

$$\frac{\partial \chi}{\partial x} = \frac{\nu}{u_0} \nabla^2 \chi \tag{12.7}$$

---

[1] On the Uniform Motion of a Sphere Through a Viscous Fluid. Phil. Mag. (6) 2, p. 112, 1911.

Substituting the values given by (12.6) into (12.2), we see that continuity is satisfied on account of (12.7), and equations (12.1) are also satisfied if, as assumed, no pressure is involved in the flow u_2, v_2, w_2. We have now to find solutions for the function χ. The simplest is

$$\chi = \frac{A}{r} e^{-k(r-x)} \tag{12.8}$$

with $$r = \sqrt{x^2 + y^2 + z^2}$$

Differentiating we find $$\frac{\partial \chi}{\partial x} = \frac{A}{r^2}\left[k(r-x) - \frac{x}{r}\right] e^{-k(r-x)}$$

$$\nabla^2 \chi = \frac{2Ak}{r^2}\left[k(r-x) - \frac{x}{r}\right] e^{-k(r-x)}$$

and hence, if (12.7) is to be satisfied, $k = u_0/2\nu$. The function χ must be discussed somewhat more closely. First, what is the geometrical locus along which $(r - x)$ is constant? We write $r = x + C$ and square, obtaining after simplification

$$y^2 + z^2 = 2Cx + C^2$$

or $$x = \frac{1}{2C}(y^2 + z^2 - C^2)$$

This is a paraboloid of revolution with the x axis as axis of revolution, and focus at the origin of coordinates. (The point of intersection with the x axis is at $x = -1/2\, C$, and the y, z plane cuts it in a circle of radius C.) The expression $e^{-k}(r - x)$ therefore gives constant values on confocal paraboloids and these values become very small as soon as the argument becomes somewhat large. Outside a certain paraboloid therefore, the function χ is practically equal to zero and the velocities u_2, v_2, w_2 derived from it also vanish. In the immediate neighborhood of the origin of coordinates $(r - x)$ is very small and therefore $e^{-k(r-x)}$ is practically equal to unity, so that here

$$\chi \approx \frac{A}{r}$$

Hence for small r $$u_2 = -A\frac{x}{r^3} + \ldots$$

$$v_2 = -A\frac{y}{r^3} + \ldots$$

$$w_2 = -A\frac{z}{r^3} + \ldots$$

The flow, which according to the preceding remarks satisfies the conditions of continuity, has a sink at the origin which must be compensated by an equally strong source of the flow u_1, v_1, w_1. Hence the potential φ must contain a term $-A/r$, and of course other terms in addition, to ensure fulfillment of the boundary conditions at the surface of the sphere. We must pass over the exact calculation. It can be found in Lamb's paper of 1911, and in § 340 of his textbook[1]. The result is that for Reynolds numbers $u_0\,a/\nu$ which are small compared to 1, the flow in the neighborhood of the sphere is almost exactly given by Stokes' calculation and the same formula for the resistance is obtained. At great distances however, the velocity distribution is quite unlike that given by Stokes' formula. Even for $u_0\,r/\nu = 1$ these velocities are already quite different. A paraboloidal domain of disturbance is found stretching behind the sphere where the velocity of flow falls off as the reciprocal of the distance. In this domain the fluid moves in a "wake" behind the body. These are the particles which have been reached by the diffusion of the rotation produced by the body. Outside this region

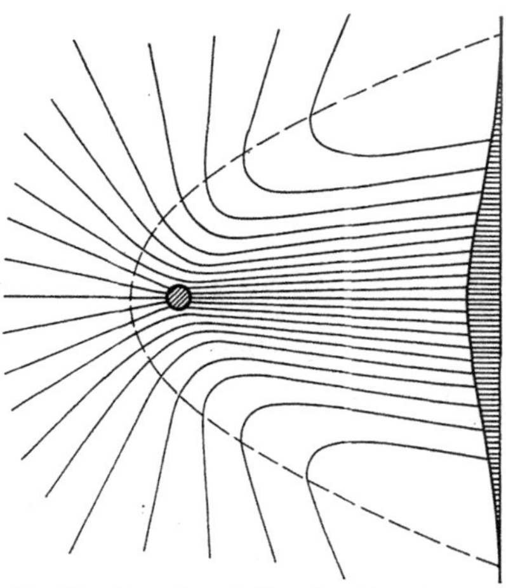

Fig. 17. Oseen's solution for the viscous flow around a moving sphere.

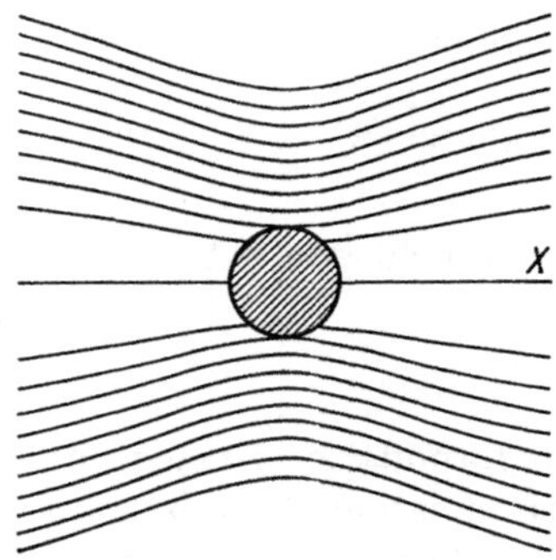

Fig. 18. Stokes's solution for the viscous flow around a moving sphere.

there is a flow as from a source, which appears to radiate from the body and to make place for the fluid which moves toward the body in the wake. Fig. 17 represents such a flow. For greater clarity a system of reference has been chosen in which the undisturbed fluid is at rest and the body is moving. The corresponding picture for the Stokes' motion is reproduced in Fig. 18. For infinitely small Reynolds numbers the paraboloid given in Fig. 17 by the broken line would move toward infinity and Fig. 17 would in fact transform into Fig. 18.

The calculations for the infinitely long cylinder were carried through by Lamb, see his paper of 1911 already mentioned, and his textbook § 343. The function χ in this case is a more complicated expression which, for large values of the radius r, can be represented by

$$\chi = \frac{A'}{\sqrt{2\,k\,r}}\,e^{-k\,(r-x)} + \dots$$

The circumstances are obviously very similar, apart from the different power of r which occurs. This knowledge will now allow us to arrive

[1] Fifth edition.

at an estimate for the resistance formula of a cylinder. We were not able to do this from the considerations of the preceding paragraph since we could give no definite velocity to the undisturbed flow corresponding to the actual flow considered. We can now say that the inertial forces produce, at a great distance, a flow of a different form corresponding to a constant velocity at infinity. As an approximation to the distance at which the Stokes' solution passes over into the solution which replaces it, we may take the distance from the origin to the vertex of that parabola for which the exponent of e becomes equal to -1. This is the case when $kr = 1/2$ or $u_0 r/\nu = 1$. Putting

$$u_0 \approx \frac{A}{2\mu} \log \frac{r}{a} + \cdots$$

in accordance with formula (11.11a), and taking the above value for r, we obtain

$$u_0 \approx \frac{A}{2\mu} \log \frac{\nu}{u_0 a}$$

and therefore

$$A \approx \frac{2\mu u_0}{\log \frac{\nu}{u_0 a}}$$

from which with (11.12) the resistance can be written:

$$D = 2\pi A l \approx \frac{4\pi \mu u_0 l}{\log \frac{\nu}{u_0 a}}$$

This formula is of course not very exact since it is based on a rough approximation. Lamb has however made the exact calculation and finds a formula which differs from the above only in having an additive term in the denominator. The correct formula is

$$D = \frac{4\pi \mu u_0 l}{\left(\log \frac{\nu}{u_0 a} + \frac{1}{2} + 2\log 2 - \gamma\right)}$$

where γ, (0.577), is Euler's constant. It may be mentioned that Stokes' formula for spheres and Lamb's formula for cylinders have been well confirmed by experimental tests (the latter however only by extrapolation from larger Reynolds numbers). They can however be applied only for Reynolds numbers which are less than 1. Numerical calculations for cases of Reynolds numbers somewhat greater than 1 have been made by A. Thom[1] and his students, and the results are in good agreement with experiment. An analytic representation of the results was however found to be impossible.

13. Boundary Layers. The other extreme, of very small viscosities, or in other words of *very large Reynolds numbers*, may now be discussed. The exposition in **9** and various examples in **10** have made it clear that

[1] THOM, A., Some Studies of the Flow Past Cylinders. Vorträge aus dem Gebiet der Aerodynamik und verwandten Gebieten, Aachen, 1929 (Julius Springer, Berlin 1930, p. 58).

the case very often arises where the flow is without rotation except for a thin layer in the neighborhood of the fixed surface, and can therefore be represented by a potential of the same kind as those which occur in the theory of ideal fluids. This state of affairs may be expected particularly when the flow commences from rest, or when some disturbing object enters a region previously at rest. We shall now investigate how the Navier-Stokes' equations can be simplified in case it is previously known that the desired solution shows deviations from the potential flow only in a thin region in the neighborhood of a fixed surface.

For the sake of clarity the problem will be formulated only for the two-dimensional flow (the so-called plane problem). We shall also assume at first that the fixed surface lies in the $x\,z$ plane. The boundary conditions therefore are: When $y = 0$, $u = v = 0$. The peculiarity of the flow is that in accordance with the previous exposition, we have practically a potential flow until comparatively near to the fixed surface, which, in accordance with previous considerations, would slide with a finite velocity past the wall. The transition from this velocity to the zero velocity demanded by the boundary conditions takes place in a thin layer whose order of magnitude can be taken from the example already mentioned in **10**. We introduce the dimension δ as a measure of this thickness, and can at first take it to mean that distance from the wall at which the transition from zero velocity to the potential flow has been completed, so that all friction effects take place inside a layer of thickness δ. The name "Boundary Layer" has been introduced for such layers. We will now see to what extent Navier-Stokes' equations can be simplified on the basis of the above exposition. If the impressed forces X, Y are left out of consideration[1], we have to deal with the following equations for the two-dimensional flows:

$$\left.\begin{aligned} \varrho\left(\frac{\partial u}{\partial t} + u\frac{\partial u}{\partial x} + v\frac{\partial u}{\partial y}\right) + \frac{\partial p}{\partial x} &= \mu\left(\frac{\partial^2 u}{\partial x^2} + \frac{\partial^2 u}{\partial y^2}\right) \\ \varrho\left(\frac{\partial v}{\partial t} + u\frac{\partial v}{\partial x} + v\frac{\partial v}{\partial y}\right) + \frac{\partial p}{\partial y} &= \mu\left(\frac{\partial^2 v}{\partial x^2} + \frac{\partial^2 v}{\partial y^2}\right) \end{aligned}\right\} \tag{13.1}$$

$$\frac{\partial u}{\partial x} + \frac{\partial v}{\partial y} = 0 \tag{13.2}$$

We next consider what takes place when the region in which the flow occurs has an extension much smaller in the y direction than in the x direction. We also introduce a characteristic length l for the x direction, and, in order to obtain dimensionless coordinates, put

$$x = l\,\xi, \qquad y = \delta\,\eta$$

In order now to relate the orders of magnitude of the velocities u and v,

[1] As previously noted, this only means the splitting away of a part of the pressure which is in equilibrium with ϱX and ϱY.

we integrate $\partial v/\partial y$, using the equation of continuity (13.2) and the boundary condition $v = 0$ for $y = 0$, and obtain for $y = \delta$,

$$v = -\int_0^\delta \frac{\partial u}{\partial x} d y$$

If we introduce ξ and η into this formula we find

$$v = -\frac{\delta}{l}\int_0^1 \frac{\partial u}{\partial \xi} d\eta$$

But $\partial u/\partial \xi$, and therefore also its integral from 0 to 1, are of normal order of magnitude. We therefore find that the order of magnitude of v is δ/l times the order of magnitude of u, or expressed in another way, $(l/\delta)\, v$ has the same order of magnitude as u. We now write (13.1) in the new coordinates and find

$$\frac{\varrho}{l}\left[l\frac{\partial u}{\partial t} + u\frac{\partial u}{\partial \xi} + \frac{l}{\delta}v\frac{\partial u}{\partial \eta}\right] + \frac{1}{l}\frac{\partial p}{\partial \xi} = \frac{\mu}{l^2}\frac{\partial^2 u}{\partial \xi^2} + \frac{\mu}{\delta^2}\frac{\partial^2 u}{\partial \eta^2}$$

If we realize that the ordinate η ranges from 0 to 1 inside the region of friction, we find that the differential coefficients $\partial u/\partial \eta$ and $\partial^2 u/\partial \eta^2$ are in general of the same order of magnitude as u itself. Similarly, by suitable choice of the length l, ξ can be made to range in a region of order of magnitude 1, so that the differential coefficients of ξ are also of normal order of magnitude. We may also assume that $\partial u/\partial t$ is at most of such an order of magnitude as to fit in suitably with the other acceleration terms. This excludes states of affairs containing very sudden accelerations produced by shocks, which in themselves are possible, but will be assumed not to occur in the present problem. (Accelerations induced by shock or a sudden blow in such cases, would of course be decisive factors in the considerations of orders of magnitude.) It is easy to see what order of magnitude may be permitted to $\partial u/\partial t$. Instead of t a new variable τ can be introduced such that

$$t = \frac{l}{u_0}\tau.$$

This means that time is related to the period taken to traverse the length l with a velocity u_0, which is characteristic of the state of affairs. Instead of $l\,(\partial u/\partial t)$ we have now $u_0\,(\partial u/\partial \tau)$, an expression of the same form as $u\,(\partial u/\partial \xi)$. Our assumptions can therefore be expressed by the statement that $\partial u/\partial \tau$ must have an order of magnitude not greater than that of $\partial u/\partial \xi$.

The velocity component v occurs in the left hand factor, multiplied by l/δ. This brings the order of magnitude of the term up to that of u. Hence all three terms in the bracket have the same order of magnitude.

One of the terms multiplied by μ is divided by l^2, and the other is divided by δ^2. According to hypothesis the viscosity μ is very small.

Since δ^2 is also very small μ/δ^2 may become finite again. It is clear however that in such case, μ/l^2 is very small, so that the expression $(\mu/l^2)\,(\partial^2 u/\partial \xi^2)$ is negligible. In what follows we shall, in fact, neglect this expression. Regarding the order of magnitude of δ, the following statement can now be made. Evidently the terms $\frac{\varrho}{l}\,u\,\frac{\partial u}{\partial \xi}$ and $\frac{\mu}{\delta^2}\,\frac{\partial^2 u}{\partial \eta^2}$ should be of the same order of magnitude. According to our discussion the two differential coefficients $\partial u/\partial \xi$ and $\partial^2 u/\partial \eta^2$ are of the same order of magnitude and can therefore be cancelled on both sides for the purpose of the present estimate. If the variable velocity u be represented by a suitably chosen value u_0 we find

$$\delta^2 \sim \frac{\mu\, l}{\varrho\, u_0}$$

which can also be written as

$$\frac{\delta}{l} \sim \sqrt{\frac{\nu}{u_0 l}} \tag{13.3}$$

The ratio of the boundary layer thickness δ to the length of the flow, measured along the wall, thus turns out to be inversely proportional to the square root of the Reynolds number. This is in agreement with the results of the exact solutions of **10**. We now consider the second equation of (13.1) in the same manner. We find first,

$$\frac{\varrho}{l}\left[u_0\frac{\partial v}{\partial \tau} + u\frac{\partial v}{\partial \xi} + \frac{l}{\delta}\,v\,\frac{\partial v}{\partial \eta}\right] + \frac{1}{\delta}\frac{\partial p}{\partial \eta} = \frac{\mu}{l^2}\frac{\partial^2 v}{\partial \xi^2} + \frac{\mu}{\delta^2}\frac{\partial^2 v}{\partial \eta^2}$$

With the former assumptions as to the accelerations, the three terms in the bracket have the same order of magnitude. In accordance with the conditions regarding δ, this is the same as that of the expression $\frac{\mu}{\delta^2}\frac{\partial^2 v}{\partial \eta^2}$. The expression $\frac{\mu}{l^2}\frac{\partial^2 v}{\delta\, \xi^2}$ is again small and can be cancelled in comparison with the other terms. The order of magnitude of the remaining terms is that of v, that is, equal to $u_0\,\delta/l$. Hence the order of magnitude of $\partial p/\partial \eta$ is $\varrho u_0^2\,(\delta^2/l^2)$. Hence the variation of p in the region of the boundary layer is so small that it may be neglected.

So far, the equations are concerned with a fixed plane surface. The results obtained can however be extended to a curved surface. In this case a coordinate system can be chosen in which the arc-length along the fixed surface can be introduced as the x coordinate, and the perpendicular distance from the surface as the y coordinate. The curvature now introduces a series of terms which are partly caused by the varying length of the coordinate x, and partly by the transverse accelerations. Since we here consider a region of small thickness δ, these first corrections can be forthwith neglected in the present approximate theory; further, if it be assumed that the smallest radii of curvature of the wall are of the order of magnitude of l, the corrections due to the transverse acceleration change the order of $\partial p/\partial \eta$ by one order δ/l, so that these are now given by $\varrho u_0^2\,\delta/l$. Nevertheless $\partial p/\partial \eta$ remains small to the first order and must be integrated between 0 and 1 in order to obtain p, so that the differences of p inside the boundary layer still remain of the order stated. In the following calculations we shall therefore take $\partial p/\partial \eta = 0$ with the consequence that p is now only a function of x. We have now finished with the second[1] equation of (13.1) and have obtained the final equations for a two-dimensional boundary layer flow as

$$\varrho\left(u\frac{\partial u}{\partial x} + v\frac{\partial u}{\partial y}\right) + \frac{d p}{d x} = \mu\frac{\partial^2 u}{\partial y^2} \tag{13.4}$$

$$\frac{\partial u}{\partial x} + \frac{\partial v}{\partial y} = 0 \tag{13.5}$$

In these equations the old coordinates x, y have been introduced once again, but care must naturally be taken that the application of these simplified equations is restricted to a region whose extension in the y direction is of the order of magnitude $\sqrt{\nu l/u_0}$. In order to satisfy the equation of continuity a stream function Ψ is usually introduced from which the velocities u and v are derived by the equations

$$u = \frac{\partial \Psi}{\partial y}, \qquad v = -\frac{\partial \Psi}{\partial x}$$

Substitution shows at once that continuity is satisfied, and we now pass on to study various solutions of the boundary layer equation.

14. The Flat Plate. One of the simplest examples is the flow along a thin flat plate (Fig. 19). Let the origin of coordinates be placed at the front edge of the plate, where the flow divides. Let the flow of the undisturbed fluid be in the direction of the x axis and have a constant velocity u_0, and, corresponding to this velocity let there be a constant pressure which will also act on the boundary layer.

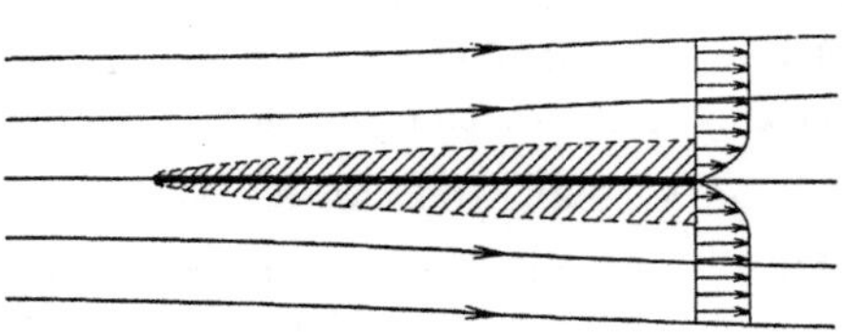

Fig. 19.

We look now for an approximate solution for values of x for which the Reynolds number $x u_0/\nu$ is already very large. This is a necessary assumption for the application of the simplified equations (13.4) and (13.5). We shall naturally be unable to do complete justice to the state of affairs at the front edge of the plate.

First of all, the following can be said from the viewpoint of dimensions: In addition to the two lengths x, y, the length ν/u_0 is at our disposal for describing the state of affairs at any point (x, y). After our previous experiences with the Navier-Stokes' equations we shall expect that the values of y here considered will have the order of magnitude $\sqrt{\nu x/u_0}$, which can be interpreted as the geometric mean of x and ν/u_0. (We can

[1] Subsequently this could be used to calculate the small value $\partial p/\partial \eta$ if that were considered necessary.

also think of the formula $\sqrt{\nu t}$ previously mentioned, by putting t equal to x/u_0 that is, equal to the time taken by a particle of fluid to flow from the front edge of the plate to the point we are considering). It may be conjectured that the velocity will here be a function of the one variable $\eta = y/\sqrt{\nu x/u_0}$. This is in fact the case. We introduce a stream function Ψ which will insure this result, namely,

$$\Psi = \sqrt{\nu x u_0}\, f(\eta) \tag{14.1}$$

This gives
$$u = \frac{\partial \Psi}{\partial y} = \frac{d\Psi}{d\eta}\frac{\partial \eta}{\partial y} = u_0 f'(\eta) \tag{14.2}$$

in accordance with requirements. In order that u should become equal to u_0 for large values of η, we must have

$$\lim_{\eta \to \infty} f'(\eta) = 1$$

With respect to the value $\eta = \infty$ it must be noted that according to the meaning of the calculation, actually infinite distances η cannot of course be considered. The limit value of η simply is large—large enough to reach into the purely potential flow. As the graphical representation shows, the curves bend very quickly toward the asymptotes. So long as we are thinking, not of the actual flow, but only of the function of η, the mathematical convention of speaking of the limiting values $\eta = \infty$ will be used.

The velocity component v perpendicular to the wall becomes

$$v = -\frac{d\Psi}{dx} = \sqrt{\nu u_0 x} \cdot f'(\eta) \cdot \frac{y}{2x}\sqrt{\frac{u_0}{\nu x}} - \frac{1}{2}\sqrt{\frac{\nu u_0}{x}}\, f(\eta)$$
$$= \frac{1}{2}\sqrt{\frac{\nu u_0}{x}}(\eta f' - f) \tag{14.3}$$

We now calculate $u\,(\partial u/\partial x) + v\,(\partial u/\partial y)$ and $\nu\,(\partial^2 u/\partial y^2)$, and equate the two expressions, since the term due to the fall in pressure drops out.

We obtain
$$-\frac{u_0^2}{2x}\eta f' f'' + \frac{u_0^2}{2x}(\eta f' - f) f'' = \nu \cdot \frac{u_0^2}{\nu x} f'''$$

or, simplified
$$f f'' + 2 f''' = 0 \tag{14.4}$$

The fact that the differential equation simplifies in this way is a sign of the correctness of our method of attack. Since we have a differential equation of the third order, three boundary conditions are necessary to determine the solution completely. One of them, $f' = 1$ for $\eta = \infty$ has already been mentioned. Further we must have u and v both zero for $\eta = 0$. This gives $f' = 0$ and $f = 0$ for $\eta = 0$. The way in which such problems are solved may be elucidated in rather more detail by this example. We may first ask for a simple solution. Of these we find only one which corresponds to the potential flow; it can however be useful and is
$$f = \eta + \text{constant}.$$

As required, $f' = 1$. Since $f'' = f''' = 0$ the differential equation (14.4) is satisfied, but not the two boundary conditions for $\eta = 0$. There are no other simple solutions. We will therefore write out a power series in order to represent the state of affairs for small and moderately large η and try to obtain in addition an asymptotic representation for large η. As far as the power series is concerned it is advisable to write it in the following form in order to simplify the differentiation:

$$f(\eta) = A_0 + A_1\eta + \frac{A_2}{2!}\eta^2 + \frac{A_3}{3!}\eta^3 + \dots$$

We can at once satisfy the boundary conditions for $\eta = 0$ by putting $A_0 = A_1 = 0$. We now calculate

$$f'' = A_2 + A_3\eta + \frac{A_4}{2!}\eta^2 + \dots$$

$$f''' = A_3 + A_4\eta + \frac{A_5}{2!}\eta^2 + \dots$$

and substitute these values in (14.4) and rearrange in powers of η. We have a solution of the differential equation only if the coefficient of every power of η vanishes by itself, since in this case the differential equation is satisfied for all values of η. We now obtain the following:

$$0 = 2A_3 + \eta \cdot 2A_4 + \frac{\eta^2}{2!}(A_2^2 + 2A_5) + \dots$$

We see at once that A_3 and A_4 must be zero, and also that $A_5 = -\frac{1}{2}A_2^2$. Since A_2 and A_3 vanish, so do A_6 and A_7. The next coefficient which is not automatically zero is that of η^5, giving

$$\frac{\eta^5}{5!}(11A_2A_5 + 2A_8)$$

from which $$A_8 = -\frac{11}{2}A_2A_5 = +\frac{11}{4}A_2^3$$

We must continue in this way till a sufficient number of terms have been evaluated. We write the whole series as follows, where A_2, which represents a constant of integration is replaced by α

$$f = \sum_{n=0}^{n=\infty}\left(-\frac{1}{2}\right)^n \frac{C_n\,\alpha^{n+1}}{(3n+2)!}\eta^{3n+2}$$

The first six coefficients have the following values

$C_0 = 1$ $\quad C_1 = 1$ $\quad C_2 = 11$

$C_3 = 375$ $\quad C_4 = 27{,}897$ $\quad C_5 = 3{,}817{,}137$

As the representation of the function f shows, the various other possible solutions differ among themselves only to the extent of a similitude transformation, as may be seen by the fact that the introduction of a new variable $\eta' = \eta\,(\alpha)^{1/3}$ produces a series without α having a factor $(\alpha)^{1/3}$ in front of the whole.

We must now look for the asymptotic solution. We already have a first approximation in the previous expression $f = \eta + \text{const.}$ We denote the constant, which is negative, by $-\beta$, thus obtaining

$$f_1 = \eta - \beta \tag{14.6}$$

We take as a second approximation $f = f_1 + f_2$ where f_2 is to be considered small relative to f_1. Instead of $f f''$ we now write $f_1 f_2''$ as an approximation. This transforms (14.4) into

$$(\eta - \beta) f_2'' + 2 f_2''' = 0,$$

or

$$\frac{f_2'''}{f_2''} = \frac{1}{2}(\beta - \eta),$$

which can be integrated giving

$$\log f_2'' = \frac{1}{2}\beta\eta - \frac{1}{4}\eta^2 + \text{constant.}$$

With an eye to what follows we write this constant equal to $-\beta^2/4 + \log\gamma$ and take the exponential of both sides, so that

$$f_2'' = \gamma e^{-\frac{1}{4}(\eta-\beta)^2}$$

This gives f_2', in which the main interest lies, as

$$f_2' = \gamma \int_{\infty}^{\eta} e^{-\frac{1}{4}(\eta-\beta)^2} d\eta \tag{14.7}$$

The lower limit of the integral is so chosen that f_2' vanishes for $\eta = \infty$. The numerical value of f_2' is therefore negative, that is, something is subtracted from the potential flow, as should be the case. We now obtain f_2 by one more integration, so that the asymptotic formula is

$$f = \eta - \beta + \gamma \int_{\infty}^{\eta}\int_{\infty}^{\eta} e^{-\frac{1}{4}(\eta-\beta)^2} d\eta \tag{14.8}$$

This approximation can of course be refined still further by putting

$$f = f_1 + f_2 + f_3,$$

where f_3 is to be small compared with f_2, and proceeding as before. Since the power series converges comparatively rapidly, this third approximation is not required in practice.

The two solutions thus found must now be harmonized. We have in all three constants of integration, α, β, and γ. The constant α occurs in the power series developed at $\eta = 0$ for which there were two boundary conditions, and β and γ appear in the asymptotic solution where there was only one boundary condition. [The number of boundary conditions, and the number of constants of integration, each equal to 3, correspond to the order of (14.4)]. The two solutions must now be so combined that they both represent the same function. We satisfy this condition by demanding that at each place where both expansions are valid, the values of f, f', f'' must be the same for both. This gives three equations for the three unknowns α, β, and γ. The solution of this problem is not in practice very convenient. It leads to the following values,

$$\alpha = 0.332, \qquad \beta = 1.73, \qquad \gamma = 0.231.$$

The above calculations are due to H. Blasius[1]. Since the solution beginning from $y = 0$ has only one constant of integration, denoting one similitude transformation, it is possible to calculate the function $f'(\eta)$ numerically from $\eta = 0$ by a step-by-step method. For any arbitrary value of α we arrive by this method at a horizontal asymptote whose height is initially unknown. The function can then be altered by changing the measures of length of abscissa and ordinate, so that the asymptotic value of f' becomes equal to unity. The problem in this form has been dealt with by C. Töpfer[2]. The above table for $f'(\eta)$ has been calculated by interpolation from the material of Töpfer's paper.

TABLE 1.

η	$f'(\eta)$	η	$f'(\eta)$
0	0		
0.2	0.0664	3.2	0.8761
0.4	0.1328	3.4	0.9018
0.6	0.1990	3.6	0.9233
0.8	0.2647	3.8	0.9411
1.0	0.3298	4.0	0.9555
1.2	0.3938	4.2	0.9670
1.4	0.4563	4.4	0.9759
1.6	0.5168	4.6	0.9827
1.8	0.5778	4.8	0.9878
2.0	0.6298	5.0	0.9916
2.2	0.6813	5.2	0.9943
2.4	0.7290	5.4	0.9962
2.6	0.7725	5.6	0.9975
2.8	0.8115	5.8	0.9984
3.0	0.8461	6.0	0.9990

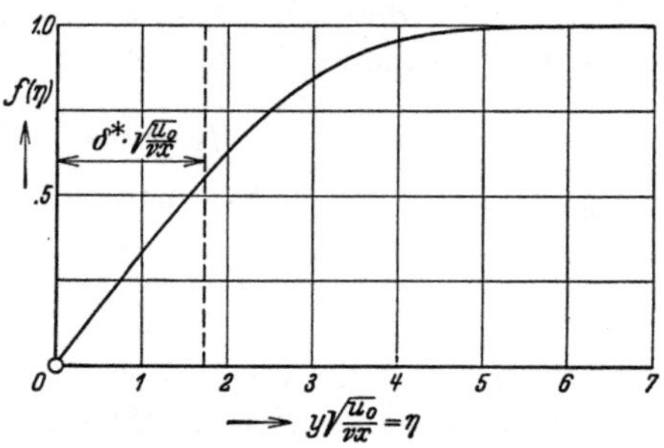

Fig. 20. Velocity distribution in the flow along a flat plate.

This function, which gives the velocity distribution is represented graphically in Fig. 20.

In order to calculate the resistance of a plate of length l and width b the shear $\tau = \mu\,(\partial u/\partial y)_0$ must be integrated. According to (14.2) $u = u_0 f'(\eta)$ and hence

$$\frac{\partial u}{\partial y} = u_0 f''(\eta)\frac{\partial \eta}{\partial y} = u_0 f''(\eta)\sqrt{\frac{u_0}{\nu x}}$$

But the power series (14.5) gives

$$f''(\eta) = \alpha - \frac{1}{2}\frac{\alpha^2\eta^3}{3!} + \dots$$

so that

$$f''(0) = \alpha$$

[1] Grenzschichten in Flüssigkeiten mit kleiner Reibung. Zeitschr. für Mathematik und Physik, Bd. 56, p. 1, 1908.

[2] Short Extract in Zeitschr. für Mathematik und Physik, Bd. 60, p. 397, 1912.

Hence the resistance of one side of the plate is

$$\frac{D}{2} = \mu \, \alpha \, b \, u_0 \sqrt{\frac{u_0}{\nu}} \int_0^l \frac{d\,x}{\sqrt{x}} = 2 \alpha \, b \, u_0 \sqrt{\mu \, \varrho \, l \, u_0}$$

since $\mu = \nu \, \varrho$. And also, since $4 \, \alpha = 1.328$[1] the total resistance is

$$D = 1.328 \, b \sqrt{\mu \, \varrho \, l \, u_0^3} \tag{14.9}$$

We can construct a resistance coefficient C_D by dividing the resistance by $1/2 \; \varrho \, u_0^2$ (the stagnation pressure) and the area of the surface. We thus obtain

$$C_D = 1.328 \sqrt{\frac{\nu}{u_0 l}} \tag{14.9a}$$

The slowing down of the flow has the effect that the main stream is thrust a little away from the plate. We can easily calculate the amount of this displacement, which will be denoted by δ^* as follows: We must have

$$\int_0^{y_1} u \, d\,y = u_0 \, (y_1 - \delta^*)$$

where y_1 is any point outside the boundary layer. This can also be written as

$$\delta^* = \frac{1}{u_0} \int (u_0 - u) \, d\,y \tag{14.10}$$

Furthermore, $$\int_0^{y_1} (u_0 - u) \, d\,y = u_0 \int_0^{\eta_1} [1 - f'(\eta)] \, d\,\eta \sqrt{\frac{\nu \, x}{u_0}}$$

$$= u_0 \, [\eta_1 - f(\eta_1)] \sqrt{\frac{\nu \, x}{u_0}}$$

This formula acquires an exact interpretation if the limiting value for large $\eta_1 \, (\eta_1 \to \infty)$ is taken. For this however the first asymptotic approximation (14.6) $f(\eta) = \eta - \beta$, can be introduced, thus furnishing the measure of the displacement which is at the same time a measure of the boundary layer

$$\delta^* = \beta \sqrt{\frac{\nu \, x}{u_0}} = 1.73 \sqrt{\frac{\nu \, x}{u_0}} \tag{14.11}$$

All the exterior stream-lines are displaced by this amount from the plate as compared with the flow of the ideal frictionless fluid, for which the constant velocity would prevail up to the surface of the plate. The thickness of the layer affected by the friction is not exactly defined since the effect of the friction falls away asymptotically as we pass into the interior of the fluid. On account of the peculiar nature of the asymptotes it does not matter much whether the point is taken where, say 95 per cent of the full velocity is attained or the point where 99 per cent is attained. One does not go far wrong in practice by taking three

[1] Blasius gives 1.327. The more exact value 1.328 is due to Töpfer.

times the measure of the displacement mentioned as the thickness of the boundary layer. We should arrive at exactly this factor 3 if we assumed the velocity to be distributed on a parabola touching the line $u = \text{const.} = u_0$ at its vertex. Fig. 20 shows in fact that the full velocity is practically reached for $\eta = 3 \times 1.73 = 5.19$. We can therefore give an approximate formula for the thickness of the boundary layer:

$$\delta = 5.2 \sqrt{\frac{\nu x}{u_0}} \tag{14.12}$$

The displacement of the stream-lines by the amount δ^* produces a slight alteration in the potential flow which was made the basis of the calculations. Instead of the simple parallel flow, the flow around a parabolic cylinder of thickness $2\,\delta^*$ should be introduced, which would slightly alter the pressure distribution. The above calculation would have to be repeated for this new pressure distribution and if necessary the process repeated on the basis of the new measure of displacement so obtained. Such calculations have so far not been performed; they would, in any case, make little difference in the regions where the calculations are usually applied in practice. They would however become necessary if the transition to smaller Reynolds numbers $u_0\,l/\nu$ were attempted. The flows here calculated have been tested experimentally by B. G. van der Hegge Zynen[1] and also by M. Hansen[2] confirming all the important details.

15. Calculation of the Boundary Layer of the Steady Flow Around Cylinders. Separation of Flow from the Wall. The investigations of the preceding section have taught us that a flow of fluid along a body gives rise to a frictional resistance, and this resistance was calculated for a very special case. The fact that they explain these small frictional resistances in this way does not however exhaust the significance of the boundary layers. In less simple cases they also exhibit a behaviour which is responsible for the remaining portion of the total resistance, which is usually much greater in magnitude. In the case of flows which are first accelerated and then retarded, there are regions of the boundary layer where a reverse flow appears in the direction opposite to that of the original flow. Reverse flows of this kind usually produce a complete change of the stream-line diagram, since the advancing flow separates from the body leaving a region usually filled with vortices of small velocity, see Fig. 21. In the next section we shall return to this phenomenon of separation in order to study it in more detail. First, however, we shall examine the steady flow after the separation is completed, and may commence with the following purely descriptive statements.

---

[1] VAN DER HEGGE ZYNEN, B. G., Thesis Delft, 1914; see also BURGERS, J. M., Proc. of the First International Congress for Applied Mathematics, Delft, p. 113, 1924.

[2] HANSEN, M., Zeitschr. f. angew. Math. u. Mech. 8, p. 185, 1928.

If velocity-profiles, such as shown in Fig. 22, be produced by some suitable arrangement of pressures, a reverse flow occurs in a region of increasing extension situated to the right of C, and the stream-lines are as drawn in the diagram. These lines are plotted by noting the fact that the amount of flow must be constant across all lines bounded by two such stream-lines. The diagram shows that at the point C, where the velocity profile touches the axis, one of the stream-lines branches and divides the fluid coming from the left and right respectively. If, for example, the fluid coming from the right hand side is colored, it would be found after some time that the color had spread over the

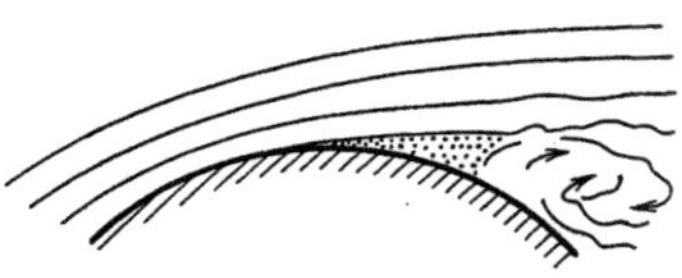

Fig. 21.

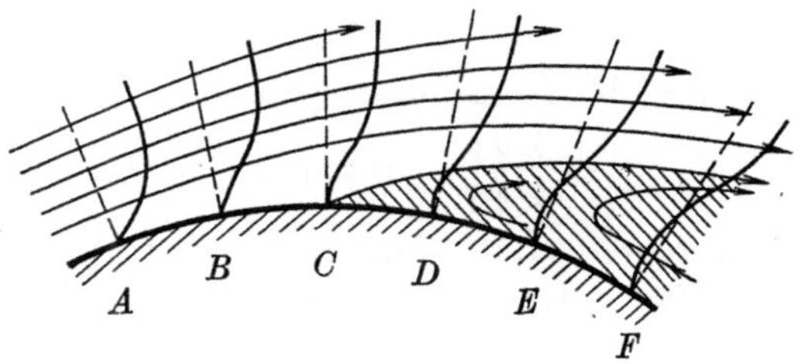

Fig. 22. Velocity profiles in the vicinity o a separation point.

region shaded in the diagram. By the differential equation of the boundary layer [see (13.4)], the curvature of the velocity-profile at $y = 0$, where $u = v = 0$, is

$$\frac{\partial^2 u}{\partial y^2} = \frac{1}{\mu} \frac{d p}{d x}$$

This shows that curvatures of the kind shown in Fig. 22, that is, for which $(\partial^2 u/\partial y^2)_0$ is positive, are associated with a positive pressure gradient dp/dx. Hence, *increasing pressure is a necessary condition for flows where separation from the wall occurs.* In the differential equations of the boundary layer, external forces (gravity, etc.) were neglected; this is always permissible for incompressible flows provided that p denotes the actual pressure minus the hydrostatic pressure due to gravity. Since the laws for frictionless fluids are applicable to the region outside the boundary layer, Bernoulli's equation allows us to replace p by $\left(\text{const.} - \frac{1}{2} \varrho u_1^2\right)$, u_1 being the local velocity. We can therefore write $-\varrho u_1 (\partial u_1/\partial x)$ instead of $\partial p/\partial x$. A positive $\partial p/\partial x$, and a positive u_1, therefore correspond to a negative $\partial u_1/\partial x$ or in other words, a decrease of velocity in the direction of flow. Hence it is established that the state of affairs represented by Fig. 22 is possible only *if the frictionless flow outside the boundary is retarded.* If the stream accelerates, the curvature of the velocity-profile in the neighborhood of $y = 0$ has the opposite sign. In this case it is certain that no separation effects will occur.

The magnitude of the retardation required to produce separation depends on the particular circumstances of each case and cannot be determined in a general manner. It will therefore be necessary to study suitable examples of solutions of the differential equations. The first example chosen is that of the flow around a cylinder whose contour is symmetrical with respect to the direction of flow. An approximate solution will be obtained in the form of a power series in x, where x is the length of arc from the front stagnation point. On account of the symmetry, the pressure p will then be an even function of x, but the velocity component u, and the fall in pressure $-dp/dx$ will be odd functions of x. The velocity of the potential flow is written as a series in the following form:

$$u_1 = a_1 x + a_3 x^3 + a_5 x^5 + \ldots \qquad (15.1)$$

Hence, by Bernoulli's equation

$$-\frac{1}{\varrho}\frac{dp}{dx} = a_1^2 x + 4 a_1 a_3 x^3 + (6 a_1 a_5 + 3 a_3^2) x^5 + \ldots \qquad (15.2)$$

To represent the flow in the boundary layer, the stream function Ψ is again introduced, so that the equation of continuity is satisfied, and

$$u = \frac{\partial \Psi}{\partial y}, \qquad v = -\frac{\partial \Psi}{\partial x}$$

Ψ is again an odd function of x and may be written as

$$\Psi = \psi_1 x + \psi_3 x^3 + \psi_5 x^5 + \ldots \qquad (15.3)$$

By substituting in (13.4) and using the condition that the coefficient of each power of x must vanish identically, the following equations are obtained for $\psi_1, \psi_3, \psi_5, \ldots$:

$$\left.\begin{aligned} \psi_1'^2 - \psi_1 \psi_1'' &= p_1 + \nu \psi_1''' \\ 4 \psi_1' \psi_3' - 3 \psi_1'' \psi_3 - \psi_1 \psi_3'' &= p_3 + \nu \psi_3''' \\ 6 \psi_1' \psi_5' - 5 \psi_1'' \psi_5 - \psi_1 \psi_5'' + 3 \psi_3'^2 - 3 \psi_3 \psi_3'' &= p_5 + \nu \psi_5''' \\ \ldots\ldots\ldots\ldots\ldots\ldots & \end{aligned}\right\} \qquad (15.4)$$

where p_1, p_3, p_5 are abbreviations for the coefficients of the series for $-1/\varrho\,(dp/dx)$ given above ($p_1 = a_1^2$, $p_3 = 4 a_1 a_3$, etc.). The first of (15.4) is a quadratic differential equation for ψ_1. The remaining differential equations are linear in ψ with the highest subscript occurring in them. Any ψ can therefore be calculated if the values of those which precede it are already known. All the equations are of the third order. The first of them is identical with that discussed in **10** c. As in that section the equation can be reduced to the form

$$f_1'^2 - f_1 f_1'' = 1 + f_1''' \qquad (15.5)$$

by replacing y by $\eta \sqrt{\nu/a_1}$ and ψ by $\sqrt{a_1 \nu}\, f_1(\eta)$. The boundary conditions are, as before,

$$f_1 = f_1' = 0 \quad \text{for} \quad \eta = 0$$

and

$$f_1' = 1 \quad \text{for} \quad \eta = \infty$$

The method of expansion in a power series can be applied for the solution of the present equation, as in the previous section. For further details reference can be made to Blasius[1]. Later, in order to obtain more exact numerical results Hiemenz used Kutta's numerical method, in which the differential equation is solved by successive small steps, each consisting of power series of degree four. This method left a constant of integration undetermined since only two boundary conditions are given at $y = 0$, and a rather long calculation, requiring 30 steps of the independent variable of magnitude 0.1, had to be repeated several times for various values of this constant before the behavior of the system at $\eta = 3$ was obtained (the asymptote is practically reached for $\eta = 3$). In contrast to the solution discussed in the previous paragraph, there is no similitude transformation in the present case, on account of the numeral 1 in (15.5), so that the simplification which was given there cannot be used here. The distribution of f_1', the factor of the velocity u, is shown in Table 2. After f_1 has been determined, substitution of $a_3 \sqrt{\nu/a_1}\, f_3(\eta)$ for ψ_3 in the differential equation for the latter leads to

TABLE 2.

η	$f_1'(\eta)$	$f_3'(\eta)$
0	0	0
0.2	0.2266	0.5004
0.4	0.4145	0.8516
0.6	0.5663	1.0752
0.8	0.6860	1.1988
1.0	0.7779	1.2500
1.2	0.8468	1.2532
1.4	0.8969	1.2280
1.6	0.9325	1.1900
1.8	0.9570	1.1484
2.0	0.9734	1.1104
2.2	0.9841	1.0784
2.4	0.9908	1.0532
2.6	0.9950	1.0348
2.8	0.9975	1.0220
3.0	0.9989	1.0132

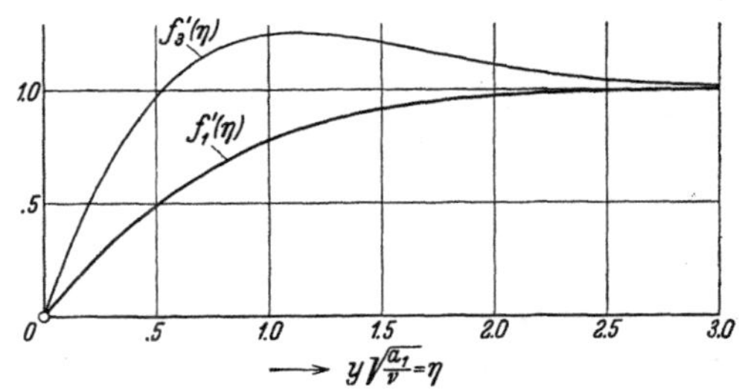

Fig. 23. Functions of the steady boundary layer flow.

$$4 f_1' f_3' - 3 f_1'' f_3 - f_1 f_3'' = 1 + f_3''' \tag{15.6}$$

The boundary conditions are again

$$f_3 = f_3' = 0 \quad \text{for} \quad \eta = 0,$$

and

$$f_3' = 1 \quad \text{for} \quad \eta = \infty$$

The solution can be obtained in a similar manner either by expansion in a power series or by the Kutta method. The results obtained by Hiemenz with the second method are also shown in Table 2. Fig. 23 shows the behavior of f_1' and f_3' which correspond to this table.

These two terms already suffice for the calculation of flows which show the separation effect. It is only necessary to take the coefficient a_3

[1] *loc. cit.* pp. 15 and 16.

negative, thus obtaining a velocity-curve with a maximum, beyond which the velocity decreases again (see Fig. 24). It can easily be seen by considering Fig. 23 that a sequence of velocity profiles similar to Fig. 22 are produced in this way.

Hiemenz has measured the distribution of pressure for a circular cylinder in a stream of water[1]. Special precautions ensured that the turbulent region behind the cylinder remained as quiescent as possible and was not oscillating. This was effected by bounding the turbulent region from beneath by still water, which was drawn into the turbulent region. In this way the flow was almost exactly steady. Hiemenz set himself the problem of obtaining, by means of the calculations described above, the boundary layer corresponding to the measured distribution of pressure as exactly as possible. It appeared that the two terms already given were not enough. He also calculated ψ_5. The solution for this term cannot be given in a general manner since, as the differential equation (14.4) shows, it depends on the ratio of a_1 to a_3. Hiemenz calculated ψ_5 by the same method from a knowledge of the functions for the pressure distribution and velocity of the potential flow in his particular case, and thus obtained the boundary layer profile. By introducing color through apertures bored in the cylinder he was able to discover the point at which separation occurred and found that the measurements and calculations agreed to within about 1 degree. It is of course not certain whether the series consisting of all the terms up to and including $\psi_5 x^5$ represents the effect of the whole infinite series with sufficient accuracy. Experiment shows that to some extent at least, this is the case. In order to improve the convergence it is natural to think of the following type of process: A new power series is begun at some suitable intermediate point by taking the velocity profile obtained by the previous calculations as initial velocity profile and so the further development can be investigated. Peculiar difficulties arise in this process because any small deviations from the exact profile due to inexactitudes in the calculation produce singularities on continuation of the procedure. Dr. S. Goldstein[2] has succeeded in overcoming these difficulties, but his expansions show that calculations of this character are very laborious, and that the continuation can be performed only

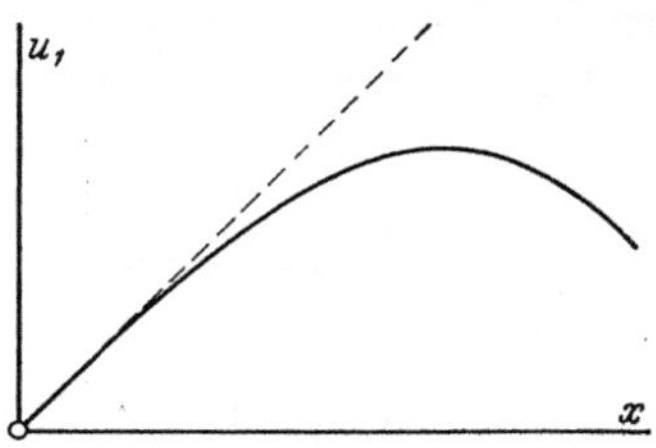

Fig. 24.

[1] Die Grenzschicht an einem in den gleichförmigen Flüssigkeitsstrom eingetauchten geraden Kreiszylinder. (Dissertation Göttingen, 1911.) Dingl. Polyt. Journal, vol. 326, p. 321, 1911.

[2] GOLDSTEIN, S., Concerning Some Solutions of the Boundary Layer Equations in Hydrodynamics. Proc. Camb. Phil. Soc., vol. XXVI, 1930.

in comparatively small steps. Nevertheless the fundamental possibility remains, that any given boundary flow can at least be numerically followed to the point where separation occurs and even a little further.

Unfortunately, the boundary layer calculations necessarily break down at some small distance beyond the separation point, since the assumption that the thickness of the layer influenced by friction is small in relation to the dimensions of the body can no longer be retained when the particles which have been set rotating by the friction move into the free fluid. Clearly, in this case, the width of the zone of rotation can no longer be taken as small. The consequence is that in practice the preceding calculations will describe the state of affairs only up to about the point where separation commences. From this point onwards the further effect of the friction can be left out of consideration as being of minor importance and the fluid may be considered as frictionless but containing vorticity. This point of view has proved useful in the theory of the aerofoil. Conditions in the boundary layers for the flow around circular cylinders have also been tested by experiment. The work of A. Thom [1] and A. Fage [2] may be referred to in this connection. An approximate theory of the steady boundary layer flow has been worked out by V. M. Falkner and S. W. Skan [3].

16. Development of the Boundary Layer with Time. Only steady boundary layer flows have been studied in the preceding sections. It is, however, also very instructive to study the development of boundary layers with respect to time. For example, we can choose the initial conditions that the system starts from rest and the fluid (or body) commences to move at the time $t = 0$. According to the simplest problem of this kind, already given under the exact solution of **10** b, boundary layers are obtained of increasing thickness proportional to $\sqrt{\nu t}$. If the stream tends to separate from the body a flow in the opposite direction will set in after some time. This will also be shown by the calculations, originally due to Blasius, which follow below. But before proceeding to this, a discussion based on considerations of energy will give a good explanation of the phenomena themselves.

In the stream outside the boundary layer the Bernoulli equation holds with sufficient accuracy of approximation, and can be interpreted as meaning that the work done by the difference of pressure on a particle, equals the change in kinetic energy of the particle. If the theoretical potential flow around a cylinder be taken as the basis of consideration,

---

[1] Thom, A., The Boundary Layer of the Front Portion of a Cylinder. Br. A.R.C. R. and M. 1177, London, 1928.

[2] Fage, A., The Airflow Around a Circular Cylinder in the Region Where the Boundary Layer Separates from the Surface. Phil. Mag., vol. VII, p. 253, 1929.

[3] Falkner, V. M., and Skan, Sylvia W., Some Approximate Solutions of the Boundary Layer Equations. Br. A.R.C. R. and M. 1314, 1930.

and the state of affairs followed from the stagnation point *A* (Fig. 25) across the velocity maximum *B* to the second stagnation point *C*, it is found that a particle sliding along in the neighborhood of the cylinder, if uninfluenced by friction, is accelerated in the region of falling pressure *A* to *B*, and retarded in the region of increasing pressure *B* to *C*. By the theory of frictionless fluids, the pressure at *C* is exactly equal to the pressure at *A*, so that by the time the particle reaches *C* it has lost exactly the kinetic energy acquired in going from *A* to *B*. We will now consider a particle of fluid which is inside the boundary layer and consequently loses energy by friction. The fall of pressure between *A* and *B* takes place in the direction of flow so that even a particle which had lost all its kinetic energy would be set in motion again in the direction of the stream. The fall of pressure between *B* and *C* takes place in the opposite direction however, so that a particle at rest in that stretch would be set into motion in the direction *C* to *B*. We see therefore that a particle which still has a measurable kinetic energy at *B* would be unable to reach *C* if its kinetic energy were smaller than that of the frictionless stream, but would come to rest previously, since the full energy of the potential flow is necessary in order to reach *C*. The particles thus brought to rest will then be set into motion in the opposite direction and will move under the stream arriving from *A*. They thus alter the whole movement, including the pressure distribution on the body, until a steady motion again becomes possible. The first phases of this transformation process will be examined in more detail. The later phases have so far proved amenable only to experimental investigation. It is found that the transition from the potential flow, which actually does exist at the start, to the final flow, proceeds in such fashion that those parts of the fluid flowing in opposite directions lock together into powerful whirlpools which grow until the transformation of the flow has been accomplished.

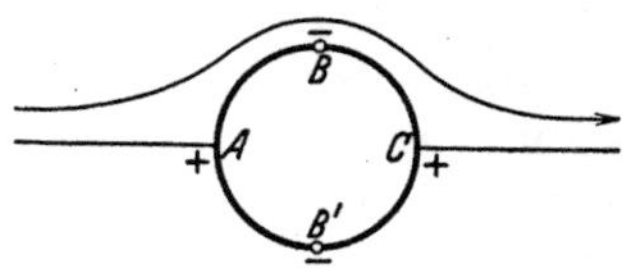

Fig. 25.

Figs. 26—37 on Plates I—V at the end of this Division show examples of this. The first example is that of a circular cylinder, the second that of a rather more elongated body. The first series of pictures consists of separate photographs, the second is taken from a cinematographic film so that the various pictures show successive stages of one and the same flow. The stream velocity was very small and the body very large, so that a relatively thick boundary layer was obtained. The various phases of the reverse flow, the formation of whirlpools (vortices) and separation of the forward flow from the body can easily be followed. It may be observed that after the formation of the first vortex, between the cylinder and the vortex the state of affairs along that part of the circumference of the cylinder is very similar to the state along the whole circumference of the cylinder at the commencement of the flow. Here too the stream starts at a stagnation point, is accelerated between the body and the vortex, and is finally retarded again. A secondary reverse flow must set in, leading to the formation of a secondary vortex. This can be seen quite clearly in Figs. 29 and 35, 36.

The analytic treatment of the commencement of the flow, as given by Blasius[1], can be summarized as follows: The differential equation of the boundary layer is,

$$\frac{\partial u}{\partial t} + u\frac{\partial u}{\partial x} + v\frac{\partial u}{\partial y} + \frac{1}{\varrho}\frac{d p}{d x} = \nu\frac{\partial^2 u}{\partial y^2} \tag{16.1}$$

At the commencement of the motion the boundary layer is very thin, and change of velocity with respect to time, of particles inside the boundary layer, very large, so that the first and last terms of the equation strongly outweigh the remaining terms. A first approximation is therefore obtained by putting

$$\frac{\partial u}{\partial t} = \nu\frac{\partial^2 u}{\partial y^2} \tag{16.2}$$

Previous considerations have shown that it is advisable to introduce a non-dimensional distance (from the wall) by means of the equation

$$\eta = \frac{y}{2\sqrt{\nu t}} \tag{16.3}$$

(The factor 2 is introduced for the sake of the special solution given later, but would be unnecessary otherwise.) The following relations must be noted in order to transform (16.2) in terms of the new variables η and t:

$$\left(\frac{\partial u}{\partial y}\right)_t = \left(\frac{\partial u}{\partial \eta}\right)_t \cdot \left(\frac{\partial \eta}{\partial y}\right)_t$$

and

$$\left(\frac{\partial u}{\partial t}\right)_y = \left(\frac{\partial u}{\partial t}\right)_\eta + \left(\frac{\partial u}{\partial \eta}\right)_t \cdot \left(\frac{\partial \eta}{\partial t}\right)_y$$

The suffix attached to the differential coefficient indicates which quantity remains constant during the differentiation. Using

$$\left(\frac{\partial \eta}{\partial t}\right)_y = -\frac{\eta}{2t}$$

and

$$\left(\frac{\partial \eta}{\partial y}\right)_t = \frac{1}{2\sqrt{\nu t}}$$

the following equation is obtained from (16.2)

$$4t\frac{\partial u}{\partial t} = 2\eta\frac{\partial u}{\partial \eta} + \frac{\partial^2 u}{\partial \eta^2} \tag{16.4}$$

If it be assumed that the velocity of the whole stream, zero at the instant $t = 0$, increases as the nth power of t for $t > 0$, the velocity of the potential flow in the neighborhood of the wall can be given by a formula

$$u_1 = t^n w_n(x).$$

---

[1] *loc. cit.* p. 20 *et seq.*

The velocity in the boundary layer may therefore be written

$$u = t^n\, w_n(x)\, \varphi(\eta)$$

with $\varphi(\eta)$ satisfying the conditions $\varphi = 0$ when $\eta = 0$, and $\varphi = 1$ when $\eta = \infty$. The abscissa x only occurs as a parameter in this first approximation. The simplified equation (16.4) now gives the following relation involving φ,

$$4\, n\, \varphi = 2\, n\, \varphi' + \varphi'' \tag{16.5}$$

With the above boundary conditions the solution is completely determined. Using this first approximation for u the process can be continued by calculating an approximate value for the expression $u\,(\partial u/\partial x) + v\,(\partial u/\partial y)$ and thence obtaining a better solution. The process can if necessary be continued and it is easy to prove that a power series in t is obtained, whose coefficients are functions of η, again containing functions of x as parameters. The series so obtained is convergent for values of t which are not too large, and gives a solution of the problem.

Blasius investigated two laws of acceleration in greater detail:

(1) The case in which movement is generated very suddenly at the time $t = 0$, *e. g.* by shock, and the motion then continues constant with respect to time. This obviously corresponds to the case $n = 0$.

(2) The case of a uniformly accelerated movement commencing at $t = 0$. Here n must be put equal to 1.

A stream function Ψ is again used to represent the complete motion which involves a velocity component v also. For the first case we write

$$\Psi = 2\sqrt{\nu t}\,[\psi_0(x_1\eta) + t\,\psi_1(x_1\eta) + \ldots] \tag{16.6}$$

where the following relations

$$\psi_0 = u_1 f_0(\eta), \qquad \psi_1 = u_1 \frac{d u_1}{d x} f_1(\eta) \tag{16.7}$$

are necessary in order to fit in with the potential flow outside the boundary layer. The boundary conditions are clearly:

$$\begin{array}{lll} \text{For } \eta = 0: & f_0 = f_0' = 0, & f_1 = f_1' = 0 \\ \text{For } \eta = \infty: & f_0' = 1, & f_1' = 0 \end{array}$$

Equation (16.5) already discussed above, is obtained for f_0, and therefore, since $u = \partial\psi/\partial y$ and $n = 0$,

$$f_0''' + 2\eta\, f_0'' = 0 \tag{16.8}$$

The equation for f_1 is:

$$f_1''' + 2\,\eta\, f_1'' - 4\, f_1' = 4\,(f_0'^2 - f_0 f_0'' - 1) \tag{16.9}$$

The solution of (16.8) has already been discussed, in essentials, in **10.** The result for f_0', in which our first interest lies, is

$$f_0' = \frac{2}{\sqrt{\pi}} \int_0^\eta e^{-\eta^2} \cdot d\eta \tag{16.10}$$

A solution can also be obtained for f_1', but it is very cumbersome. It may be found in the article by Blasius or in that by Tollmien[1].

For the second case (uniform acceleration) the series must be written

$$\Psi = 2\sqrt{\nu t}\,(t\,\psi_1(x,\eta) + t^3\,\psi_3(x,\eta) \qquad (16.11)$$

Writing the velocity in the potential flow as

$$u_1 = t\,w_1(x)$$

we have, $\psi_1(x,\eta) = w_1 f_1(\eta)$, and $\psi_3(x,\eta) = w_1 \frac{d w_1}{d x} f_3(\eta)$ (16.12)

This reduces the differential equations for f_1 and f_3 to

$$f_1''' + 2\eta f_1'' - 4 f_1' = -4$$

and

$$f_3''' + 2\eta f_3'' - 12 f_3' = 4(f_1'^2 - f_1 f_1'' - 1)$$

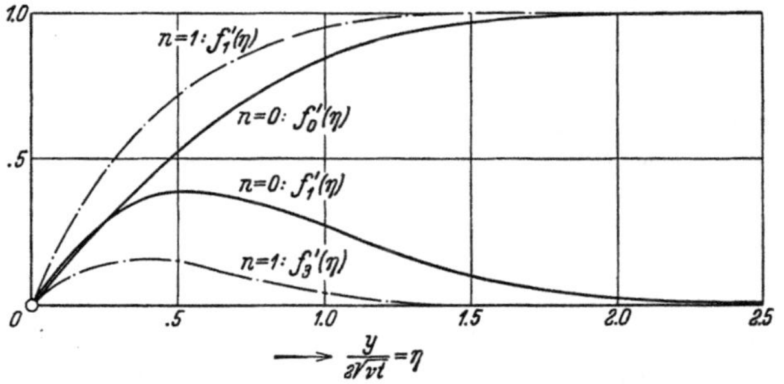

Fig. 38. Functions of the accelerated boundary layer flow.

The boundary conditions are the same as in the first case. The solution for f_1' is

$$f_1' = 1 + \frac{2}{\sqrt{\pi}}\left[\eta\, e^{-\eta^2} + (1 + 2\eta^2)\int_{\infty}^{\eta} e^{-\eta^2} d\eta\right]$$

As before f_3' can be obtained in the form of integrals, but the expression is quite unwieldy. It can be found in the article by Blasius[2]. The behavior of the four functions is shown in Fig. 38; the dotted curves refer to $n = 1$, and the full lines to $n = 0$.

If only two terms of the series (16.6) and (16.11) respectively are taken, the separation effect is clearly seen in the case of a stream which is sufficiently retarded. The results obtained from these two-termed series will naturally be trustworthy only when t is small. It would be desirable to have more terms of the series but the requisite calculations for the succeeding terms are too lengthy. Comparison with Fig. 22 shows that the condition for the point at which separation occurs is

$$\left(\frac{\partial u}{\partial y}\right)_{y=0} = 0$$

[1] Handbuch der Experimentalphysik, Bd. IV, Part I, p. 277.

[2] *loc. cit.* p. 29.

We therefore have to find the values of f_0'', f_1'' etc. at $y = 0$. The expressions thus obtained, in the case of motion starting from rest, are very simple. When $y = 0$, we get

$$f_0'' = \frac{2}{\sqrt{\pi}}\left(1 + \frac{4}{3\pi}\right)$$

$$f_1'' = \frac{2}{\sqrt{\pi}}$$

so that the following equation for the time t_a, required for separation to commence is obtained:

$$u_1 + \left(1 + \frac{4}{3\pi}\right) t_a u_1 \frac{d u_1}{d x} = 0$$

and on inserting numerical values we have

$$t_a = \frac{-0.702}{\frac{d u_1}{d x}}$$

Qualitatively, the relations are quite similar for the case of uniformly accelerated motion. The corresponding calculation gives a time of separation t_a where

$$t_a = \frac{1.53}{\sqrt{-\frac{d w_1}{d x}}}$$

In addition Blasius carried through a second approximation involving a quadratic equation for t_a^2. For separation to occur at the hindermost point the final value obtained was

$$t_a = \frac{1.445}{\sqrt{-\frac{d w_1}{d x}}}$$

From this approximation it is seen that the time taken for the separation to occur is even smaller than that given by the first approximation, so that it may be taken as quite certain that this calculation ensures the reality of the separation effect. The formulae for the separation times in the two cases considered have this in common, that separation times only occur for negative du_1/dx and dw_1/dx respectively. The point where a reverse flow first occurs is characterized by a minimum value of t_a, obviously corresponding to the maximum of $-du_1/dx$ and dw_1/dx respectively. For circular cylinders and elliptic cylinders with the major axis parallel to the stream, this maximum occurs at the hindermost point. With elliptic cylinders whose major axes are perpendicular to the direction of the stream and whose semi-axes satisfy the relation $b/a > \sqrt{4/3}$, the separation begins at two points symmetrically situated with respect to the direction of the stream, which moves farther away from the rear point as the ellipse becomes flatter. For ellipses of constant major axis in a stream of constant velocity at infinity the separation times becomes smaller as the ellipse becomes flatter. On proceeding to the limiting case of a plate placed at right angles to the stream, a separation time of zero is obtained, or in other words, the formation of vortices at the edge commences at the first instant of motion. For the rest, the separation times for very thin ellipses placed at right angles to the stream are already very small, being approximately proportional to the third power of the ratio a/b in the case of sudden generated motion with constant velocity.

It may also be noted that E. Boltze, in his dissertation[1], performed the corresponding calculations for the flow around bodies of revolution in the special case of sudden motion from rest. The calculations are more complicated, but the application of the results obtained to the movement of a sphere show exactly the same qualitative effects as have been described for the two-dimensional case. Boltze worked out the power series expansion up to terms of the 3rd degree so that his results are particularly reliable. He obtains a cubic equation for the time taken for the separation effect to reach a point lying at an angle β from the foremost stagnation point. Writing, for brevity, $t_a \frac{u_\infty}{r} = z$,

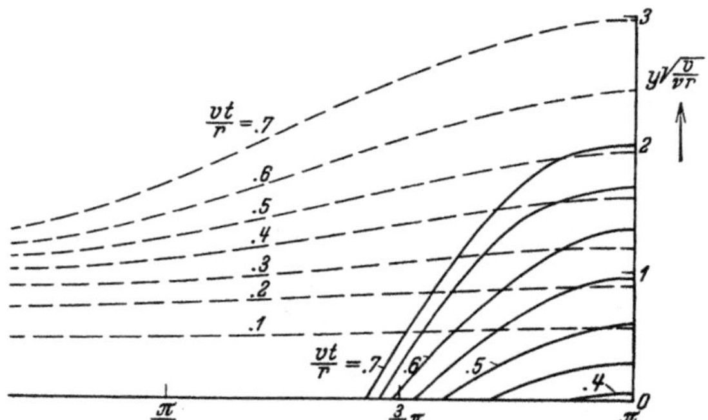

Fig. 39. Boundary layer thickness of a sphere depending on time (Boltze).

the equation becomes

$$1 + 2.360\, z \cos\beta - z^2 (0.662 \cos^2\beta - 0.180) + z^3 (0.010 \cos^3\beta - 0.017 \cos\beta) = 0$$

It is seen that the coefficient of z^3 is very small so that on neglecting this term we introduce no large error. The results can be summarized as follows: The separation begins at the rear point after a time

$$t_a = 0.392 \frac{a}{u_\infty} \qquad (a = \text{radius of sphere})$$

The values of the numerical factor for the occurrence of the backward flow at various values of β are,

for $\beta =$	160°,	140°,	120°,	110°
factor $=$	0.407,	0.521,	0.853,	1.505

[1] Grenzschichten an Rotationskörpern und Flüssigkeiten mit kleiner Reibung, Göttingen, 1908.

Boltze also computed the measure of displacement δ^*, and further the position of the stream-line $\Psi = 0$ after the separation, obtaining the diagram which is reproduced in Fig. 39. A diagram of the velocity distribution, and of the stream-lines showing the conditions within the vortex forming at $t = 0.6\, a/u_\infty$ is reproduced in Fig. 40. The radial measurements of the boundary layer have been much magnified in the diagram and the scale is the same as that of Fig. 39. It can be seen that the vortex, at this stage, exhibits only very small velocities and the highest values of rotation are found outside it, approximately at a distance δ^*.

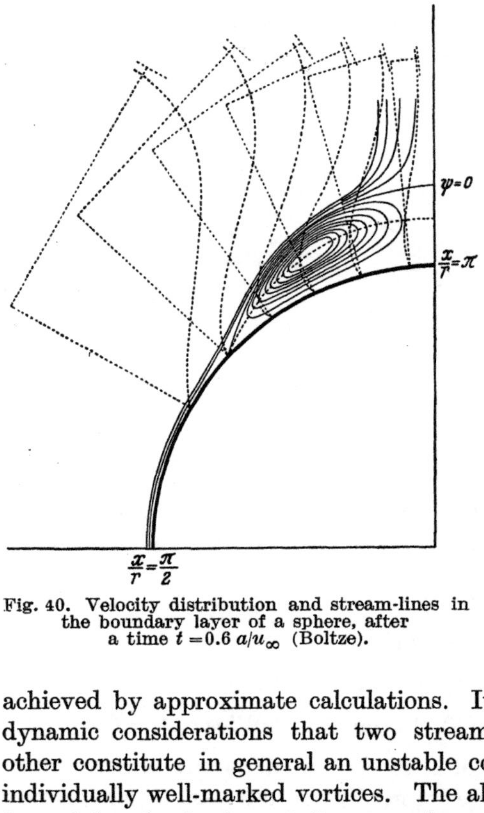

Fig. 40. Velocity distribution and stream-lines in the boundary layer of a sphere, after a time $t = 0.6\, a/u_\infty$ (Boltze).

In the succeeding stages which cannot be correctly given by the calculation without introducing higher powers of t, particles of fluid streaming in the reverse direction pile up more and more in the space between the wall and the advancing fluid, so that the shape of the exterior flow becomes changed. An analytic investigation of these occurrences has so far not been attempted and would present very great difficulties. Nevertheless something could be achieved by approximate calculations. It is known from other hydrodynamic considerations that two streams of fluid sliding over each other constitute in general an unstable configuration and tend to form individually well-marked vortices. The almost motionless body of fluid formed by the backward flow together with the more rapidly moving forward flow, constitutes such an unstable configuration. The gradual formation of vortices on account of this instability is shown by Figs. 34 to 37.

17. Theorem of Momentum and Kármán's Approximate Theory. As the preceding paragraphs have shown, the mathematical methods for handling boundary flows analytically are rather complicated and difficult, and are insufficient to dispose of all the problems whose investigation is required. It is therefore desirable to estimate the required results by means of approximate methods. Such approximations can be obtained by abandoning the attempt to satisfy the hydrodynamic differential equations for each particle and by choosing some plausible velocity distribution, merely taking care that the momentum relation obtained by integrating the differential equations, is satisfied. A simple example is the flow along a flat plate. Application of the theorem of momentum to a section AA (Fig. 41) will express the fact that the time rate of decrease in the momentum in this cross-section, as compared with any cross-section in front of the plate, where all the particles of fluid have the undisturbed velocity u_0, is equal to the sum of the frictional resistances of the plate from its beginning to the position AA. The loss per second of momentum flowing across the section AA can be calculated as follows: For a surface element of width 1 and height dy, the mass of fluid crossing per second is equal to $\varrho u\,dy$. This mass had previously a velocity u_0. Its velocity is now u. The loss of momentum per second is therefore $\varrho u (u_0 - u)$—which we shall call briefly the "momentum loss". If τ_0 is the shearing force along the plate, the resistance per unit width of one side must be

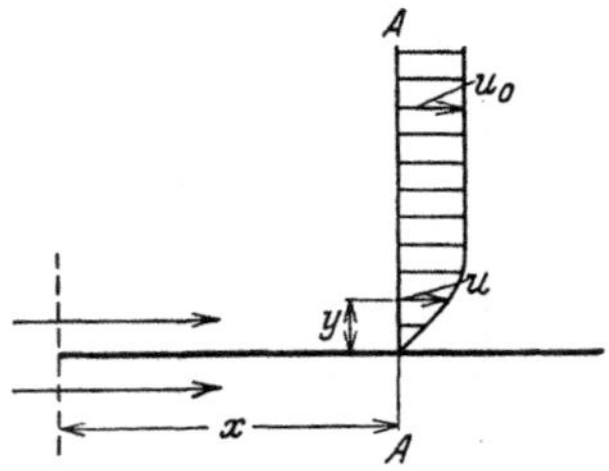

Fig. 41.

$$\frac{D}{2b} = \int \tau_0\,dx$$

which, by the above theorem, must equal the loss of momentum $\int_0^h \varrho u (u_0 - u)\,dy$. This integral ought, strictly, to extend to ∞ but since the boundary layer has the property of vanishing completely for practical purposes within a distance which is some fairly small multiple of $\sqrt{\nu x/u_0}$, a distance h of the same order of magnitude will suffice for the upper limit of the integral. The contributions of elements dy to the integral obviously vanish as soon as u becomes equal to u_0 so that the exact value of h is not important. It must only be chosen sufficiently large for u to be practically equal to u_0. The relation obtained above can be differentiated with respect to x in order to obtain the surface friction itself. This gives

$$\tau_0 = \frac{d}{dx}\int_0^h \varrho u (u_0 - u)\,dy \tag{17.1}$$

We may regard h as being independent of x, but the result would be unchanged if h were to be variable, since the integrand vanishes for $y = h$. If any suitable velocity-profile be chosen for the boundary layer, say

$$u = u_0 f\left(\frac{y}{\delta}\right) = u_0 f(\eta) \tag{17.2}$$

where δ, the thickness of the boundary layer, is to remain indeterminate for the present, the value of the momentum integral can at once be calculated. It becomes

$$\int_0^h \varrho\, u\,(u_0 - u)\, d\,y = \varrho\, u_0^2\, \delta \int_0^{h/\delta} (f - f^2)\, d\,\eta \tag{17.3}$$

Hence, once the form of f has been established, the integral, which simply furnishes a numerical value, can be completely evaluated. This numerical value will be called α, so that the momentum loss becomes equal to $\alpha\,\varrho\,u_0^2\,\delta$. The surface friction at the wall is however

$$\tau_0 = \mu \left(\frac{\partial u}{\partial y}\right)_0 = \frac{\mu\, u_0}{\delta} f'(0) \tag{17.4}$$

Putting $f'(0)=\beta$, we have from (17.1)

$$\beta \cdot \frac{\mu\, u_0}{\delta} = \alpha\, u_0^2 \frac{d\,\delta}{d\,x}$$

or

$$\delta \frac{d\,\delta}{d\,x} = \frac{\nu\,\beta}{\alpha\, u_0}$$

This equation can be integrated, giving

$$\frac{1}{2}\delta^2 = \frac{\nu\,\beta}{\alpha\, u_0} x$$

or

$$\delta = \sqrt{\frac{2\nu\,\beta\, x}{\alpha\, u_0}} \tag{17.5}$$

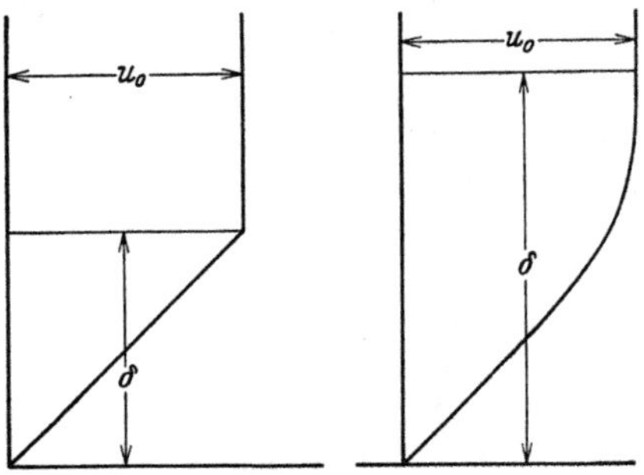

Fig. 42.

Hence the surface friction becomes

$$\tau_0 = \sqrt{\frac{\alpha\,\beta}{2} \cdot \frac{\mu\,\varrho\, u_0^3}{x}} \tag{17.6}$$

and

$$\int_0^x \tau_0\, d\,x = \sqrt{2\,\alpha\,\beta\,\mu\,\varrho\, u_0^3\, x}$$

so that the resistance for both sides of the plate, taken together, is

$$D = 2\,b \sqrt{2\,\alpha\,\beta\,\mu\,\varrho\, u_0^3\, x} \tag{17.7}$$

As a numerical example let us make the very crude assumption that the velocity distribution inside the boundary layer can be represented by a straight line: $f(\eta) = \eta$ for $\eta < 1$, and $f(\eta) = 1$ for $\eta > 1$ (see left hand diagram of Fig. 42). This gives

$$\alpha = \int_0^1 (\eta - \eta^2)\, d\,\eta = \frac{1}{2} - \frac{1}{3} = \frac{1}{6},$$

and β becomes equal to 1. Hence

$$\delta = 2\sqrt{3}\sqrt{\frac{\nu\, x}{u_0}} = 3.464\sqrt{\frac{\nu\, x}{u_0}},$$

and

$$D = \frac{2}{\sqrt{3}}\sqrt{\mu\,\varrho\, u_0^3\, x} = 1.155\sqrt{\mu\,\varrho\, u_0^3\, x}$$

It is also easy to use a function of the third degree, as represented by the right hand diagram of Fig. 42. Here can be written

$$f(\eta) = \frac{3}{2}\eta - \frac{1}{2}\eta^3$$

which must again be replaced by $f(\eta) = 1$ for $\eta > 1$. The numerical calculation, which will be omitted here, gives

$$\delta = 4.64 \sqrt{\frac{\nu x}{u_0}}$$

and $$D = 1.29 \sqrt{\mu \varrho u_0^3 x}.$$

It is also of interest to obtain the amount of displacement δ^* which is given by the general expression

$$\delta^* = \delta \int_0^{h/\delta} (1 - f)\, d\eta$$

The value of the integral is 1/2 for the first example, and 3/8 for the second. Hence, in the first case

$$\delta^* = 1.732 \sqrt{\frac{\nu x}{u_0}},$$

and in the second $$\delta^* = 1.740 \sqrt{\frac{\nu x}{u_0}}$$

The exact value is $1.73 \sqrt{\nu x / u_0}$, so that in spite of crude assumptions the value of the displacement has been obtained with astonishing exactitude, which must naturally be regarded as more or less accidental. The two examples give the values 1.155 and 1.29 respectively for the numerical constant occurring in the resistance formula, instead of 1.328. Nevertheless, when the simplicity of the calculation is considered, the result must be considered very useful.

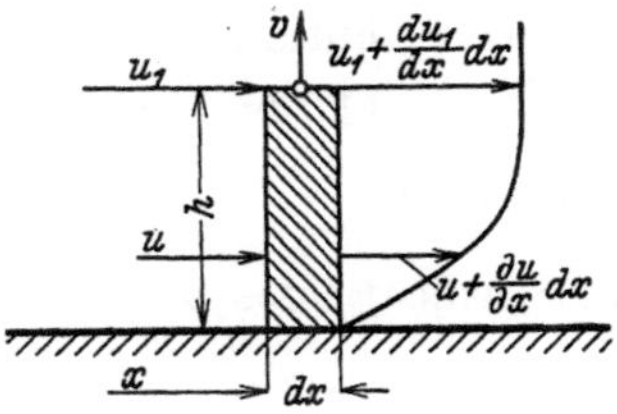

Fig. 43.

Th. v. Kármán[1] has shown that this process can be extended to cover the general case of boundary flows with fall in pressure. The calculation can be performed in the following way: The momentum relation is formulated at once in such a way as to correspond to (17.1), that is, two surfaces separated by an infinitesimal distance dx are taken and the momenta flowing in and out of them are compared. The result could also be obtained by integrating the differential equations of the boundary layer with respect to y, but it is simpler to formulate the momentum condition in this way. We consider the region which is represented in detail in Fig. 43. The height h will again be so chosen that the particles

[1] Kármán, Th. v., Zeitschrift f. angew. Math. u. Mech., vol. I, p. 235, 1921.

between $y = 0$ and $y = h$ include practically all those influenced by friction. Further, h is chosen so as to be independent of x and t so that when differentiating integrals between 0 and h, the limits need not be differentiated.

The various contributions to the x component of "momentum loss", that is to the change of momentum in the x direction, will now be described in turn. The interior of the parallelepiped of base area $h\,dx$, and unit width perpendicular to the plane of the diagram contributes $dx\int_0^h \varrho\,(\partial u/\partial t)\,dy$ to the change in the momentum. The left side contributes $\int_0^h \varrho u^2\,dy$, and on the right side an amount

$$\int_0^h \varrho\,u^2\,dy + dx\cdot\frac{d}{dx}\int_0^h \varrho\,u^2\,dy$$

is carried away. The difference is therefore $dx\cdot(d/dx)\int_0^h \varrho u^2\,dx$. In conformity with the derivation of the momentum theorem, in which we fix our attention on a fluid surface composed of the same fluid particles, the momentum carried away is considered positive, and that brought in, negative.

Turning now to the horizontal boundaries of length dx, nothing is introduced or removed at $y = 0$, but at $y = h$ a mass $\varrho v\,dx$ is removed, the x velocity being equal to the velocity of the potential flow (u_1), which, as before, can be taken independent of y in the small region h. The contribution at this place is accordingly $\varrho u_1 v\,dx$. We wish further to here express the velocity v in terms of u. By the equation of continuity

$$\frac{\partial v}{\partial y} = -\frac{\partial u}{\partial x}$$

and hence, since $v = 0$ at $y = 0$

$$v = -\int_0^h \frac{\partial u}{\partial x}\,dy$$

The time rate of change in the momentum in the x direction resulting from these contributions is now to be equated to the x component of the resulting force. Two sets of forces have to be taken into account:

1) the differences of pressure on the sides.

2) the frictional forces on the wall.

The pressure will be taken as being independent of y inside the domain h, as before. There is therefore a contribution ph from the left hand side, acting in the positive direction of the x axis. From the right hand side there is $\left(p + \frac{dp}{dx}\,dx\right)h$, acting in the negative direction of the x axis. The difference is accordingly $(-dp/dx)h\,dx$. The surface friction at $y = 0$ gives a contribution $\tau_0\,dx$ over the base of area $(dx \times 1)$. This force acts on the fluid in the negative direction of the x axis and is therefore to be introduced with a negative sign. At the upper end $(y = h)$ there is a potential flow, and the shear is practically zero. Hence, after division by dx the "momentum relation" reduces to

$$\int_0^h \varrho \frac{\partial u}{\partial t}\,dy + \frac{d}{dx}\int_0^h \varrho\,u^2\,dy - \varrho\,u_1 \int_0^h \frac{\partial u}{\partial x}\,dy = -h\frac{dp}{dx} - \tau_0 \quad (17.8)$$

On account of the invariance of h with respect to t and x the differential operators in the first and third terms of this equation can naturally be taken outside the integral. It is however, advisable to replace the pressure p by the appropriate expression involving the velocity of the potential flow as given by the Bernoulli equation. Since there is no friction, and gravitational forces may be neglected (see the remarks in 7) this equation is $\dfrac{dp}{dx} + \varrho\left(\dfrac{\partial u_1}{\partial t} + u_1\dfrac{du_1}{dx}\right) = 0$

[the term $v\,(du_1/dy)$ can be neglected on account of the smallness of v in the neighborhood of the wall in the region considered]. If the value of dp/dx thus obtained be substituted in (17.8) it can be so arranged that the integrands of the three integrals vanish identically for $y = h$, which shows that the choice of the length h is unimportant provided that it is greater than the thickness of the region influenced by friction. In order to put the equations into the correct form, the term $hu_1\,(du_1/dx)$ which must stand on the right hand side of (17.8) with a positive sign, must occur with a factor two in the second integral and be subtracted with a factor one in the third integral. This leads to the new equation

$$\tau_0 = \frac{d}{dt}\int_0^h \varrho\,(u_1 - u)\,dy + \frac{d}{dx}\int_0^h \varrho\,(u_1^2 - u^2)\,dy - \varrho\,u_1\frac{d}{dx}\int_0^h (u_1 - u)\,dy \quad (17.9)$$

which is the required extension of (17.1). If a fixed velocity u_0 be substituted for the velocity u_1 which varies with x and t, and in addition a steady flow be taken, (17.9) at once reduces to (17.1).

The case of steady flow with a velocity u_1 varying with respect to x [so that the first integral of (17.9) vanishes] can be interpreted as follows: Denote the "momentum loss" for the flat plate, in accordance with the considerations at the beginning of this section, by M, so that

$$M = \varrho \int_0^h u\,(u_1 - u)\,dy$$

The integrand can be written as

$$u\,(u_1 - u) = u\,u_1 - u^2 = (u_1^2 - u^2) - u_1\,(u_1 - u)$$

This makes

$$\frac{dM}{dx} = \frac{d}{dx}\int_0^h \varrho\,(u_1^2 - u^2)\,dy - \varrho\,u_1\frac{d}{dx}\int_0^h (u_1 - u)\,dy - \varrho\frac{du_1}{dx}\int_0^h (u_1 - u)\,dy,$$

and hence

$$\tau = \frac{dM}{dx} + \varrho\frac{du_1}{dx}\int_0^h (u_1 - u)\,dy$$

or, by introducing the displacement δ^*, [see (14.10)],

$$\tau = \frac{dM}{dx} + \varrho\,u_1\frac{du_1}{dx}\delta^* \tag{17.10a}$$

$$= \frac{dM}{dx} - \delta^*\frac{dp}{dx} \tag{17.10b}$$

Hence it is seen that in addition to the contribution given by (17.1) the fall in pressure acting on the displaced mass of fluid supplies an additional term.

In order to interpret the magnitude M a special length can be introduced by the equation $M = \varrho u_1^2\,\delta^{**}$

(δ^{**} is a measure of the momentum change, for example, for the straight line velocity profile in the approximate calculation for the flat plate, $\delta^{**} = \frac{1}{6}\delta = \frac{1}{3}\delta^*$). Introducing δ^{**} into (17.10a) it is easily found that

$$\tau = \varrho\,u_1\frac{du_1}{dx}(\delta^* + 2\,\delta^{**}) + \varrho\,u_1^2\frac{d\delta^{**}}{dx} \tag{17.11}$$

Later use will be found for this equation.

In accordance with a suggestion due to v. Kármán, the theorem of momentum, written as in (17.8) or (17.9), is now used in conjunction with the boundary layer equation [see (13.4)] in which we put $y = 0$. Here, in the immediate neighborhood of the wall, the velocities u and v are both zero and the equation reduces to

$$\left(\frac{d^2u}{dy^2}\right)_0 = \frac{1}{\mu}\frac{dp}{dx} = -\frac{u_1}{\nu}\frac{du_1}{dx} \tag{17.12}$$

Now let us choose some function suitable for representing the distribution of velocities in boundary layers. This function must satisfy the boundary condition $u = 0$, the equation just obtained, and must also permit the transition to the velocity u_1 given by the potential flow. The thickness of the boundary layer is left as an undetermined parameter. If now this first approximation towards a velocity profile is also made to satisfy Kármán's momentum condition, the following result is obtained: the hydrodynamic equations are satisfied in the immediate neighborhood of the wall and also outside the boundary layer, where a potential flow has been assumed. In the boundary layer itself the differential equation is not satisfied—but an integral of the equation is satisfied. It is therefore to be expected that useful approximations can in general be obtained by this method. The usefulness depends on the suitability of the functions chosen to represent the velocity profile which actually occurs in the boundary layer. At v. Kármán's suggestion, K. Pohlhausen[1] took a polynomial of the fourth degree as an approximation to the profile,

$$u = a\,y + b\,y^2 + c\,y^3 + d\,y^4 \tag{17.13}$$

This already satisfies the boundary condition $u = 0$ for $y = 0$. For $y = 0$, $du/dy = a$ and hence $\tau_0 = \mu a$. Also from (17.12)

$$b = -\frac{u_1}{2\nu}\frac{d u_1}{d x}$$

The following are the conditions to be satisfied at $y = \delta$

$$u = u_1$$

$$\frac{\partial u}{\partial y} = 0$$

$$\frac{\partial^2 u}{\partial y^2} = 0$$

In this manner a very good transition to the constant velocity u_1 is obtained. The velocity profile passes over at a point of inflection into the line of constant velocity u_1.

The values of the quantities a, b, c, d have now to be calculated. For simplicity, differentiations with respect to x will be denoted by primes (thus u_1' for du_1/dx, etc.). Pohlhausen introduces the non-dimensional magnitude $\lambda = u_1'\,\delta^2/\nu$ and obtains

$$\left.\begin{aligned} a &= u_1\frac{(12+\lambda)}{6\,\delta} & b &= -u_1\frac{\lambda}{2\,\delta^2} \\ c &= -u_1\frac{(4-\lambda)}{2\,\delta^3} & d &= u_1\frac{(6-\lambda)}{6\,\delta^4} \end{aligned}\right\} \tag{17.14}$$

A long calculation, here omitted, is required to produce the integrals in the momentum theorem (17.9). As can be seen from (17.11) the differential coefficient $d\,\delta^{**}/dx$ occurs, which naturally gives an expression for $d\,\delta/dx$. The momentum theorem therefore supplies a differential equation of the first order in δ by means of which δ may be obtained. This differential equation, according to Pohlhausen is

$$\frac{dz}{dx} = \frac{0.8\left[-9072 + 1670.4\,\lambda - \left(47.4 + 4.8\frac{u_1 u_1''}{u_1'^2}\right)\lambda^2 - \left(1 + \frac{u_1 u_1''}{u_1'^2}\right)\lambda^3\right]}{u_1\left[-213.12 + 5.76\,\lambda + \lambda^2\right]} \tag{17.15}$$

where

$$z = \frac{\lambda}{u_1'} = \frac{\delta^2}{\nu}$$

Since $\lambda = u_1'\,z$ this is a differential equation of the first order in z dependent on x, and can therefore in general only be solved numerically

[1] Pohlhausen, K., Zur näherungsweisen Integration der Differentialgleichung der laminaren Grenzschicht. Zeitschr. f. angew. Math. u. Mech., Vol. I, p. 252, 1921.

or graphically if the velocity u_1 and its first two differential coefficients are given as functions of x. The most suitable is the so-called "isoclinic" method of solution, in which the curves $dz/dx =$ constant are plotted and in which the direction of the integral curves is marked on these lines. This can be arranged by first calculating the value of

$$k = \frac{dz}{dx} = \frac{P(x, z)}{Q(x, z)}$$

for a series of values of z, from which the lines $k =$ constant can then be interpolated. Finally the directions $\tan \alpha = k$ can be marked on the

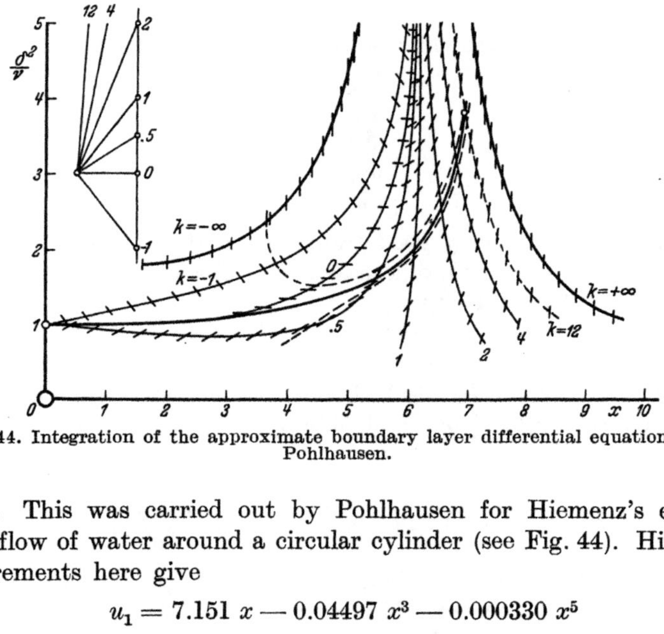

Fig. 44. Integration of the approximate boundary layer differential equation after Pohlhausen.

curves. This was carried out by Pohlhausen for Hiemenz's example of the flow of water around a circular cylinder (see Fig. 44). Hiemenz's measurements here give

$$u_1 = 7.151\,x - 0.04497\,x^3 - 0.000330\,x^5$$

where the length x, measured along the arc, and u_1 are measured in cm. and cm. per sec. respectively. The kinematic viscosity is taken as 0.01 corresponding to water at 20° C.

The following detailed comments may be made on the differential equation (17.15): At a stagnation point, that is, where $u_1 = 0$, dz/dx becomes equal to infinity, with the exception of those values of λ for which the numerator of the fraction also vanishes. Thus dz/dx becomes indeterminate for the real roots of the cubic equation

$$-9072 + 1670.4\,\lambda_0 - 47.4\,\lambda_0^2 - \lambda_0^3 = 0$$

which gives $$\lambda_0 = 7.052$$

As may be seen from Fig. 44, it is precisely the solution going through this singular point which is of interest. It is shown by the continuous curve in Fig. 44. Two other solutions are indicated by dotted lines. Bearing in mind the interpretation of λ, a boundary layer thickness is obtained at the stagnation point of

$$\delta_0 = \sqrt{\frac{\nu \lambda_0}{u_1'}} = 2.65 \sqrt{\frac{\nu}{u_1'}}$$

The point at which separation occurs was previously shown to be given by $(\partial u/\partial y)_0 = 0$. With the polynomial of degree four at present under discussion this condition obviously implies that the coefficient a is zero. From the above discussion, this is the case for $\lambda = -12$. The curve for $\lambda = -12$ is drawn in Fig. 32. The results of the approximate calculation agree well with Hiemenz's calculations for about three quarters of the way from the stagnation point to the separation point. For the last quarter of the way however, deviations arise, since the boundary layer thickness in Pohlhausen's approximation increases more rapidly than in Hiemenz's. The position of the point of separation however, agrees with that given by Hiemenz to within less than 1 per cent.

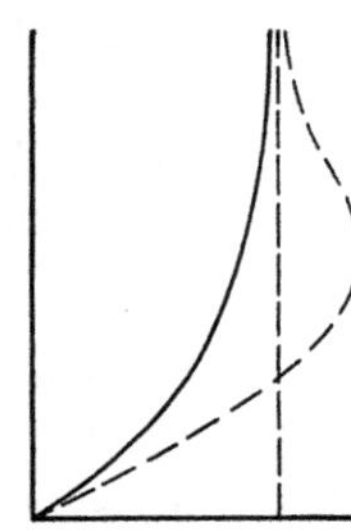

Fig. 45.

Although the Kármán-Pohlhausen calculations work very well in this, as in some other cases which have been tested, it may be pointed out that there are also cases in which the method fails. This occurs just in those cases in which physical considerations may lead us to expect that, concerning the further development of the boundary layer, all should be quite in order, as for instance when a boundary layer, developed over a long approach, suddenly receives a very strong acceleration in a short distance (flow toward an opening or something similar). In this case the condition

$$\left(\frac{\partial^2 u}{\partial y^2}\right)_0 = -\frac{1}{\nu} u_1 u_1'$$

in the Pohlhausen polynomials produces velocity distributions similar to the dotted line of Fig. 45, whereas the actual boundary layer has the character of the continuous line of Fig. 45. In this case other functions must be introduced instead of the Pohlhausen polynominal. A suitable process has, however, not yet been found.

The criterion for the failure of the Pohlhausen method is, as the foregoing shows, the occurrence of values of u such that $|u| > |u_1|$. If a Taylor expansion is taken at the point $y = \delta$, the terms with $\partial u/\partial y$ and $\partial^2 u/\partial y^2$ vanish in accordance with the conventions agreed upon. Hence the sign of $\partial^3 u/\partial y^3$ is decisive in determining whether $|u|$ is greater or less than $|u_1|$. If $d^3 u/\partial y^3$ has the same sign as u_1 then $|u| < |u_1|$ for $y < \delta$. It is easy to verify that if this is to be the case λ must be less than 12. Equation (17.15) gives $dz/dx = \infty$ when $\lambda = 12$ so that when searching for the solution, attention is at once drawn to this condition of failure.

Kármán's momentum method is, as (17.8) or (17.9) at once shows, also suited to non-steady flows. If the velocity of the potential flow u_1 is given as some function of x and t, the method leads to a partial differential equation of the first order for the auxiliary quantity z, which can be solved by a suitable graphical process. In those cases where it can be assumed that the velocity u_1 becomes steady after a certain time, it is found that the boundary layers inside the region stretching up to the separation point tend towards this steady value. But beyond the separation point the thickness of the boundary layer is found to increase indefinitely. An investigation of this kind has been made by W. Tollmien but is as yet unpublished. Naturally the lack of agreement shown in Fig. 45 can also appear here, so that care must be taken in the investigation. The small regions where the method breaks down can often be bridged by some reasonable guess as to the variation of the thickness of the layer.

18. Prevention of the Separation Effect. In practical cases of fluid motion the problem of passing from a low pressure to a higher one often arises. In such cases it is very desirable that the stream should not separate from the wall since that would interfere with the intended increase of pressure. The aerofoil is a good example of this state of affairs: In order to produce a lift on an aerofoil its upper side (the suction side) must be a surface of low pressure; but passing along to the trailing edge, the pressure must there rise to that of the undisturbed pressure at the edge itself. Another example of the same nature is the motion in a canal of gradually increasing section, a flow, the purpose of which is to transform kinetic energy of the stream into pressure.

We may at first investigate how such increases of pressure are possible with ordinary boundary flows. The approximate method of Kármán-Pohlhausen can here be applied. We first ask what retardation laws are permissible if no separation is to occur. We shall assume that the boundary layer moves with some velocity derived from a potential almost up to the separation point and that the form of the velocity-profile through the boundary layer remains constant from that point onward. This simply means that the parameter λ remains constant from this point onward. The profile at the separation point corresponds to $\lambda = -12$. The profile of the boundary layer may, for example, be allowed to continue until $\lambda = -10$ and then be made to remain at this value. From a previous equation λ is equal to $u_1' z$, so that with λ equal to -10 we have $z = -10/u_1'$, and hence

$$\frac{dz}{dx} = +\frac{10\, u_1''}{u_1'^2}$$

If the factor u_1 is taken out of the denominator of (17.15) and put on the other side of the equation, we find with the above value of dz/dx that the expression $u_1 u_1''/u_1'^2$ appears on both sides of the equation. Writing σ as an abbreviation for this expression and evaluating the various numbers, we find

$$10\,\sigma = \frac{(-23600 + 416\,\sigma)}{(-170.7)}$$

or

$$\sigma = 11.13$$

In order to simplify the calculation, which after all is only an approximate one, we put σ equal to 11 and obtain

$$\frac{u_1''}{u_1'} = 11\,\frac{u_1'}{u_1}$$

or, on integration, $log\ u_1' = 11\ log\ u_1 + log\,(-C_1')$

or

$$\frac{u_1'}{u_1^{11}} = -C_1'$$

which, when integrated, becomes

$$\frac{1}{10}\,u_1^{-10} = C_1'\,x + C_2$$

The constant C_2 is so determined that at $x = 0$, that is, at the point where the imposed condition shall begin, $u_1 = u_0$. In that case, obviously

$$C_2 = \frac{1}{10}\,u_0^{-10}$$

and therefore, putting $u_0^{10}\,C_1' = C_1$, we have

$$u_1 = \frac{u_0}{(1 + 10\,C_1\,x)^{0.1}} \tag{18.1}$$

In order to calculate how the boundary layer thickness varies it may be observed that

$$\lambda = \frac{u_1'\,\delta^2}{\nu} = -10$$

so that

$$\delta = \sqrt{\frac{-10\,\nu}{u_1'}}$$

Since

$$u_1' = -\frac{C_1\,u_0}{(1 + 10\,C_1\,x)^{1.1}}$$

it follows that

$$\delta = \sqrt{\frac{10\,\nu}{C_1\,u_0}}\,(1 + 10\,C_1\,x)^{0.55} \tag{18.2}$$

If the thickness of the boundary layer is put equal to δ_0 for $x = 0$, then

$$C_1 = \frac{10\,\nu}{u_0\,\delta_0^2}$$

and we have the relation

$$\delta = \delta_0\left(1 + 100\,\frac{\nu\,x}{u_0\,\delta_0^2}\right)^{0.55} \tag{18.3}$$

As a numerical example we may consider the case in which half the pressure at the stagnation point is to be regained. This is equivalent to stating that a velocity $u_1 = u_0/\sqrt{2}$ must be attained. Writing the denominator of (18.1) as $(1 + p)^{0.1}$, we have

$$(1 + p)^{0.1} = \sqrt{2}$$

Hence $$p = 2^5 - 1 = 31$$

This gives the required length

$$x_1 = \frac{31\, u_0\, \delta_0^2}{100\, \nu} = 0.31\, \delta_0 \frac{u_0\, \delta_0}{\nu}$$

This shows that the initial boundary layer thickness plays a very important part. If it be doubled, x_1 is quadrupled. The thickness of the boundary layer at the end of the length x_1 is also of interest, and obviously is equal to

$$\delta_1 = \delta_0\,(1 + p)^{0.55} = \delta_0 \cdot 32^{0.55} = 6.72\, \delta_0$$

The thickness of the boundary layer has therefore increased almost seven fold. If on the other hand it is desired to use three-quarters (instead of a half) of the initial stagnation pressure, the same process must be continued under less favorable conditions—which are due to the new value for δ_0. We should then require a length 32 times as great as the one obtained, and the boundary layer thickness at the end of this length would be $$\delta_2 = 45.2\, \delta_0$$

This shows that in practice great increases of pressure are not to be obtained by this method.

A good standard of comparison for possible increases of pressure is furnished by the non-dimensional ratio of the increase of pressure over a length along the x axis equal to the boundary layer thickness, to the stagnation pressure produced by the velocity due to the potential flow just outside the boundary layer. The formula for this ratio is $\left[\delta \frac{d\,p}{d\,x} \Big/ \frac{1}{2} \varrho\, u_1^2\right]$ which may also be written as $[2\, \delta\, u_1'/u_1]$ since $dp/dx = -\varrho\, u_1\, u_1'$. Substituting the values of δ, u, u' and using the value of the constant C_1,

we have $$\frac{\delta \dfrac{d\,p}{d\,x}}{\dfrac{1}{2}\varrho\, u_1^2} = \frac{20\, \nu}{\delta_0\, u_0} (1 + 10\, C_1\, x)^{-0.45}$$

This can also be written in a somewhat different form. It is easy to verify that the Reynolds number for the boundary layer at any point x is

$$R_\delta = \frac{u_1\, \delta}{\nu} = \frac{u_0\, \delta_0}{\nu} (1 + 10\, C_1\, x)^{0.45}$$

from which the non-dimensional ratio is given as

$$\frac{\delta \dfrac{d\,p}{d\,x}}{\dfrac{1}{2}\varrho\, u_1^2} = \frac{20}{R_\delta} \tag{18.4}$$

This is a very simple and clear result[1]. The numerical factor 20 is naturally connected with the previous assumption that $\lambda = -10$. If instead we were to take the separation profile itself, *i. e.* $\lambda = -12$, almost the same numerical value is obtained for σ, namely 11.09. Within the limits of accuracy here achieved the exponents therefore remain unaltered, but C_1 becomes equal to $12\,\nu/u_0\,\delta_0^2$, which may be explained as due to a somewhat larger δ_0, other conditions being equal. The non-dimensional ratio now has the value $24/R_\delta$ instead of that given by (18.4). Since values of R_δ of the order of magnitude 1,000 easily occur for fluids of slight viscosity, it can be seen how small is the capacity of laminar boundary layers to bear increases of pressure. If at some point the value of δ does not produce the difference of pressure desired, then the increased value of δ at a point farther along is still less favorable to its realization.

It is very fortunate for technical applications that with higher Reynolds numbers R_δ, the flow no longer remains as the previous calculations would indicate; instead, the phenomena denoted by the name *turbulence*, and consisting of an irregular mixing movement, appear. This mixing with the exterior flow results in momentum being continually brought to the wall and conversely, in the retarded liquid at the wall being carried away into the free fluid and there accelerated by mixture with surrounding particles. It is this phenomenon that makes possible the increases of pressure actually observed in technical application. This will be discussed in some detail in the following section. If the velocity of such a flow be continually reduced, the turbulence may disappear. If it does, we are left with a flow of a kind that has been described above, a flow where very slight increases of pressure are possible and where the fluid very soon separates from the body. Turbulence, which may appear undesirable on account of the fact that it causes an increase of frictional resistance, is, however, useful from the point of view that it makes possible considerable increases of pressure, and is thus a very necessary phenomenon for many technical applications.

A few methods will now be discussed, the purpose of which is to avoid the separation of the stream in the case of increase of pressure:

1) By the special application of energy in a comparatively narrow boundary region, the velocity can be increased so that with the same increase of pressure the non-dimensional ratio $\delta \frac{d p}{d x} \Big/ \frac{1}{2} \varrho\, u_1^2$ falls below the value given in (18.4). One means of producing such an increase is to allow a jet of fluid to flow out under pressure (Fig. 46). If the width of the jet of increased velocity be small the energy involved is unimportant. In practice the gain with this method is not large since

[1] See also Prandtl, L., On the Rôle of Turbulence in Technical Hydrodynamics. Proc. World's Engineering Congress, vol. V, Tokyo, 1929.

the band of fluid having increased velocity produces an unstable flow and is therefore soon destroyed by the resulting vortices.

2) Another method, specially applicable to aerofoils, but also to other types of flow, consists in arranging one or more slits in the body, as in Fig. 47; a flow which is practically unretarded can pass through these slits. The effect, in contrast with that of an unbroken surface (from A to D in Fig. 47), is that the boundary layer formed from A to B is

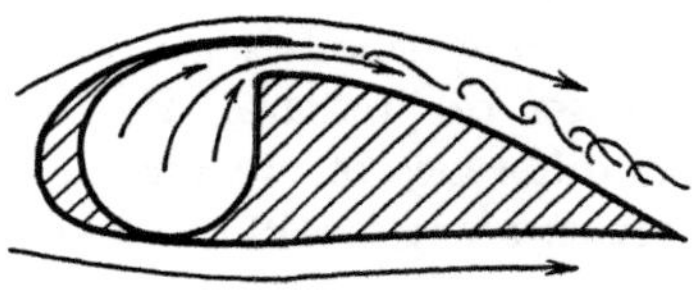

Fig. 46.

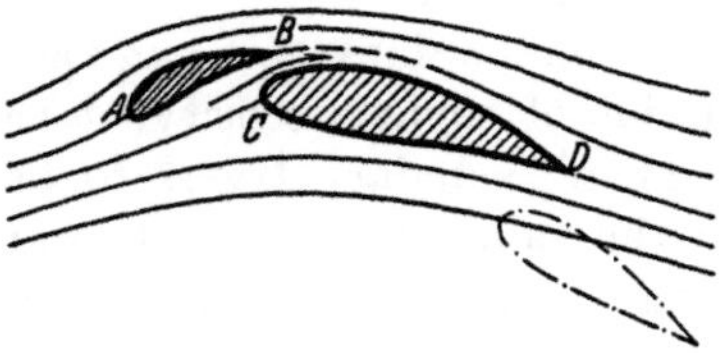

Fig. 47.

brought into the free fluid before it separates from the body, and so becomes harmless. At C a new boundary layer is formed. This is however, very thin at first and can flow on to D without separating. If necessary it can be arranged that a new section of the aerofoil begins at D. This procedure has also proved successful with bodies of other forms, *e. g.* with the Townend Ring and the related N. A. C. A. Cowling. Very good deflections around blunt bodies are obtained by means of this method (see Fig. 48). Instructive experiments on this type of flow have been made by G. Flügel[1].

3) A third and very effective method capable of producing a further continuous increase of pressure (see Fig. 49) consists of sucking that part of the boundary layer retarded by friction, into the interior of the

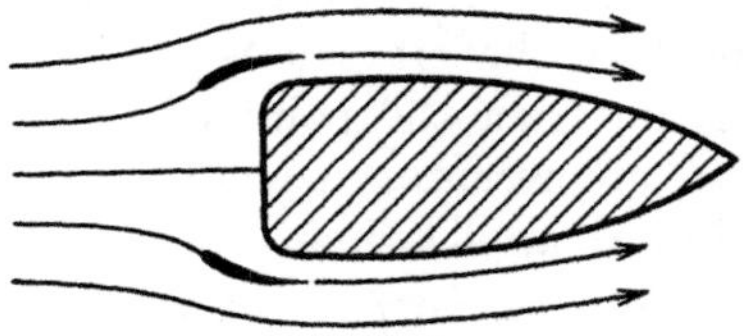

Fig. 48.

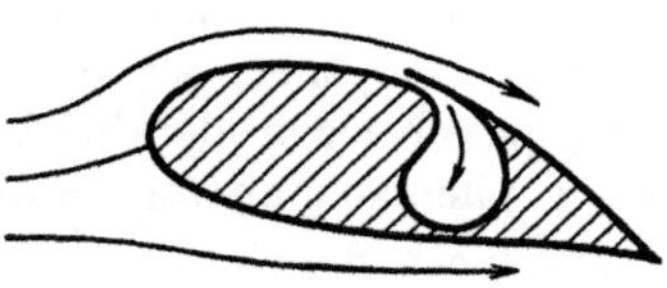

Fig. 49.

body, at a point forward or slightly to the rear of the point of separation. No appreciable reverse flow can then occur because the boundary layer material is diverted, and a new boundary layer commences beyond this point, which again permits another pressure increase (see Fig. 49). If necessary, several slits can be arranged one behind the other. If the slit is made as in Fig. 50 so that the wall behind the slit is a continuation of the wall in front of it, and does not project as in Fig. 49, the result is to produce a disturbance in the external stream equal to that which

[1] Jahrbuch der Schiffbautechnischen Gesellschaft, p. 87, 1930.

would be produced in the potential flow by a sink of appropriate strength situated at the slit. This, of course, somewhat alters the pressure distribution in the neighborhood of the slit and therefore also the boundary layer flow. There is a fall of pressure on the upstream side of the slit and a stagnation point immediately behind with higher pressure. It is seen that this local pressure distribution in the neighborhood of the slit favors very much a pressure increase without separation, especially when the suction is rather strong. A pressure distribution produced by suction strong enough to cause parts of the potential flow to be drawn into the slit is shown in Fig. 50 by the continuous line. If it be compared with the pressure distribution of a potential flow without suction it can be seen that a very considerable increase in pressure can be obtained as a result of the occurrence of the stagnation point P at the right edge of the hole. There is no danger of separation, since the body is touched at that point by a stream-line which until then has been quite uninfluenced by friction. No pressure disturbance of this kind occurs in the flow of Fig. 49 if the suction strength is suitably chosen, and in accordance with the previous exposition a measurably favorable effect is to be expected also in this case, even when the suction is much weaker. But neglecting the greater difference of pressure required to produce the suction of the flow in Fig. 50 it can at once be seen that greater effects can be produced here by the pressure field than is the case in Fig. 49.

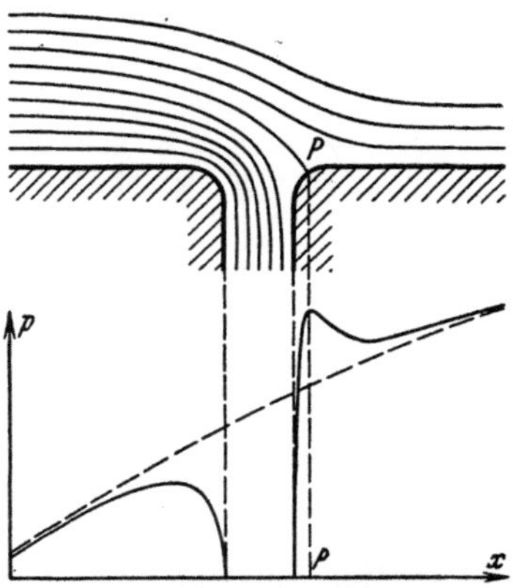

Fig. 50. Pressure distribution near a suction slit.

A very considerable effect can also be obtained by making the wall porous over the whole surface and thus producing a weak but uniformly distributed suction. This can be verified by the following calculations:

In order to make the problem as simple and as clear as possible it will be assumed that the suction is so powerful all over the region of increase of pressure that the boundary layer thickness δ remains constant. In formulating the momentum relation in the manner shown in the previous paragraph, it must be noticed that in addition to the momenta there considered there is a further term in the present case corresponding to the fluid mass absorbed into the wall with velocity component $u = 0$, and replaced by an equal mass with velocity u_1 entering the region under consideration at $y = \delta$. If the velocity of suction be denoted by c so that the mass of fluid absorbed per unit surface area is equal to $\varrho\, c$, the additional momentum is $\varrho\, c u_1$. Instead of (17.11) the correct formula

is here
$$\tau_0 = \varrho c u_1 + \varrho u_1 \frac{d u_1}{d x}\left(\delta^* + 2\delta^{**}\right) + \varrho u_1^2 \frac{d\delta^{**}}{d x}, \tag{18.5}$$
where
$$\tau_0 = \mu\left(\frac{\partial u}{\partial y}\right)_0$$

We will assume that the velocity profile throughout the system is that at the separation point (see Fig. 51), and therefore, in accordance with the previous paragraph, and because $\lambda = -12$, we have
$$u = u_1\left[6\left(\frac{y}{\delta}\right)^2 - 8\left(\frac{y}{\delta}\right)^3 + 3\left(\frac{y}{\delta}\right)^4\right]$$
from which it follows that
$$\delta^* = \frac{2}{5}\delta \quad \text{and} \quad \delta^{**} = \frac{4}{35}\delta$$
so that
$$\delta^* + 2\delta^{**} = \frac{22}{35}\delta$$

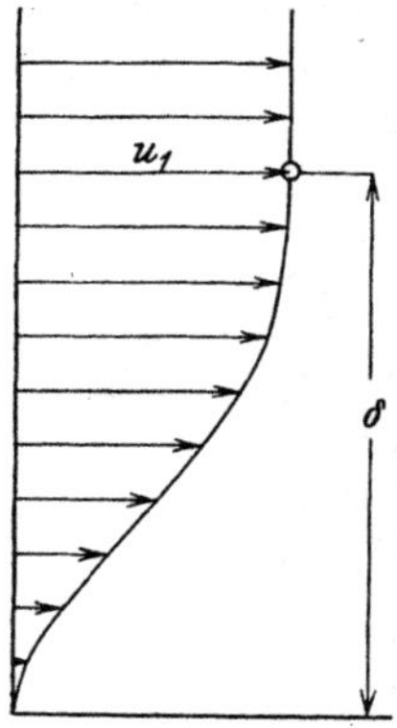

Fig. 51. Separation profile after Pohlhausen.

Since $(\partial u/\partial y)_0 = 0$, τ vanishes; and since it has been postulated that δ should be constant $d\delta^{**}/dx$ vanishes, so that the momentum relation reduces to
$$c = -\frac{22}{35}\delta\frac{d u_1}{d x} \tag{18.6}$$

In accordance with the Kármán-Pohlhausen method we have yet to obtain the relation given by the differential equation of the boundary layer for $y = 0$. At this point $u = 0$, and $v = -c$, so that
$$-c\left(\frac{\partial u}{\partial y}\right)_0 - u_1\frac{d u_1}{d x} - \nu\left(\frac{\partial^2 u}{\partial y^2}\right)_0 \tag{18.7}$$
In the present case the first term vanishes since $(\partial u/\partial y)_0 = 0$. Further,
$$\left(\frac{\partial^2 u}{\partial y^2}\right)_0 = \frac{12 u_1}{\delta^2},$$
which gives
$$\delta = \sqrt{\frac{12\nu}{-(d u_1/d x)}} \tag{18.8}$$
and from (18.6) and (18.8)
$$c = 2.18\sqrt{-\nu\frac{d u_1}{d x}} \tag{18.9}$$

Formula (18.8) shows that these assumptions are particularly applicable to an example for which $-du_1/dx$ is constant (as for example in the neighborhood of a stagnation point). The formula shows clearly that, for example, in the case of flow around the rear stagnation point of a cylinder, freedom from separation can be obtained by suction over the whole surface.

If an arbitrary law be given for du_1/dx it would be necessary to revert to (18.5) and the solution could then be found by methods similar to those adopted in Pohlhausen's example of the previous section. In all cases it can be proved that an arbitrary potential flow can be generated by the use of suitable suction methods. This is of special importance in experimental investigations where such flows are required.

4) For the sake of completeness another process for the elimination of separation must be mentioned, which though very effective can be applied only in rare cases. If the walls of the canal, or the surfaces of the rigid body as the case may be, are set into motion so that they move with the velocity of the potential flow, or with a somewhat greater velocity, the friction of the fluid nowhere decreases the kinetic energy below that of the potential flow. It follows then that there is no reverse flow and no separation. If, for example, two rotating cylinders, as

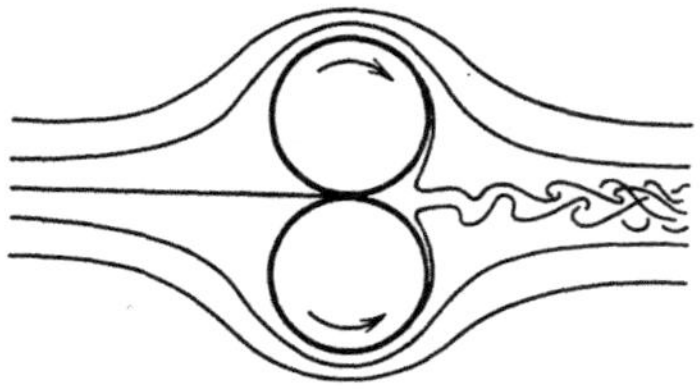

Fig. 52.

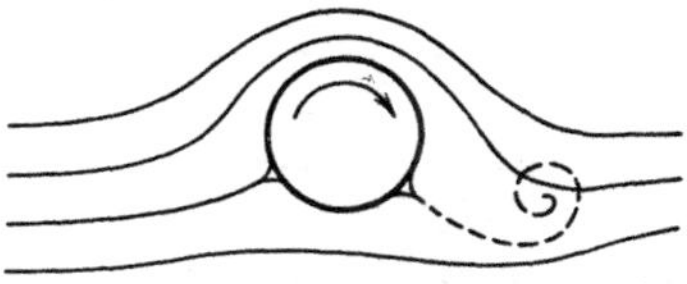

Fig. 53.

represented in Fig. 52 are in a stream, all separation will be absent if the velocity of rotation is sufficiently great. The velocity of the cylinders in the neighborhood of the rear stagnation point produces a jet of fluid which is unstable in the remaining fluid and generates small vortices. If the flow could be correctly produced at the cylinders, a result difficult to achieve, it would exhibit no resistance, but rather furnish a slight forward thrust which nevertheless would only be a fraction of the effort involved in rotating the cylinders. When a stream flows around a single rotating cylinder, no separation occurs on the side of the cylinder moving in the same sense. A vortex therefore forms only on the other side, see Fig. 53. The flow around the cylinder thereby constitutes a circulation and gives rise to a corresponding lateral thrust. This phenomenon is called the Magnus[1] effect, after its discoverer.

The behavior of boundary layers in the case of rotating cylinders has been further studied by W. Tollmien[2].

19. The Facts about Turbulent Flow. Actual flows at high Reynolds numbers are characterized by a peculiar phenomenon that has been

[1] Magnus, Abhandlungen d. Berliner Akad., 1852, or Pogg. Annalen, Vol. 88, 1853, p. 1, and further Prandtl, L., Magnuseffekt und Windkraftschiff. Naturwissenschaften, 13. Jahrg., Part 6, p. 93, 1925.

Also, Ackeret, Das Rotorschiff und seine physikalischen Grundlagen. Göttingen, 1925.

[2] See his dissertation, Göttingen 1924 (manuscript); further, Handbuch der Experimentalphysik, vol. IV, part 1, p. 241.

named "Turbulence". If we consider first of all a long straight tube with constant cross-section and smooth walls, the previous exposition gives a flow in which every particle of fluid advances uniformly and rectilinearly, the exterior layers moving more slowly than the interior ones on account of the viscosity of the fluid. In order to preserve the motion, a fall of pressure along the axis of the tube is required of magnitude given by the Hagen-Poiseuille formula. If flows of various maximum velocities are compared, the fall in pressure should be proportional to the first power of this velocity according to the formula. This however, as a rule, is not observed in practice. The fall of pressure is much more nearly proportional to the square of the velocity, and at any rate increases with

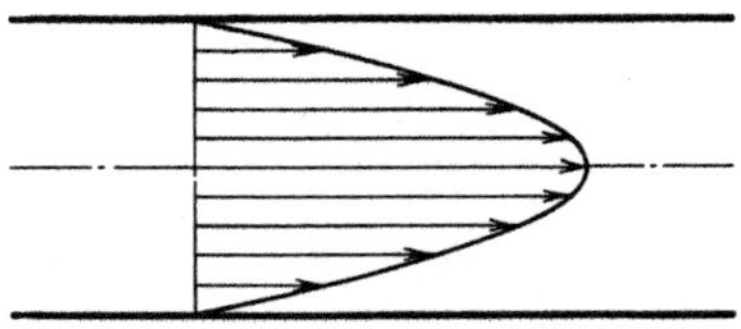

Fig. 54. Velocity distribution of the laminar flow in a tube.

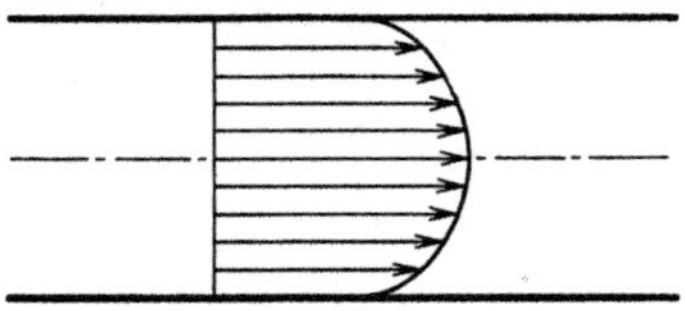

Fig. 55. Velocity distribution of the turbulent flow in a tube.

increasing velocity much more rapidly than as indicated by the Hagen-Poiseuille formula. If the experiment be performed in a glass tube and some of the fluid particles are marked, *e. g.* by introducing color into the fluid, or by suspending small bodies, it is also found that the flow is by no means rectilinear. Instead of this, the suspended particles are whirled about, and colored fluid, introduced through a small tube is torn into shreds and after a short distance is uniformly mixed with the whole fluid. If, however, the velocity is sufficiently small, or, to be more precise, the Reynolds number is sufficiently small, the results of the Hagen-Poiseuille equation are completely verified and the suspended particles or colored fluid introduced show a rectilinear flow. We conclude that the increased resistances are fundamentally connected with the confused movement which is observed. If the fluid particles are mixed the momenta are averaged out since each particle takes its own momentum with it. Hence it is not surprising that the velocity is more uniformly distributed in the tube where turbulent motion occurs than in the case of motion without mixture (laminar motion). The velocity distributions for the two types of flow, as given by observations, are reproduced in Figs. 54 and 55 (both figures are drawn for the same amount of fluid flowing past per second). It can immediately be seen that the shearing forces at the wall become considerably greater when the velocities near the wall are increased. It can therefore be understood why the pressure gradient along a tube with turbulent flow is so much greater than that with laminar flow with the same mass of fluid per unit time. The first experiments on the resistances to flow in tubes were all concerned with the turbulent state. Especially well known are those of Darcy[1], carried out in connection with the design of a large water supply for the city of Paris. The first quantitative observations of the laminar state that are known are those of Hagen[2]. He also described the transition from the laminar to the turbulent state. The first systematic investigation of the phenomenon was however that of Osborne Reynolds[3] in his famous paper of 1883. Reynolds observed the relation between the decrease of pressure in the tube and the mass of fluid passing through it, and established the details of the transition from one state to the other as exactly as possible. He also attacked the problem from the theoretical standpoint and in this connection formulated the law of similitude which bears his name. He found that for tubes of various diameters the transition from one state to the other always occurred at approximately the same value of the Reynolds number ud/ν. The actual numerical values were in the neighborhood of 2,000. They were, however, materially higher if the water in the supply tank was kept calm and if the tube had a well-rounded inlet. Ekman[4] later repeated the experiments with Reynolds original apparatus, and was able to obtain laminar motion up to Reynolds numbers $ud/\nu = 24{,}000$, by allowing the supply of water to remain undisturbed for several hours before the experiments were performed. For technical applications however, interest is more or less concentrated on the lowest values, which are obtained when arbitrary disturbances are produced, whether by irregularities in the inflowing stream or by vortices forming at an insufficiently rounded inlet.

Reynolds also experimented with the introduction of colored fluid, and was able to show that in the laminar state a straight thread of colored fluid passes through the entire tube. When turbulent flow occurred, however, the thread of fluid could be recognized only for a short distance and then quickly became mixed with the rest of the fluid. In the transition region, turbulent mixture and laminar movement alternated more or less regularly.

Reynolds found that for a turbulent flow the resistance was proportional to the 1.73 power of the mean velocity.

In recent times it has been discovered that the flow along the surface of a body, a flow which we have thus far known as boundary layer flow, may also become turbulent. First of all it was found that

---

[1] DARCY, Mémoires des Savants étrangers. Vol. XV, 1858.

[2] Pogg. Ann., vol. 46, p. 423, 1839.

[3] REYNOLDS, OSBORNE, Philos. Trans. Roy. Soc., 1883, or Collected Papers, vol. II, p. 51.

[4] EKMAN, W. V., Arkiv. för Math. Astr. och. Fys, vol. VI, No. 12, 1911.

calculations of the frictional forces in laminar boundary layers[1] do not agree, either in value or in the index with observations, *e. g.* those of Froude[2]. The theory gave the 1.5 power of the velocity, but experiments showed roughly the 1.85 power, so that the assumption of turbulence was suggested. A clear decision was given by Eiffel's discovery (1913)[3] in connection with the behavior of the air-resistance of spheres. It appeared that instead of the air-resistance of spheres increasing continuously with increase in velocity, the value fell off suddenly at a certain point, so that the resistance coefficient (air-resistance divided by $(1/2)\,\varrho v^2$ times the projection of the surface area) decreased to 2/5 and less, of its earlier value. The explanation was found in 1914 by Prandtl[4] who showed that the phenomenon is connected with the fact that the laminar boundary layer becomes turbulent forward of the point of separation, so that the region where separation occurs is shifted backward and the whole vortex region becomes very much smaller. The experimental proof was as follows: A ring of wire 1 mm. in diameter was put around a sphere about 28 cm. in diameter, with the result that the reduced value of the resistance occurred at still smaller velocities. The wire ring was attached to the sphere at a short distance in front of the separation point for laminar motion, so that vortices must be formed as a result of this slight obstacle to the boundary layer, which was also approximately 1 mm. thick. The effect was, as already mentioned, that approximately the same small resistance coefficient was obtained as appeared of itself at the higher velocities. That the point of separation is shifted backward can be seen from the fact that, through the mixing, fluid with greater velocity is mixed with the more slowly moving flow near the wall and accelerates it. Hence, only much farther to the rear of the body, a backflow occurs and, corresponding to the smaller vortex region, there is a smaller resistance. As already explained in the previous section, the phenomenon of the turbulence of boundary layers is of great importance for all cases where a flow with a continuous increase of pressure is required. The critical Reynolds numbers vl/ν (v = velocity, l = length of the boundary flow) are of the order of magnitude 300,000 for very steady (undisturbed) flows, and correspondingly smaller values are obtained for violently disturbed fluids. These numbers should not be compared directly with the Reynolds numbers of the tube, but rather the magnitudes for comparison are, for the tube, the radius r, and for the submerged body, the thickness δ of the boundary

---

[1] Prandtl, L., Verhandlungen des III. intern. Mathematiker-Kongresses zu Heidelberg, 1904, Leipzig, 1905.

[2] Froude, Experiments on the Surface Friction. Brit. Ass. Rep., 1872.

[3] Eiffel, G., Compt. Rend., vol. 155, p. 1597, 1912.

[4] Prandtl, L., Über den Luftwiderstand von Kugeln. Göttinger Nachrichten, p. 177, 1914.

layer before the development of turbulence. Taking for example a plate in a stream in which the flow is parallel to the plane of the plate, the thickness of the boundary layer, by previous results is

$$\delta \approx 5 \sqrt{\frac{\nu l}{v}}$$

and hence

$$\frac{v\,\delta}{\nu} \approx 5 \sqrt{\frac{v l}{\nu}}$$

Taking $vl/\nu = 250{,}000$, this becomes

$$\frac{v\delta}{\nu} = 5 \times 500 = 2{,}500$$

This number is to be compared with $u_{max}\, r/\nu$, or the number $u_{mean}\, d/\nu$ which, on account of the paraboloidal distribution of pressure ($u_{mean} = 1/2\, u_{max}$, and $d = 2r$) equals $u_{max}\, r/\nu$. Hence for purposes of comparison we must put

$$\frac{u_{mean}\, d}{\nu} = 2{,}500$$

For a state of flow through a tube corresponding to the above state of flow around a plate, the Reynolds number ud/ν would be slightly higher, but the results agree to the order of magnitude considered.

Note: The above results for the value of the critical number vl/ν for a flat plate can be used with good approximation to determine the length measured from the inlet (assumed to be well-rounded) traversed by the flow inside the tube before turbulence appears. If l is this length and u the mean velocity, ul/ν can in this case be put equal to 300,000, providing that non-turbulent inflow can be assumed. This would, for example, give a length $l = 20\, d$ for a Reynolds number $ud/\nu = 15{,}000$, but only $0.2\, d$ for $ud/\nu = 1{,}500{,}000$.

20. Older Theories of Turbulence. The first detailed theoretical investigations of turbulence are due to O. Reynolds. In connection with this work there are several questions which must be cleared up. On the one hand we have to show by considering the hydrodynamic differential equations, that the spatial and temporal oscillations of the velocity occurring in the mixing movement affect the average motion in the sense of a resistance to deformation, that is, they have an effect like an internal friction. A second group of theoretical considerations is concerned with the investigation of the conditions under which a disturbing movement, such as occurs in turbulent motion, can increase with time. Starting from small initial disturbances, whose temporal growth or decline is studied, a kind of stability theory was obtained, which it was hoped would yield the value of the critical Reynolds numbers. It soon appeared however, that the mathematical development of this problem led to very complicated calculations and could in many cases give no practical result. Quite lately however, calculations have been successfully performed with comparatively satisfying results. These theories regarding the formation of turbulence will be described in **26**. Certain other considerations, which deal with developed turbulence will be briefly explained here.

In accordance with experimental facts, the motion subject to fluctuations in space and time is resolved into a mean value which is assumed constant with respect to time, and with oscillations about this mean. Various authors differ in the choice of this mean value. Some consider a spatial mean, that is values averaged over a certain definite space of moderate extension; others a temporal mean, that is, values averaged for a single point of space over a long time, and finally, also mixed spatial-temporal mean values have been used. In the following only temporal mean values will be employed. The velocity components, pressure, etc., will be resolved into their temporal mean values (care being observed that the average is taken over a sufficiently long period of time at the point of space considered) and into fluctuations about these mean values. If, for example, the velocity component in the x direction is u, the mean value $\bar{u}$ is constructed as follows

$$\bar{u} = \frac{1}{T}\int_{t_0}^{t_0+T} u\,dt$$

The time T is taken so large that $\bar{u}$ is independent of T and t_0 (strictly the limit for $T \to \infty$ should be taken). Motions for which such a mean value independent of time exists will be called motions steady at the mean values.

The fluctuation about the mean value is obviously $u - \bar{u}$ which will be denoted by u'. By definition, the mean value of u' is zero, or, more explicitly

$$\frac{1}{T}\int_{t_0}^{t_0+T} u'\,dt = \frac{1}{T}\int_{t_0}^{t_0+T} u\,dt - \frac{\bar{u}}{T}\int_{t_0}^{t_0+T} dt = \bar{u} - \bar{u} = 0$$

The velocity components u, v, w are therefore compounded of the mean values and the fluctuations, and we write

$$\left.\begin{aligned} u &= \bar{u} + u' \\ v &= \bar{v} + v' \\ w &= \bar{w} + w' \end{aligned}\right\} \qquad (20.1)$$

The corresponding expression for the pressure is

$$p = \bar{p} + p'$$

These values can now be introduced into the Navier-Stokes' equation and into the equation of continuity in order to construct the mean values of the expressions which occur in them. The terms which in this way come to contain only "barred" members (*e. g.*, $\bar{u}$) remain unaltered when the mean value is taken since all such members are constant with respect to time. Where a fluctuation member, alone or in conjunction with an expression constant with respect to time, occurs, it vanishes by definition when the mean value is taken. The quadratic terms of the Navier-Stokes' equations however, give rise to terms containing two fluctuations multiplied together. These products in general do not vanish identically when the mean value is taken and it is these terms which in fact produce the frictional effect of the mixing movement.

Instead of performing this process on the hydrodynamic differential equations, for which reference may be made to Reynolds[1] or also to a very elegant account by H. A. Lorentz[2], we may here consider the result obtained from the momentum theorem. Some plane surface is taken, its normal being chosen as the x axis, and two axes in the plane as y and z axes. The corresponding velocities are taken as u, v, w. The amount of fluid passing in any small time dt through the surface of area dA is $(dA \cdot \varrho \cdot u \cdot dt)$. The x component of the momentum transported is then $dA \cdot \varrho \cdot u^2 \cdot dt$. The y and z components are $dA \cdot \varrho uv \cdot dt$ and $dA \cdot \varrho uw \cdot dt$. The mean values can now be obtained by finding the

values $\quad \frac{1}{T} dA \int_0^T \varrho u^2 dt; \quad \frac{1}{T} dA \int_0^T \varrho u v dt; \quad \frac{1}{T} dA \int_0^T \varrho u w dt;$

where for simplicity the arbitrary t_0 has been put equal to zero. The expressions of (20.1) are now introduced, bearing in mind that $\bar{u}$, $\bar{v}$, $\bar{w}$ are constant, and that for T sufficiently large expressions like $\frac{1}{T}\int_0^T u'\, dt$ are approximately zero with an accuracy which increases as T increases. If the last mentioned expression be neglected, then since

$$(\bar{u} + u')^2 = \bar{u}^2 + 2\,\bar{u}\,u' + u'^2$$

the x component becomes

$$\frac{1}{T} dA \int_0^T \varrho u^2 dt = dA \left(\varrho \bar{u}^2 + \frac{1}{T} \int_0^T \varrho u'^2 dt \right)$$

Writing the last mean value as $\varrho \overline{u'^2}$, the x component of the rate of change of momentum on the surface dA becomes[3] $dA\,(\varrho \bar{u}^2 + \varrho \overline{u'^2})$. Similarly the y and z components are $dA\,(\varrho\,\bar{u}\,\bar{v} + \varrho \overline{u'\,v'})$ and $dA\,(\varrho\,\bar{u}\,\bar{w} + \varrho \overline{u'\,w'})$. The formulae are written for the case of an incompressible fluid. When variations of density occur the average must be taken

[1] Reynolds, O., Phil. Trans. Roy. Soc., 1895 or, Collected Papers, vol. II, p. 355.

[2] Lorentz, H. A., Abhandl. über theoret. Physik, vol. 1, pp. 43—71, 1907.

[3] $\overline{u'^2}$ is the mean of the squares of u' and not the square of the mean u' which would be indicated by $\bar{u}'^2$. They are not the same thing, the latter being zero according to the statements made above.

also over the variations in density. This complicates matters, since terms which are linear in the velocities alone, when combined with the densities, no longer vanish. This case however is quite definitely to be excluded and we may therefore restrict the entire considerations to fluids of constant density. In this simpler case, however, it must be noted that the components of momentum change contain terms which are made up of products and squares of the fluctuations in addition to those which come from the mean values of the velocities. The latter terms (those to the left in the above expressions) are exactly the same as those obtained by applying the theorem of momentum to the ordinary case of steady motion. The terms due to the fluctuations (those to the right in the above expressions) are now terms introduced by the turbulence. The above expressions for momentum change have the dimensions of force acting on the element dA. Dividing by dA therefore gives force per unit area, or stresses. The term $\varrho\,\overline{u'^2}$ corresponds to a normal stress. The terms $\varrho\,\overline{u'\,v'}$ and $\varrho\,\overline{u'\,w'}$ denote tangential stresses in the y and in the z directions. It can therefore be clearly seen that the temporal mean values of the velocities have exactly the same effect as stresses in solid bodies or in strongly viscous fluids. With respect to the sign of the terms it must be verified that the amounts written out above do actually denote the time rate of momentum transport. A time rate of momentum transport can always be interpreted as a force acting from the surroundings on the mass system considered. In this view, however, we must reverse the signs. The following "apparent stresses" are then obtained, acting on the cross-section considered:

$$\left.\begin{aligned} \sigma_x &= -\varrho\,\overline{u'^2} \\ \tau_{xy} &= -\varrho\,\overline{u'\,v'} \\ \tau_{xz} &= -\varrho\,\overline{u'\,w'} \end{aligned}\right\} \qquad (20.2)$$

These apparent stresses, are to be treated in exactly the manner prescribed in the general theory of stress of **3**.

Exactly similar expressions are obtained for surface elements in the directions of the other axes, and it is superfluous to write them down here. The stresses on surface elements which make some angle with the axes can also be immediately obtained by using the general rules for compounding forces, but these details will also be omitted.

The above relations are very valuable as an explanation of the existence of apparent stresses in a turbulent flow, but since the magnitude of the fluctuations and above all, the magnitude of the mean values of the products in question are unknown, it is at first not possible to utilize them. In order to overcome this difficulty it is assumed, following Boussinesq[1], that the mixing movement is simply equivalent

[1] Boussinesq, T. V., Mem. pres. par. div. Sav., vol. XXIII Paris, 1877, Théorie des l'écoulement tourbillant, Paris, 1897.

to an increase in the viscosity effect, so that the viscosity to be taken into account is much greater and may vary from case to case. The magnitude which corresponds to the kinematic viscosity is considered as a measure of the turbulence and is denoted by ε. The magnitude corresponding to the viscosity μ is then equal to $\varrho\varepsilon$. Hence we put

$$\left.\begin{aligned} \sigma_x &= 2\,\varrho\,\varepsilon\,\frac{\partial \bar{u}}{\partial x} \\ \tau_{xy} &= \varrho\,\varepsilon\left(\frac{\partial \bar{u}}{\partial y} + \frac{\partial \bar{v}}{\partial x}\right) \end{aligned}\right\} \tag{20.3}$$

How large ε is to be taken, in any individual case, is to be decided by experiment. In more exact discussions ε has been assumed to vary with the space coordinates. Since this makes the calculations very difficult however, most writers content themselves with introducing a mean value supposed constant with respect to location.

It may be noted at this point that the mixing movement also produces other effects. Thus in a flow exhibiting differences of temperature between its parts, these differences are equalised. Differences in chemical composition or in the concentration of mixtures are also evened out by the turbulent flow. The assumption that the magnitude ε which occurs in the expression for the turbulent apparent friction (τ_{xy}, etc.) also determines the increased heat conductivity or diffusion, has also been applied with some success. For further consideration of these questions, in so far as they affect interchanges in a free atmosphere, reference may be made to the text by Wilhelm Schmidt[1].

Further results as to the connection of the quantity ε with the velocities of the mean flow $\bar{u}$, $\bar{v}$, $\bar{w}$ will be found in the following section.

21. Newer Theory. Mixing Length and Velocity Distribution. In order to proceed further from Boussinesq's starting point (replacement of the kinematic viscosity ν by turbulence strength ε) it is necessary to connect the quantity ε with the values occurring in the mean flow. The discussion which follows, starts from the case in which the mean flow has the same direction at all points but different values on different streamlines, a case which because of its simplicity is of special interest for us. Taking the direction of flow as the x axis and the direction of greatest difference of velocity as the y axis, the equations in the simplest case are

$$\bar{u} = \bar{u}\,(y), \qquad \bar{v} = \bar{w} = 0$$

Only one component of the apparent stresses now appears, and is

$$\tau_{xy} = -\varrho\,\overline{u'v'} = \varrho\,\varepsilon\,\frac{d\,\bar{u}}{d\,y} \tag{21.1}$$

which will be called *apparent friction* and denoted for brevity by τ without

[1] SCHMIDT, W., Der Massenaustausch in freier Atmosphäre und verwandte Erscheinungen. Hamburg, 1925.

suffixes. The formula shows that $|\tau|/\varrho$ is the square of a velocity, which will be denoted by v_*. Hence

$$v_* = \sqrt{\frac{|\tau|}{\varrho}} = \sqrt{|\overline{u' v'}|} \qquad (21.2)$$

It follows that the turbulence strength ε must have the dimensions (length $\times$ velocity). If the velocity be regulated, *e. g.* put equal to v_* say, it is always possible by choice of some length, l say, to obtain all values for the turbulence strength. Replacing τ by ϱv_*^2 and ε by $l v_*$ (21.1) becomes

$$v_* = l \frac{d\bar{u}}{dy} \qquad (21.3)$$

an equation which will occur again in an intuitive derivation of the apparent friction. Equation (21.3) now gives

$$\varepsilon = l v_* = l^2 \frac{du}{dy} \qquad (21.4)$$

The only unknown quantity which remains is the length l which, by comparison with similar considerations in the kinetic theory of gases, may be supposed to be something like a mean "free path" length. This is in fact the kernel of the whole matter; but l can also be considered as a gross dimension of what may be called "bodies" of fluid having a common velocity at any time. We will temporarily denote the diameter of such a body of fluid by d and ask what distance l, such a body, having relative to the rest of the fluid, some velocity v' say, must traverse, before its kinetic energy is practically destroyed. We may assume that the resistance to this body's movement is similar to the resistance to a solid body, that is, that it is proportional to the cross-section and to the stagnation pressure $(1/2)\,\varrho v'^2$. The work done by the resistance is then equal to (number) $d^2 \cdot (\varrho v'^2/2) \cdot l$ and the kinetic energy to be disposed of is (number) $\varrho d^3 (v'^2/2)$. Equating these two quantities gives $l =$ (number) d as was asserted.

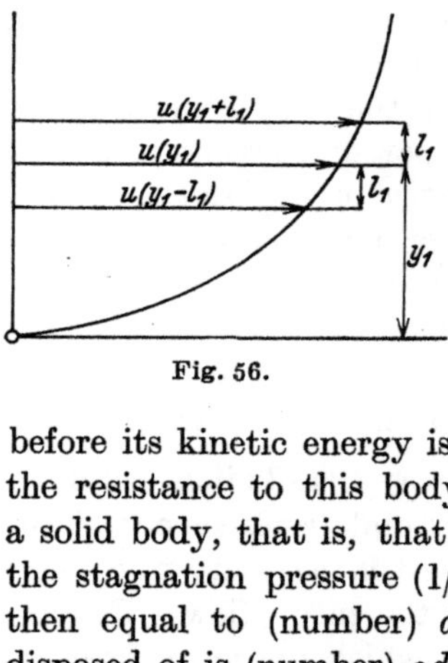

Fig. 56.

We will now try to realize in more detail what happens when such a body of fluid travels along a distance l_1 perpendicular to the direction of flow. Let the velocity increase as y increases. We will assume that as a result of circumstances whose nature is of no further interest in this connection, a body of fluid which originally belonged to a layer $y = y_1 - l_1$ and which had a velocity equal to the mean velocity of this layer, is set moving at right angles to the general direction of motion and reaches the layer y_1 (see Fig. 56). Assuming that the body has lost none of its velocity in the x direction, it will be moving more slowly than the mean velocity of flow in the new layer, by an amount

$$-u_1' = \bar{u}(y_1) - \bar{u}(y_1 - l_1)$$

Conversely a body of fluid coming from the layer $y = y_1 + l_1$ will have a greater velocity than the mean at y_1. If it loses no part of this excess velocity on the way it will show a deviation

$$u_2' = \bar{u}\,(y_1 + l_1) - \bar{u}\,(y_1)$$

Expanding $\bar{u}\,(y - l_1)$ and $\bar{u}\,(y + l_1)$ in Taylor series and neglecting terms of order higher than the second, it is easily found that the mean value of the amounts u_1' and u_2' is given by

$$\overline{|u'|} = \frac{1}{2}\,(|u_1'| + |u_2'|) = l_1 \left|\left(\frac{d\,\bar{u}}{d\,y}\right)_{y_1}\right| \tag{21.5}$$

This now gives an interpretation of the meaning of the length l_1, which however we do not wish to identify completely with the length l previously mentioned. The length l_1 will denote the distance which a particle of fluid having the mean velocity of the layer it belongs to must move perpendicular to the direction of flow in order that the difference between its velocity and the mean velocity of the new layer should equal the mean deviation of velocity in the turbulent flow. It is left open in this definition whether particles entering a new layer from layers above or below it at a distance l_1 actually had the mean velocities of their previous layers or whether the state of affairs was such that they had travelled much greater distances and acquired some part of the velocity of the layers through which they had passed. It might be assumed for example that such a particle travels through all the layers, its velocity, if it comes from the region of smaller velocities, being at any instant less than the mean velocity of the layer in which it happens to be by an amount of the order of magnitude u', and conversely, that the velocity of particles proceeding from a region of higher velocities exceeds the mean velocity of the layer in which they happen to be by the same amount. This alternative representation of the circumstances is permissible provided nothing further is known concerning the actual mechanism of turbulence. Nevertheless (21.5) above remains here applicable; from the second standpoint however, l_1 is merely a quantity representing the connection between the mean value of $|u'|$ and the gradient of the mean velocity $d\bar{u}/dy$. The first standpoint naturally gives a more intuitive representation.

Information concerning the mean value of v' is now required. The following remarks will supply this. We assume that two bodies of fluid coming from the layers $y_1 + l_1$ and $y_1 - l_1$ respectively meet at the layer y_1 in such a way that one body stands in the line of flow of the other one. If the faster one is behind, it will approach the other with a velocity $2u'$; if the slower one is behind, the two will separate with the same velocity. In both cases the fluid between them will be set into motion in a perpendicular direction, in the first case away from the layer y_1 and in the second case, towards y_1. The transverse velocities may perhaps be smaller than the velocities u' but will at any rate be of the same order of magnitude, so that we may write

$$\overline{|v'|} = (\text{number})\,\overline{|u'|} = (\text{number})\,l_1\,\frac{d\bar{u}}{dy} \qquad (21.6)$$

In order to get back to (21.1) it is necessary to consider the mean value of $u'v'$ in more detail. After the previous discussion it can easily be seen that for the particles arriving at the layer y_1 with a positive velocity v', (*i. e.* particles which come from below in Fig. 56), the velocity u' is negative. The product $u'v'$ is then negative. The particles coming from above have v' negative and u' positive so that the product is again negative. Hence the mean value of the product is different from zero and negative. Since in practice the mechanism is probably rather more complicated than here assumed, the mean value of the product will be put equal to, not simply the product of two mean values, but this value multiplied by a numerical factor, which is naturally less than one if values of opposite sign occur in the product $u'v'$. Hence

$$\overline{u'v'} = -\,(\text{number})\,\overline{|u'|}\cdot\overline{|v'|}$$

which in conjunction with (21.5) and (21.6) gives

$$\overline{u'v'} = -\,(\text{number})\,l_1^2\left(\frac{d\bar{u}}{dy}\right)^2 \qquad (21.7)$$

where the second "number" differs from the first. Comparison with (21.1) and (21.4) gives however

$$\overline{u'v'} = -\,l^2\left(\frac{d\bar{u}}{dy}\right)^2 \qquad (21.8)$$

from which the connection between l and the length l_1 used in the previous discussion follows. The two lengths are in fact proportional. Since the number occurring in (21.7) is not known, only (21.8) will be used. The following relation obviously follows from this equation.

$$\tau = \varrho\, l^2\left(\frac{d\bar{u}}{dy}\right)^2$$

If it be observed that the sign of $d\bar{u}/dy$ must change as the sign of τ changes, this formula can be more correctly written

$$\tau = \varrho\, l^2\left|\frac{d\bar{u}}{dy}\right|\cdot\frac{d\bar{u}}{dy} \qquad (21.9)$$

A name will now be introduced for l. We shall call it the *mixing-length*, observing however that this name would be even more appropriate to the length l_1.

It may be asked what has been gained by these considerations since l is again an unknown quantity—and apparently no actual progress beyond (21.1) has been made.

The following may serve as an answer: If certain secondary effects of viscosity may be neglected, the turbulent apparent friction is, like all hydraulic resistances, proportional to the square of the velocity. Formula (21.9) is so constructed that this quadratic dependency on the velocity is obtained by taking a mixing length l independent of the velocity. A length however, can be more easily imagined than for example, a turbulence strength ε, and can in a rather intuitive fashion be brought into relation with the lengths which characterize each individual problem (diameter, distance from the wall, etc.). It can therefore be expected that application of experimental results will give comparatively simple laws of distribution of the length l as a function of position, and that arguments by analogy from flows already known, should enable us to obtain with fair accuracy the value of the mixing length for each point of a flow as yet uninvestigated. In specially simple cases, where the problem contains only a single length measurement which can be brought into connection with the length l, we shall be entitled on the basis of dimensional analysis, to consider these two lengths proportional. It can at once be seen in connection with the distribution of values of l in the neighborhood of the wall, that l must tend to zero as we approach a *smooth* wall, but towards a relatively small finite value if the wall is somewhat rough. The limiting value in the second case is of the same order of magnitude as the roughness grain of the rough wall. This is so because the paths of the particles transverse to the flow are bounded by the wall and tend to zero in the neighborhood of a *smooth* wall, while they will be of the order of magnitude of the depths of the irregularities in a rough wall.

It would be very important if a rule could be given showing how long the mixing length must be at every point for any given case. A very elegant advance in this direction, which nevertheless does not seem to represent the final solution, has been obtained by considerations due to Th. v. Kármán[1] who assumes that the internal mechanism of the turbulence at any two places differs only to the extent of a change of the units of length and time. Kármán takes as his starting point the differential equation of turbulent motion, but the essential result may be obtained by a simple application of dimensional analysis. Instead of units of length and time we can obviously choose units of length and velocity. As far as velocity is concerned the correct unit is already given by the velocity v_* of formula (21.2). The unit of length must still be found. This will give us the magnitude of the mixing length l. Kármán develops the velocity[2] $u = u(y)$ in a Taylors series and ends it at the second term: $u = u(y_1) + (y - y_1)\left(\frac{du}{dy}\right)_1 + \frac{1}{2}(y - y_1)^2\left(\frac{d^2u}{dy^2}\right)_1$

[1] Kármán, Th. v., Mechanische Ähnlichkeit und Turbulenz. Nachr. d. Ges. d. Wiss. zu Göttingen, Math.-Phys. Klasse, p. 58, 1930.

[2] For sake of simplicity we will now omit the bar of $\bar{u}$, so that u means the temporal mean at the point y.

The velocity at the point y_1 is without importance in finding the value of l since by Newton's relativity relation the addition of a constant velocity is without influence on the motion of a system, *i. e.* the amount of the mean velocity can have no effect on the value of l. Hence only the terms $(du/dy)_1$ and $(d^2u/dy^2)_1$ remain as characteristic data for the velocity distribution and as of essential influence in the mechanism of turbulence. The only length which can be constructed from these two quantities is the quotient $\frac{du}{dy}\Big/\frac{d^2u}{dy^2}$ which must now be proportional to the mixing length. We write, in accordance with Kármán's notation

$$l = \varkappa \left| \frac{du}{dy} \Big/ \frac{d^2u}{dy^2} \right| \tag{21.10}$$

It is clear that continuing the Taylor series by another term would furnish a second length, *viz.* $\frac{d^2u}{dy^2}\Big/\frac{d^3u}{dy^3}$ so that the mixing length could be treated as a suitable function of these two lengths. This shows that the proceedings are not free from an arbitrary element. This can be avoided by the following alternative procedure, the field of application of which is however much more restricted.

Strict similarity of the circumstances at the various points of a turbulent flow, as assumed above, can only be expected when the main flow itself obeys the same conditions, that is, the main flow, apart from an additional uniform velocity, must differ at various points only to an extent removable by a different choice of the units of length and time. This can be formulated in the following way: If $u = f(y)$ then we must have

$$f(y + a) \equiv b\, f(cy) + d$$

Here b is the factor by which the unit of velocity must be altered, c the factor for altering the unit of length and d the alteration of the system of reference. It is easy to see that this condition can be satisfied only by functions of the form

$$f(y) = A\,(y + B)^n + C \tag{21.11}$$

If consideration is restricted to velocities that satisfy this relation, the characteristic length of the problem is the length $y + B$, that is, the distance from the axis of similarity, and it must be expected that the mixing length l is somehow proportional to this length. In fact we have

$$\frac{du}{dy}\Big/\frac{d^2u}{dy^2} = \frac{1}{n-1}\,(y + B),$$

and

$$\frac{d^2u}{dy^2}\Big/\frac{d^3u}{dy^3} = \frac{1}{n-2}\,(y + B)$$

so that the Kármán definition and the definition arbitrarily added furnish no new length for reference, but only the distance from the axis of similarity multiplied by an arbitrary numerical factor. If the axis of similarity is chosen as the x axis, $B = 0$ and hence

$$l = \varkappa_1 y \tag{21.12}$$

We will now consider the special case where the shear is constant throughout the whole region. Then, by (21.2), the velocity v_* is also constant. Connecting (21.3) and (21.12) we have

$$v_* = \varkappa_1 \, y \frac{d\,u}{d\,y}$$

or
$$\frac{d\,u}{d\,y} = \frac{v_*}{\varkappa_1 \, y}$$

Hence by integration
$$u = \frac{v_*}{\varkappa_1} \log y + \text{constant} \tag{21.13}$$

This form of the velocity distribution in no way contradicts (21.11). It is on the contrary the well known limiting value obtained by making n tend to zero, A to ∞, and C to $-\infty$, in (21.11). Equation (21.13) now gives
$$\frac{d\,u}{d\,y} \Big/ \frac{d^2 u}{d\,y^2} = -\,y$$

so that (21.10) and (21.12) give exactly the same result, and

$$\varkappa = \varkappa_1$$

We can now attack the practical problem of what law is to be found if regard be given to the viscosity effective in a narrow region along the wall. There will be a laminar layer directly next the wall and further, at small distances from it, a turbulence "mechanism" influenced by the viscosity. From the standpoint of dimensions the two influences can be expressed by stating that in addition to the length y measured from the centre of similarity, an additional length ν/v_* comes into consideration (ν has the dimensions [length $\times$ velocity], so that ν/v_* has the dimensions of a length). It is only natural to take v_* as the velocity of reference when we start with the assumption that the shear τ is constant over the region under consideration.

So far as we do not attempt to express the deviations in the velocity distribution produced by the influence of the viscosity on the turbulence, we may use formula (21.13) and shall have simply to determine the constant of integration in such manner that the viscosity effect is included. It can easily be seen from (21.13) that when $y = 0$, that is on the axis of similarity, u becomes equal to $-\infty$. The axis of similarity must therefore lie on the other side of the wall. Let the distance of the axis from the wall be called y_0, then the velocity u must equal zero for $y = y_0$. Hence, from (21.13)

$$u = \frac{v_*}{\varkappa} (\log y - \log y_0) \tag{21.13a}$$

We now put y_0 (so far undetermined) as proportional to the length ν/v_*

thus
$$y_0 = \beta \frac{\nu}{v_*}$$

from which
$$u = \frac{v_*}{\varkappa} \left(\log \frac{y\,v_*}{\nu} - \log \beta \right) \tag{21.14}$$

This equation contains two empirical constants $\varkappa$ and β which may be adjusted to experimental results. This will be examined more closely in the sequel. For the present it may be observed that this formula has proved very useful for large Reynolds numbers. The value of the particularly important numerical factor $\varkappa$ is given by the best experiments as 0.40. Fig. 57 gives a graphical representation of the behavior of the function defined by (21.14).

For smaller Reynolds numbers—not far beyond the critical value—a less summary consideration of the effect of viscosity is desirable. A general formulation of the problem can be obtained quite simply. We may start from the mixing length. We shall then be able to retain formula (21.12) but $\varkappa_1$ must now be a function of the ratio of the two lengths y and ν/v_*, that is,

$$l = y \cdot f\left(\frac{y\, v_*}{\nu}\right) \tag{21.15}$$

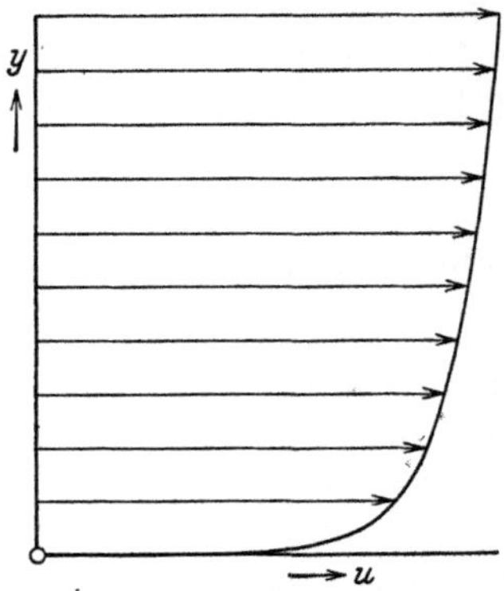
Fig. 57. Logarithmic velocity distribution.

It is seen that formula (21.15) represents the most general method of constructing a third length from two given lengths. From (21.3) we have

$$\frac{du}{dy} = \frac{v_*}{l} = \frac{v_*}{y f\left(\frac{y\, v_*}{\nu}\right)}$$

and hence, again for $v_* = \text{const.}$,

$$u = v_* \int_{y_0}^{y} \frac{dy}{y f\left(\frac{y\, v_*}{\nu}\right)} \tag{21.16}$$

where y_0 has the same meaning as before. The function f can be so arranged that it formally includes the usual viscosity effect[1], and then the lower limit of the integral can be put $y_0 = 0$. If this is not desired, the previous result for y_0, but with an altered β, may be used.

The formulae (21.14) to (21.16) suggest writing the relations in non-dimensional form. One of these non-dimensional quantities is $v_* y/\nu = \eta$ which can now be called a "dimensionless distance" from the wall. It is obviously a kind of Reynolds number. Another non-dimensional quantity is u/v_* which may be called a "dimensionless velocity", and a third is $l/y = f(\eta)$. We may now replace dy/y by $d\eta/\eta$ in (21.16), so that

$$\frac{u}{v_*} = \int_{\eta_0}^{\eta} \frac{d\eta}{\eta f(\eta)} = \varphi(\eta) \tag{21.17}$$

[1] It is only necessary to put $f(\eta) = 1/\eta$ for very small $\eta = y\, v_*/\nu$, where the flow is laminar, giving $u = (v_*^2/\nu)\, y = (\tau/\mu)\, y$.

This shows that for the case of constant internal shear, dimensional analysis lead to the result that the non-dimensional velocity u/v_* can be represented as a function of the non-dimensional distance from the wall $\eta = v_* y/\nu$. The form of $\varphi(\eta)$ which is determined by observation will be discussed in greater detail in the following section, on the basis of experimental results. For the present it may be observed that the form of $\varphi(\eta)$ obtained for large Reynolds number (that is, large η) in (21.14) can now be written

$$\varphi(\eta) = A \log \eta + B \tag{21.18}$$

We shall likewise obtain a very useful formula for smaller Reynolds numbers in the form

$$\varphi(\eta) = C\eta^n \tag{21.19}$$

22. The Laws of Surface Friction from Experiments on Flows in Tubes. Resistance Formulas. Effect of Roughness. The relations for the shear inside a straight tube of circular cross-section have already been discussed in 2. The equilibrium between a difference of pressure $(p_1 - p_2)$ and the shearing forces on the enclosing cylindrical surface of radius y was considered and a relation obtained which will here be taken as a point of departure for further development. Formula (2.3) gave the shear as

$$\tau = \frac{p_1 - p_2}{L} \cdot \frac{y}{2} \tag{22.1}$$

This formula can be used, as it stands, for turbulent flows. The only difference is that τ no longer represents the "true shear" due to the viscosity of the fluid, but the sum of the true shear and the "apparent shear", the latter being determined by the turbulent interchange. The part played by the true shear, except in layers directly next the wall, is small enough to be neglected in comparison with the apparent shear, so that in essentials only the latter need be considered. Two results of different natures may be deduced from the above formula. On the one hand, that in the same cross-section, τ is proportional to the distance y from the axis of the tube; and, on the other hand, that its value can be determined by experimental observation of the pressure difference $(p_1 - p_2)$ between two sections distant L from each other. Considering in particular the shear τ_0 at the boundary, where $y = r$,

$$\tau_0 = \frac{(p_1 - p_2)\, r}{2\, L} \tag{22.2}$$

This permits us to test experimentally the statements of the previous paragraph in an approximate manner (only approximately because the shear in the region near to the wall of the tube is not exactly constant, but linear in the distance from the wall, in accordance with (21.1).

Many formulae for the decrease of pressure in a tube have been published in the course of time. Since however the calculation of the flow in a tube is not an end in itself, so far as we are here concerned, we shall apply only some few formulae of the more recent literature on the subject which have been tested with special thoroughness. In addition it should be noted that only these more recent formulae satisfy the Reynolds similarity conditions. One well-known formula of this kind is due to Blasius[1] but can only be used for Reynolds numbers $\bar{u}\, d/\nu < 100{,}000$.

If a resistance coefficient λ for the flow in tubes be defined by the equation

$$\frac{p_1 - p_2}{L} = \frac{\lambda}{d}\, \frac{\varrho\, \bar{u}^2}{2} \tag{22.3}$$

where d is the diameter of the tube and $\bar{u}$ the mean velocity[2] in the tube, Blasius gives λ as

$$\lambda = 0.3164 \left(\frac{\bar{u}\, d}{\nu}\right)^{-1/4} \tag{22.4}$$

Comparing (22.2) and (22.3) and using $d = 2r$, we have the following equation for the shear at the wall:

$$\tau_0 = \frac{\lambda}{8}\, \varrho\, \bar{u}^2 \tag{22.5}$$

and therefore on taking (22.4) into account,

$$\tau_0 = 0.03955\, \varrho\, \bar{u}^{\,7/4}\, \nu^{\,1/4}\, d^{-1/4}$$

Since we require a formula using the radius r instead of the diameter, the above number must be divided by $(2)^{1/4} = 1.19$, giving

$$\tau_0 = 0.03325\, \varrho\, \bar{u}^{7/4}\, \nu^{\,1/4}\, r^{-1/4} = \varrho\, v_*^2 \tag{22.6}$$

In this formula the velocity v_* known to us from the previous paragraph has again been introduced. In the succeeding account v_* will always be supposed to be connected with the surface friction τ_0. If v_*^2 be split into two factors $v_*^{7/4}\; v_*^{1/4}$, it is easily found that

$$\left(\frac{\bar{u}}{v_*}\right)^{7/4} = \frac{1}{0.03325} \left(\frac{v_* r}{\nu}\right)^{1/4}$$

or

$$\frac{\bar{u}}{v_*} = 6.99 \left(\frac{v_* r}{\nu}\right)^{1/7} \tag{22.7}$$

This equation is very similar in construction to (21.19). However, a mean velocity stands on the left hand side, and the radius of the tube instead of a distance from the wall, on the right hand side. If we realize that in turbulent flows the increase in velocity is chiefly concentrated in the parts near the wall, whereas the differences of velocity are small in the interior of the tube, it is evident that the behavior of layers near the wall is most important. We can therefore see from the construction of (22.4) or (22.6), and (22.7) which was derived from it, that they depend almost entirely on the state of affairs in the neighborhood of the wall. We shall therefore have mastered the main features of this flow if we can find a law exhibiting these characteristic features of the behavior

[1] Blasius, H., Forschungsheft 131 des Vereins Deutscher Ing., 1911.

[2] This notation differs from that in **20** and **21**, in so far that here $\bar{u}$ denotes not the temporal mean at any point, but the mean value for any cross-section of the tube.

near the wall. In this region however, as previously mentioned, the shear is approximately constant and equals τ_0, so that the relations at the end of the last section may be applied as approximations. In order to obtain a result it will now be assumed that a power formula of the form given in (21.19) holds from the wall to the middle of the tube. It is certain that this will not be exact for the middle portions of the tube, but as already seen, this does not matter very much. First of all the transition may be made from the mean velocity $\bar{u}$ to the velocity at the axis of the tube u_1, by putting $\bar{u}$ equal to 0.8 u_1, thus giving

$$\frac{u_1}{v_*} = 8.74 \left(\frac{v_* r}{\nu}\right)^{1/7}$$

Now this formula must hold for all distances from the wall, in accordance with what has been said above, so that

$$\frac{u}{v_*} = 8.74 \left(\frac{v_* y}{\nu}\right)^{1/7} \tag{22.8}$$

or

$$\varphi = 8.74\, \eta^{1/7} \tag{22.8a}$$

Hence, starting from a law for the fall in pressure given by (22.4) we have obtained a relation for the velocity distribution, by assuming that the specific character of the pressure law is produced entirely by the velocity distribution and does not depend on other factors. The result obtained is that the velocity at different distances y is proportional to the 7th root of the distance from the wall. In accordance with a remark made at the commencement of this paragraph the Blasius formula holds only for a limited range of Reynolds numbers. It is however to be regarded, for this domain, as a very good and reliable interpolation formula. The derived formulae (22.8) and (22.8a) are valid only for the corresponding region, that is up to $\eta = 700$ but have proved to be very useful for representing the results of experiments in this region. In order to prove this, the 7th power of the velocity can be plotted as a function of y, the distance from the wall, and an almost straight line is usually obtained. It is surprising to find that points near the middle of the tube also fit very well to this line, which however only shows that the influence of the weakened shear toward the center is compensated by some other influences. According to our reasoning we should have expected the regularity described to hold only in the neighborhood of the wall, but it is naturally very convenient to know that formula (22.8) can be employed even to the middle of the tube. If v_* be calculated backwards from formula (22.8), the result is

$$v_* = 0.150\, u^{7/8} \left(\frac{\nu}{y}\right)^{1/8} \tag{22.9}$$

which gives

$$\tau_0 = \varrho\, v_*^2 = 0.0225\, \varrho\, u^{7/4} \left(\frac{\nu}{y}\right)^{1/4} \tag{22.10}$$

This formula which naturally agrees very closely in its construction with formula (22.6), from which we started, has proved very useful for a series of problems, as for example, the calculation of the frictional resistance of a plate (see the following sections). If we take Reynolds numbers outside the limits mentioned, it is found that the 7th power no longer gives a straight line, but that the 8th or, for even higher Reynolds numbers, the 9th or 10th power must be taken. This agrees with the statements that the resistance numbers λ now deviate from

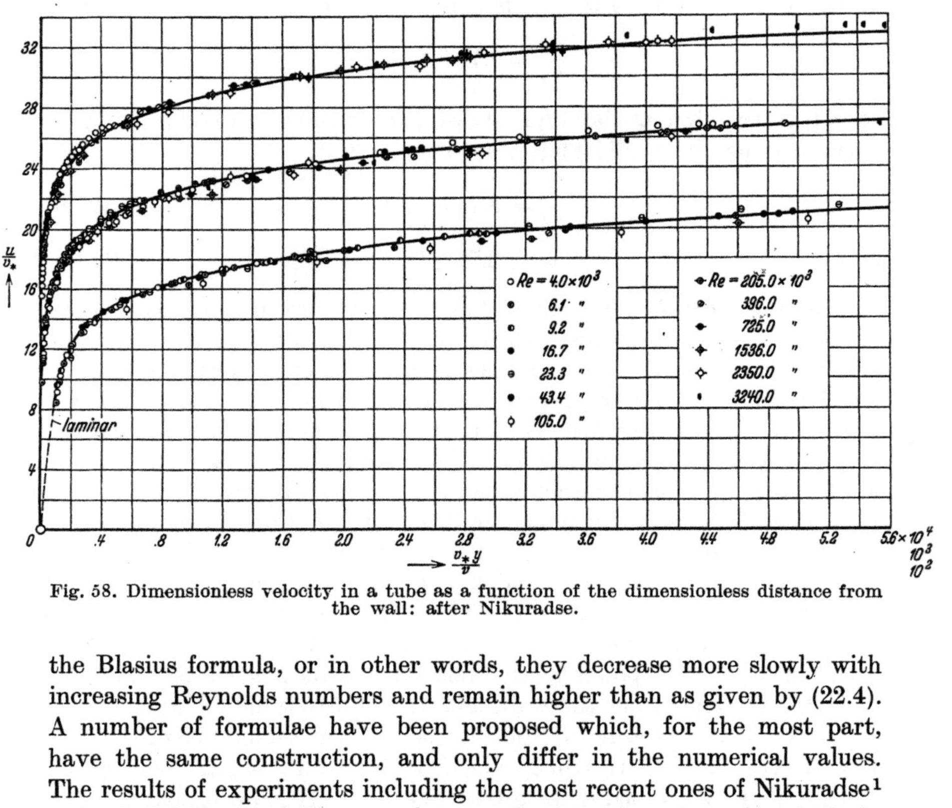

Fig. 58. Dimensionless velocity in a tube as a function of the dimensionless distance from the wall: after Nikuradse.

the Blasius formula, or in other words, they decrease more slowly with increasing Reynolds numbers and remain higher than as given by (22.4). A number of formulae have been proposed which, for the most part, have the same construction, and only differ in the numerical values. The results of experiments including the most recent ones of Nikuradse[1] and of Schiller and Hermann[2] agree best with a formula which in our notation is

$$\lambda = 0.00540 + 0.397 \left(\frac{v d}{\nu}\right)^{-0.3} \qquad (22.11)$$

The problem of deducing a formula for the velocity distribution from the above resistance law could be performed exactly as in the example

[1] Vorträge aus dem Gebiete der Aerodynamik, p. 63, Aachen, 1929.

[2] Schiller, L., Vorträge aus dem Gebiete der Aerodynamik, p. 69, Aachen, 1929. — Hermann, Dissertation, Leipzig, 1930. There $\lambda_1 = 0.0027 + 0.161 (v\,r/\nu)^{-0.3}$ referred to the radius.

of the Blasius law. The difficulty arises however that it is not possible to invert in a finite form the function occurring in (22.11) so that a graphical representation or table of values would have to be used. It is therefore more expedient, since direct measurements of the velocity distribution have been obtained, to obtain a function $\varphi = \varphi(\eta)$ directly from experiment, by plotting the quantity u/v_* on $v_* y/\nu$ for various points, after the values of τ_0 and hence $v_* = \sqrt{\tau_0/\varrho}$ are calculated. This has been done by Nikuradse[1]. It can naturally only be expected that the points

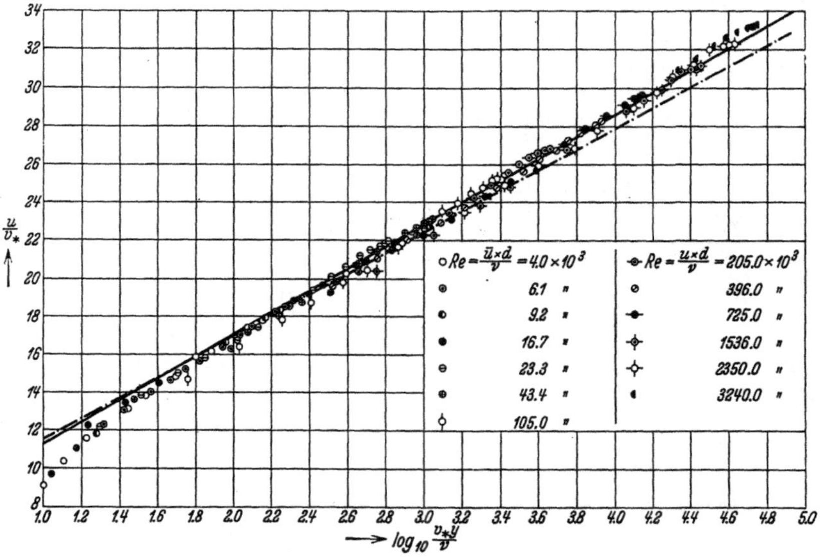

Fig. 59. Logarithmic representation of Fig. 58. The straight lines represent formula (22.12) (— · — · —) and (22.13) (————).

near the wall will give a satisfactory curve. The result, represented in the ordinary manner in Fig. 58, and with the logarithms of $v_* y/\nu$ as abscissa in Fig. 59, shows however that even the points lying near to the middle of the tube do not lie too far from the curve given by points near the wall; so that we again have the advantage of being able to apply the law for the particles near the wall with approximate accuracy up to the middle of the tube.

The representation with logarithmic abscissae furnishes the very noteworthy result that the points, excluding the smallest values of η, (less than 10) lie almost exactly on a line. A line like this has however already been obtained by the considerations of similitude in (21.14), (21.18). We are now obviously in a position to supply numerical values in (21.18).

[1] Nikuradse, J., Forschungsheft 356 des Vereins Deutscher Ingenieure, 1932.

If the points near the wall, for which the law is supposed to hold accurately, be taken, we obtain[1]

$$\varphi = 2.4 \log \eta + 5.8 \tag{22.12}$$

If, however, an approximate formula is desired, which is to be applicable over the whole region from the neighborhood of the wall to the middle, a straight line can be so arranged among the experimental points that for small values of η it represents the state near the wall and for large η the state near the center. The line so obtained can be expressed in round figures as

$$\varphi = 2.5 \log \eta + 5.5 \tag{22.13}$$

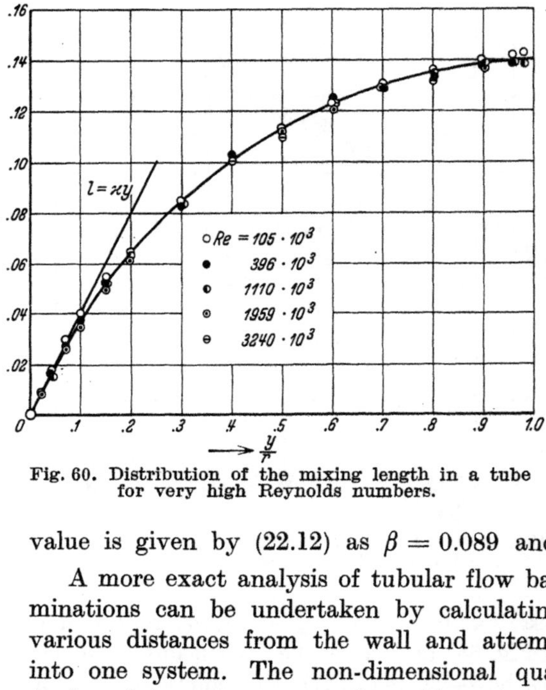

Fig. 60. Distribution of the mixing length in a tube for very high Reynolds numbers.

Comparison with (21.14) shows first of all that Kármán's universal constant $\varkappa$ of (21.10) which is identical with $\varkappa_1$ of (21.12), has the following numerical values:

Equation (21.12) above gives $\varkappa = 0.417$; (22.13) gives $\varkappa = 0.400$. It should be mentioned that Kármán himself, on the basis of experimental results of a different kind found $\varkappa$ to lie between 0.37 and 0.38. Of further interest is the quantity β of (21.14). Its value is given by (22.12) as $\beta = 0.089$ and by (22.13) as $\beta = 0.111$.

A more exact analysis of tubular flow based on experimental determinations can be undertaken by calculating the mixing length l for various distances from the wall and attempting to bring the results into one system. The non-dimensional quantities which we consider to be of importance are the ratios y/r, and $v_* r/\nu$ or $v_* y/\nu$. It is immaterial which of the two latter quantities is chosen since $v_* y/\nu = (y/r)\, v_* r/\nu$, so that the alternative choice only changes the analytic form of the function. As the behavior of the portions near to the wall already indicate, the viscosity influences vanish for large Reynolds numbers, that is, large $v_* r/\nu$, which implies that ν no longer occurs in the formulae. Hence in this case a function of y/r alone is to be expected. Experiments by Nikuradse do in fact show this result. The function l/r for these high Reynolds numbers is represented in Fig. 60.

[1] It may be noticed that in (22.12) and (22.13) "log" means the logarithm to the base e.

It can be expressed with good approximation by the following equation:

$$\frac{l}{r} = 0.14 - 0.08\left(1 - \frac{y}{r}\right)^2 - 0.06\left(1 - \frac{y}{r}\right)^4 \qquad (22.14)$$

where y always denotes the distance from the wall, and r the radius of the tube. Developing l as a series, the first terms are

$$l = 0.4\,y - 0.44\,\frac{y^2}{r} + \ldots \qquad (22.15)$$

Hence for positions near the wall (21.12) is satisfied with $\varkappa = 0.4$. This equation can be applied with good approximation from $\eta = v_* y/\nu = 2{,}500$ onward. Rather higher values for l are found with lower values of η, see Fig. 61.

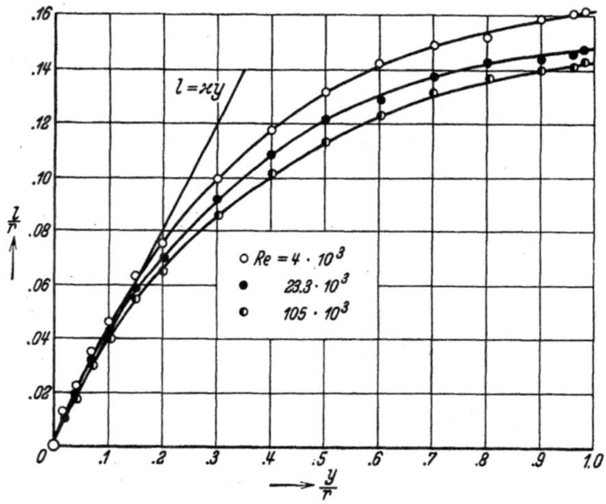

Fig. 61. Distribution of the mixing length in a tube for moderate Reynolds numbers.

The foregoing shows that on closer examination the circumstances are considerably more complicated than (22.13) appears to indicate. It should be noted first of all that in accordance with the arguments leading to (21.14), this equation, and hence also (22.12) above, only hold for the region where the viscosity effects are restricted to a thin layer at the wall, so that they no longer appear in the equation for l. According to the previous argument this would have been the case for $\eta \geqslant 2{,}500$, and according to the same arguments (22.12) and (21.14) do not hold for layers near the wall. It is however a particularly fortunate circumstance that as the result of the combination of effects, the nature of which is unknown up to the present time, (22.12) is also applicable for values of η which are one hundred times smaller than was expected, and (22.13) which differs only slightly, can be used right up to the middle of the tube.

In determining l it was necessary to differentiate an empirical curve. In calculating the velocity distribution from the behavior of l we had to perform an integration [see (21.16)]. It is known that small irregularities in a curve represent considerable deviations in the differential coefficient while conversely, considerable variations of a curve to be integrated, provided that they balance one another somewhat, lead only to minor differences in the integral curve. Hence it can be understood why the very different statements as to the mixing length obtained on the one hand from (22.14), and on the other by differentiating (22.13), can produce final velocity distributions varying so slightly. The most important practical result is that (22.13) gives an approximate representation of the velocity distribution in the turbulent state, dependent on the shear at the wall.

Since the value of l is independent of the viscosity for large values of η and is then a function of y/r alone, it follows that the value of $(u_{max} - u)$ is then a function only of the shearing force at the wall that is, of v_* and of y. The equation is

$$u_{max} - u = v_* F\left(\frac{y}{r}\right) \tag{22.16}$$

Although it had already been enunciated in the same sense by Darcy[1], the first research worker to measure the velocity distribution, this relation was soon forgotten. It was rediscovered by von Kármán in 1929[2].

With the help of the approximate formula (22.13) above it can be written

$$u_{max} - u = 2.5\, v_* \log \frac{r}{y} \tag{22.17}$$

When the function F in (22.16) is known it is possible to obtain the mean velocity $\bar{u}$ by calculating the mean value of $u\,(y)$ for the cross-section. Using the value of F in (22.17) in this fashion, the formula obtained as a first approximation is

$$\bar{u} = u_{max} - 3.75\, v_* \tag{22.18}$$

If more strict regard is paid to the form of the function F it is found that more accurate values are given by[3]

$$\bar{u} = u_{max} - 4.07\, v_* \tag{22.19}$$

Equations (22.17) and (22.18), which are mutually consistent, should be used in conjunction with each other, while the more exact equation (22.19) is to be preferred for application to experiments in which u_{max} is measured.

Equations (22.17) and (22.18) furnish a method for finding an approximate value of λ, the usual resistance coefficient. It is customary to write

$$-\frac{dp}{dx} = \frac{\lambda}{d} \cdot \frac{\varrho \bar{u}^2}{2}$$

From (22.2) we have

$$-\frac{dp}{dx} = \frac{2\tau_0}{r} = \frac{2}{r} \varrho v_*^2$$

---

[1] Darcy, Mémoires des Savants étrangers., Vol. XV, p. 141, 1858.

[2] Kármán, Th. v., Nachr. d. Ges. d. Wiss. zu Göttingen, Math.-Phys. Kl., p. 58, 1930.

[3] Nikuradse, J., V.D.I.-Forschungsheft 356, Berlin, 1932.

and since $d = 2r$ we find by equating these two values,

$$\lambda = 8\left(\frac{v_*}{\bar{u}}\right)^2 \tag{22.20}$$

Also, from (22.13) $u_{max} = v_*\left(2.5\,log\,\frac{r\,v_*}{\nu} + 5.5\right)$

which, in conjunction with (22.18) leads to

$$\bar{u} = v_*\left(2.5\,log\,\frac{r\,v_*}{\nu} + 1.75\right) \tag{22.21}$$

It is now possible to introduce the Reynolds number ud/ν by writing

$$\frac{r\,v_*}{\nu} = \frac{1}{2}\,\frac{\bar{u}\,d}{\nu}\cdot\frac{v_*}{\bar{u}} = \frac{\sqrt{\lambda}}{4\sqrt{2}}\cdot\frac{\bar{u}\,d}{\nu}$$

This gives, with reference to (22.20) and (22.21),

$$\lambda = \frac{8}{\left[2.5\,log\left(\frac{\bar{u}\,d}{\nu}\sqrt{\lambda}\right) - 2.5\,log\,4\sqrt{2} + 1.75\right]^2}$$

On simplifying this equation and transforming the logarithms to base 10 we get

$$\lambda = \frac{1}{\left[2.035\,log_{10}\left(\frac{\bar{u}\,d}{\nu}\sqrt{\lambda}\right) - 0.91\right]^2} \tag{22.22}$$

This formula shows that the graph of $1/\sqrt{\lambda}$ as a function of $log\,(\bar{u}\,d/\nu)\sqrt{\lambda}$ is a straight line. Measurements by Nikuradse have verified the linearity of this relationship very accurately. This property is found to hold with good approximation for small Reynolds numbers, $\bar{u}\,d/\nu$, beyond the range where the theory can properly be applied, see Fig. 62. The following empirical equation has been deduced from experiments on the resistance in pipes and holds with fair accuracy for the entire domain:

$$\frac{1}{\sqrt{\lambda}} = 2.0\,log\left(\frac{\bar{u}\,d}{\nu}\sqrt{\lambda}\right) - 0.8 \tag{22.23}$$

so that the numerical coefficients obtained in practice differ only slightly from those obtained in the above theory. The value of λ for any measured value of $\bar{u}\,d/\nu$ can always be obtained by successive approximation thus: first substitute any convenient value for $\sqrt{\lambda}$ in the right hand side of (22.22) or (22.23) and calculate a first approximation to the value of λ; use this value to repeat the procedure if the initial value substituted for λ differs too greatly from the first approximation.

The calculations in the theory described above are related in certain respects to similar calculations in the paper by Professor von Kármán already quoted, and owe their inspiration to that paper.

A few remarks on rough walls may be conveniently inserted at this point. It has already been pointed out that account can be taken of any degree of roughness by choosing a suitable value for y_0 in (21.13a). As a first approximation y_0 can be regarded as proportional to k, the diameter of the grain of roughness.

Hence putting $y_0 = \alpha k$ we get

$$u = \frac{v_*}{\varkappa}\left(\log\frac{y}{k} - \log\alpha\right)$$

The corresponding equation obtained as the result of a series of experiments at Göttingen in which roughness was produced by attaching grains of sand to the inner tubular surface was

$$u = v_*\left(2.5\log\frac{y}{k} + 8.5\right) \tag{22.24}$$

giving $\alpha = 1/30$.

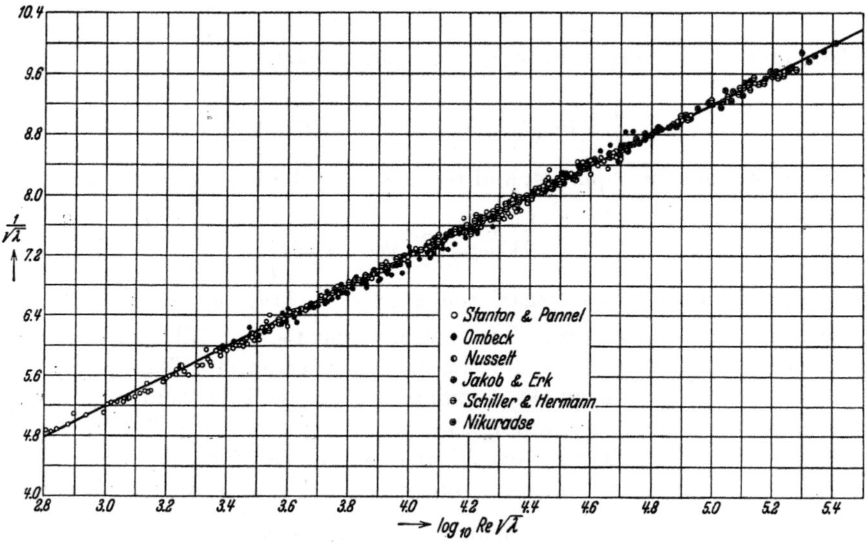

Fig. 62. Logarithmic representation of the coefficient of resistance for smooth tubes. The straight line represents formula (22.23).

Since (22.17) and (22.18) refer to the state of affairs in the interior of the tube they can be immediately applied to the case of the rough tube. The coefficient of resistance for the rough tube can therefore be obtained by repeating the corresponding calculations, as explained above, for the case of smooth tubes. It must be observed however that in their present form these equations are only correct if the Reynolds number is sufficiently large. The effect of the deviations from these equations when the Reynolds number is too small is that the resistance of the rough tube approaches the value for the smooth tube, and these values coincide for roughness sufficiently small. Naturally similar conditions prevail for the velocity distribution. Hence the constant $-\log\alpha$ must be replaced by a variable magnitude, and it is easy to see the quantities upon which this variable must depend. We must construct a Reynolds number from the magnitudes which characterize the flow at the wall, *i. e.* k, the diameter of the grain of roughness must be taken as the length and v_*, the velocity of shear, as the necessary velocity. The expression $(u/v_* - 2.5 \log y/k)$ has the value 8.5 for the completely developed flow with roughness. Let this expression which is equal to $-2.5 \log \alpha$ be introduced as a new variable χ. This value gives a very simple expression for the smooth tube. Using (22.13) we have

$$\chi = 2.5 \left(\log \frac{v_* y}{\nu} - \log \frac{y}{k}\right) + 5.5 = 2.5 \log \frac{v_* k}{\nu} + 5.5$$

which is, as expected, a function of the kind of Reynolds number as required.

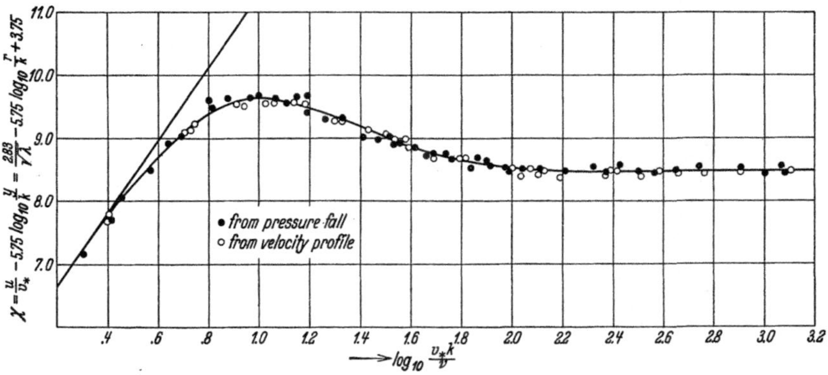

Fig. 63. Roughness function.

Experimental values obtained from various rough tubes and plotted to show χ as a function of $\log (v_* k/\nu)$ do in fact show this state of affairs (see Fig. 63). Thus the variable χ has the value 8.5 for large $v_* k/\nu$, coincides with the above function for small values and has a characteristic transition curve for intermediate values. A value of χ, dependent on $\log v_* k/\nu$, can also be obtained from the measurements of pipe resistance. The distribution of these values agrees very well with those obtained from the velocity profiles. Fig. 64 shows the manner in which the coefficients of resistance of rough and smooth tubes depend on the Reynolds numbers.

23. Turbulent Flow Along a Wall with Special Reference to the Frictional Resistance of Plates. With the help of the facts established in the preceding paragraphs, turbulent flows along a wall can now be treated in the same manner as the laminar boundary layers of **16** and **17**. Since a solution involving the differential equations is not possible, the methods to be used can only supply approximations to the required values, as in the laminar case of **17**.

The first flow to be discussed will be the comparatively simple case in which the pressure is constant; this is approximately true for a flow past a thin plate set parallel to the direction of the stream.

As in the laminar case it is necessary to distinguish between two portions of the flow: an essentially linear flow at some distance from the plate and having the given velocity u_1, and a retarded boundary layer in the immediate neighborhood of the plate. Under the cumulative action of friction at the wall the boundary layer which commences with practically zero thickness continually increases in thickness as the

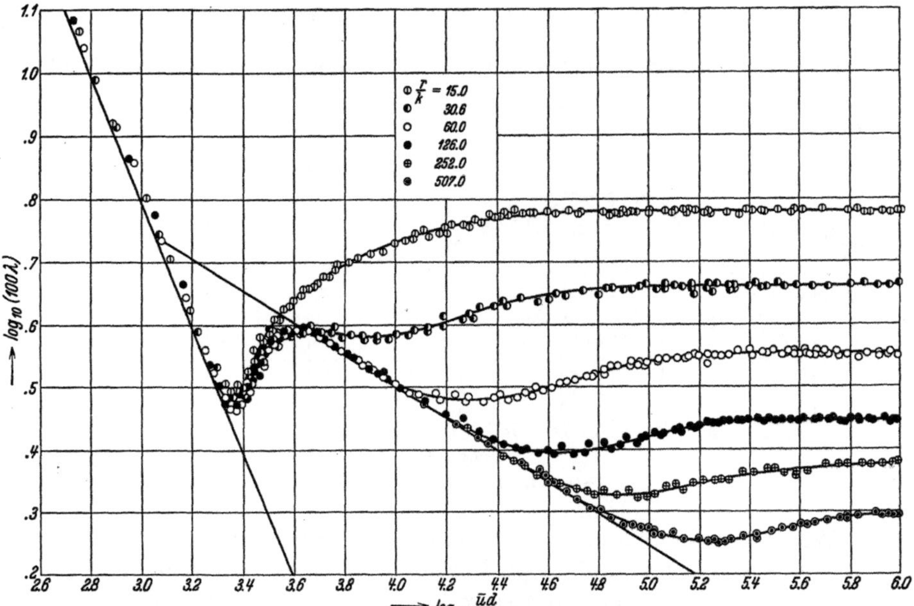

Fig. 64. Coefficient of resistance of rough and smooth tubes dependent on Reynolds number.

distance from the leading edge increases. Here again it is possible to use the theorem of momentum: the loss of momentum at a distance x from the leading edge is equal to the resistance produced by the surface friction from the leading edge to the point distant x from it. If this resistance is measured for unit width the appropriate formula is therefore

$$D(x) = \varrho \int_0^\delta u\,(u_1 - u)\,d\,y \tag{23.1}$$

In order to simplify the calculation of the momentum integral let it be assumed that the velocity distribution in the boundary layer is of the same kind as in the case of tubular flow. Since we have only to find the total momentum no serious error will be introduced when using the velocity distributions characterized by (22.8) and (22.13) respectively, if the transition from boundary layer to undisturbed flow is contrived by simply connecting the undisturbed flow and the velocity profile of

(22.8) or (22.13). This produces a slight discontinuity (see Fig. 65) which will of course be rounded off in reality. Since this modification cannot alter the total momentum in any significant degree, the approximation is reliable.

The assumption that the velocity, inside the boundary layer is proportional to the seventh root of the distance from the wall, y, [see (22.8)] is found to give a good approximation for moderate Reynolds numbers. A simple integration of the momentum integral shows that its value is then $(7/72)\,\varrho u_1^2\,\delta$. This can be connected with the shearing force by using the fact that the shearing force τ is equal to the increase in drag per unit length, *i. e.* $\tau = dD/dx$. Thus substituting the value of D obtained above we have

$$\tau = \frac{7}{72}\,\varrho\,u_1^2\,\frac{d\,\delta}{d\,x} \qquad (23.2)$$

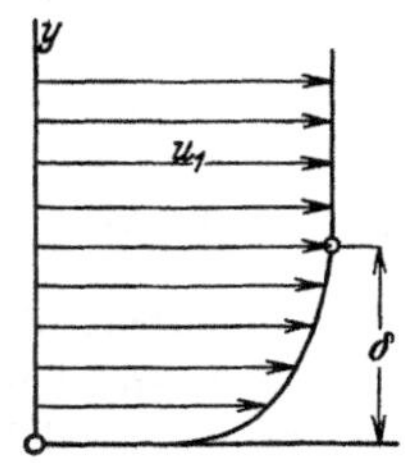

Fig. 65.

On the other hand, from (22.10)

$$\tau = 0.0225\,\varrho u_1^{7/4}\,(\nu/\delta)^{1/4} \qquad (23.3)$$

Integrating the differential equation obtained by equating these two values of τ gives $\quad \frac{4}{5}\,\delta^{5/4} = \left(\frac{72}{7}\right) 0.0225 \left(\frac{\nu}{u_1}\right)^{1/4} \cdot x$

and thence, solving for δ

$$\delta = 0.37 \left(\frac{\nu}{u_1}\right)^{1/5} x^{4/5} \qquad (23.4)$$

Substituting this value in $D = (7/72)\,\varrho u_1^2\,\delta$ gives

$$D = 0.036\,\varrho\,u_1^2\,x \left(\frac{\nu}{u_1\,l}\right)^{1/5}$$

A coefficient of friction for plates C_f is obtained by dividing this value of D by the stagnation pressure $(1/2)\,\varrho u_1^2$, and by the area of the plate, which in this case equals x, on account of the unit width assumed. In terms of a Reynolds number defined by

$$R_x = \frac{u_1\,x}{\nu}$$

the value of C_f is therefore,

$$C_f = 0.072\,R_x^{-1/5} \qquad (23.5)$$

For larger Reynolds numbers it is advisable to use the logarithmic law of velocity distribution as expressed in (22.13). The calculations are considerably more involved, for the total momentum of the boundary layer cannot be expressed as simply as in the other case. The procedure will be clearer if the law of velocity distribution is first used in a generalized form. Let us put $\varphi = \varphi\,(\eta)$ where, as before,

$$\varphi = \frac{u}{v_*} \quad \text{and} \quad \eta = \frac{v_*\,y}{\nu}$$

Let $\eta = \eta_1$ and $\varphi = \varphi_1$ when $u = u_1$. From $\varphi_1 = u_1/v_*$ it follows that $v_* = u_1/\varphi_1$, and hence we have the following equation for the general velocity

$$u = v_* \varphi = u_1 \frac{\varphi}{\varphi_1}$$

Substituting the above value for v_* in $\eta = v_* y/\nu$ gives

$$y = \frac{\nu \varphi_1}{u_1} \eta$$

and hence it may be noted in passing,

$$d y = \frac{\nu \varphi_1}{u_1} d \eta_1$$

The next stage is to write

$$\tau = \frac{d D}{d x}$$

and to substitute ϱv_*^2 for τ, and the expression of (23.1) for D. Since v_* and δ, and hence also η_1 which equals $v_* \delta/\nu$, are functions of x, $d/d x$ can be replaced by $(d \eta_1/d x)(d/d \eta_1)$.

Hence, after cancelling a number of factors

$$\frac{u_1}{\varphi_1^2} = \nu \frac{d \eta_1}{d x} \cdot \frac{d}{d \eta_1} \int_0^{\eta_1} \left(\varphi - \frac{\varphi^2}{\varphi_1} \right) d \eta \tag{23.6}$$

In differentiating the integral it is necessary to differentiate once with respect to the upper limit; but this obviously gives zero, since $\varphi = \varphi_1$, for $\eta = \eta_1$. Next, in differentiating within the sign of integration, it should be observed that φ_1 is a function of η_1 whereas φ is constant during the differentiation. Hence the derivative of the integrand is $\varphi^2/\varphi_1^2 \cdot d \varphi_1/d \eta_1$. To abbreviate the notation it is convenient to write

$$F(\eta_1) = \frac{d \varphi_1}{d \eta_1} \int_0^{\eta_1} \varphi^2 \, d \eta$$

so that the following relation holds

$$d x = \frac{\nu}{u_1} F(\eta_1) \, d \eta_1$$

If it is assumed that this formula holds from the leading edge of the plate ($x = 0$) and that here $\eta_1 = 0$, then by integration we have

$$x = \frac{\nu}{u_1} \int_0^{\eta_2} F(\eta_1) \, d \eta_1 = \frac{\nu}{u_1} \Phi(\eta_2) \tag{23.7}$$

If however this assumption is unjustified the presence of an initial laminar region will necessitate adding a suitable constant of integration. This will not be taken into account for the present. It may be observed that the method by which $\Phi(\eta_2)$ was obtained shows it to be a function of zero dimension. It can be seen that the remaining quantities occurring in (23.7) can also combine to form a non-dimensional quantity, $R_x = u_1 x/\nu$ which has already been encountered. The final result is therefore

$$R_x = \Phi(\eta_2) = \int_0^{\eta_2} \frac{d\varphi_1}{d\eta_1} \left(\int_0^{\eta_1} \varphi^2 \, d\eta \right) d\eta_1 \tag{23.8}$$

It still remains to calculate the resistance either by using (23.1) or by means of the relation $D = \int \tau \, dx$, where

$$\tau = \varrho \, v_*^2 = \frac{\varrho \, u_1^2}{\varphi_1^2}$$

As a result of either method

$$D = \varrho \, \nu \, u_1^2 \int_0^{\eta_2} \frac{F(\eta_1)}{\varphi_1^2} d\eta_1 = \varrho \, \nu \, u_1 \, \Psi(\eta_2) \tag{23.9}$$

where $\Psi(\eta_2)$ is another non-dimensional function. The Reynolds number R_x and the resistance D in this exposition are functions of the parameter η_2, which denotes the dimensionless thickness of the frictional layer at the point x. If x is taken as the entire length of the plate, so that D is now the total resistance per unit width of one side of the plate, the following value for C_f, the coefficient of frictional resistance, is easily obtained

$$C_f = \frac{D}{1/2 \, \varrho \, u_1^2 \, x} = \frac{2\nu}{u_1 x} \Psi(\eta_2) = \frac{2\Psi(\eta_2)}{\Phi(\eta_2)} \tag{23.10}$$

The next stage consists in replacing the function φ in its undetermined form by one of the special values given in the last section. If the power formula (22.8) is taken as a starting point, (23.4) and (23.5) reappear.

On the other hand, the logarithmic formula (22.13) used in its present form gives the velocity zero—not where $\eta = 0$, but at a small positive distance away. For when $\varphi = 0$

$$\log \eta = -\frac{5.5}{2.5} = -2.2$$

so that

$$\eta = e^{-2.2} \approx \frac{1}{9}$$

It is therefore advisable to transform the coordinates in such a way that $\varphi = 0$ when $\eta = 0$. This can obviously be achieved by writing

$$\varphi = 2.5 \log(1 + 9\eta)$$

This formula is practically identical with the previous one for all sufficiently large values of η. The numerical coefficients in this equation are approximations, so that the following generalized form of the equation may be written:

$$\varphi = a \log(1 + b\eta)$$

If $\varphi(\eta)$ is assumed to be of this form and $(1 + b\eta)$ is replaced by z to abbreviate the notation, integration furnishes the following values

$$F(\eta) = a^3\left(\log^2 z - 2\log z + 2 - \frac{2}{z}\right),$$

$$\Phi(\eta) = \frac{a^3}{b}\,(z\log^2 z - 4z\log z - 2\log z + 6z - 6),$$

$$\Psi(\eta) = \frac{a}{b}\left(z + 1 - \frac{2(z-1)}{\log z}\right)$$

Sets of numerical values have been calculated for $a = 2.495$ and $b = 8.93$, values which differ only slightly from those above. The following table is taken from the "Ergebnisse der Aerodynamischen Versuchsanstalt", Part IV, p. 25.

$\frac{\eta}{10^3}$	$\frac{R_l}{10^6}$	$C_f \cdot 10^3$
0.500	0.337	5.65
1.00	0.820	4.75
2.00	1.96	4.05
3.00	3.25	3.71
5.00	6.10	3.34
12.0	17.7	2.81
20.0	32.5	2.57
50.0	96.5	2.20
100.	217.5	1.96
500.	1401.	1.55

Since the relation between C_f and R_x furnished by the above set of formulae is inconvenient for calculations, it is advisable to use some interpolation formula which agrees sufficiently well with the tabulated values for the range occurring in practice. It has been found by Dr. H. Schlichting that this can be achieved satisfactorily by using the formula

$$C_f = \frac{0.472}{(\log R_x)^{2.58}} \tag{23.11}$$

It should be mentioned that calculations very similar to the above have been made by von Kármán. His results are in good agreement with those given above. They were announced by him in a lecture at the International Congress for Applied Mechanics, Stockholm, 1930, and are given in detail in the printed proceedings of that congress[1] and in a report[2] at the Conference on Hydromechanical Problems Connected with the Propulsion of Ships, Hamburg, 1932[3].

Fig. 66 shows several sets of values. It shows the numbers given in tabular form above, which in practice differ very little from those calculated from the approximate formula, together with the values obtained from (23.5) and also some values obtained experimentally. Wieselsberger's experiments[4] were performed on plates with blunted leading edges in the wind tunnel and consequently turbulence began

[1] Verhandlungen des III. Internationalen Kongresses für Technische Mechanik, p. 85, Stockholm, 1931.

[2] See Kempf and Foerster, Hydromechanische Probleme des Schiffsantriebs, Hamburg, 1932.

[3] Kármán's calculations have the advantage of priority over those of the present writer, but so far his articles contain no tables of numerical values for practical applications.

[4] „Ergebnisse der Aerodynamischen Versuchsanstalt", Pt. I., pp. 121—126.

quite near to the leading edge; Gebers[1] on the other hand used well sharpened plates in the water tank thus producing a laminar flow near the edge.

Kempf's experiments[2] which were also carried out in the water tank involve such large Reynolds numbers that the laminar portion of the flow has practically no influence on the calculations.

Allowance can be made for the laminar flow at the leading edge of a flat plate by comparing the states of affairs in two flows: (1) when the flow commences by being laminar and becomes turbulent only after

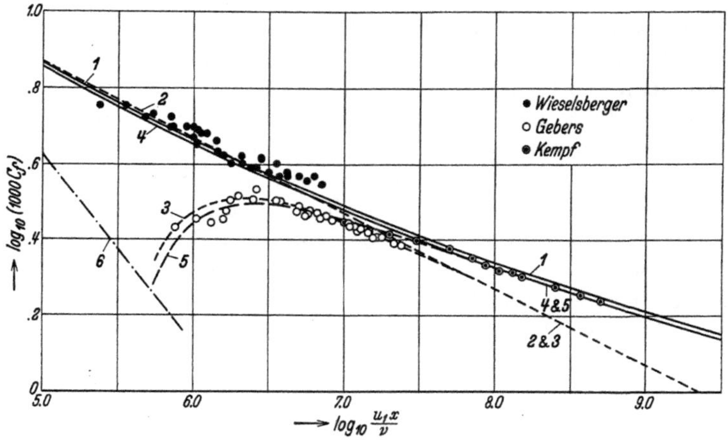

Fig. 66. Coefficient of frictional resistance of smooth plates. The lines of the diagram correspond with the following equations: (1), (23.11); (2), (23.14); (3), (23.15); (4), (23.16); (5), (23.17); (6), (23.18).

a definite Reynolds number R_x is reached, and (2) when turbulent flow occurs immediately at the leading edge. For the purpose of such a comparison the plate in (1) is taken to be longer than the plate in (2) by a length l just sufficient to ensure the same value for the resistance in each case. Since the resistances are equal, it follows from the theorem of momentum that δ, the final thickness of the frictional layer, is the same in both cases, and hence that the state of affairs throughout the two entire turbulent portions can be compared. Hence a plate of length x with initial laminar flow has the resistance of a plate of length $x - l$ as calculated from the formulae so far obtained. If symbols referring to the case of turbulent flow beginning from the leading edge are distinguished for convenience by an asterisk, it follows that

$$D = D^* = \frac{1}{2} \varrho u_1^2 (x - l) C_f^*$$

[1] „Schiffbau", Vol. IX, 1908.

[2] Werft, Reederei, Hafen, p. 234 and p. 247, 1929. The values given refer to the local coefficient of friction and have been integrated by the present writer.

And also, $$D = \frac{1}{2} \varrho u_1^2 x \cdot C_f$$

from which it follows that $C_f = C_f^* \cdot \frac{x - l}{x}$

This relation can be put into a slightly different form by multiplying both terms of the fraction by u_1/ν. For $u_1 x/\nu = R_x$, and $u_1 l/\nu$ is a Reynolds number characterizing the initial laminar flow and may be denoted by R_0. The final relation is therefore

$$C_f = C_f^* \left(\frac{R_x - R_0}{R_x} \right) \tag{23.12}$$

This relation can naturally be applied to (23.5) as well as to (23.10) and (23.11). In such a connection it must be remembered that (23.12) is based upon the assumption that at some definite point the laminar profile suddenly transforms into the turbulent profile. In actual fact the transition is accomplished in a region of appreciable length and moreover experiments show that the position of the point when turbulence commences oscillates with time. These two facts somewhat lengthen the transition between the laminar and turbulent flows. If the transition is not complete by the time the end of the plate is reached the above formula cannot be expected to agree with experimental results and in such a case C_f will in reality be somewhat higher. In the domain where transition occurs, the formula

$$C_f = C_f^* - \frac{const.}{R_x} \tag{23.13}$$

has proved to be closer to the facts but it is not possible to deduce it from theoretical considerations. The same formula can also be applied beyond the transition domain since the correction is of little account when large Reynolds numbers are involved. The Reynolds number R_0 in (23.12) as well as the constant in (23.13) naturally depend on the sharpness of the plate. They also depend on the degree of freedom from turbulence of the approaching flow, or otherwise on the smoothness of the plate's motion, as the case may be. In Gebers' experiments R_0 can be taken as 500,000 and the constant of (23.13) as 1,700.

The numerical coefficients of (23.5) and (23.11) were deduced from measurements on tubular flows. Since the velocity profile for a flow past a plate is not completely identical with that for a tubular flow it is advisable to improve the numerical values in accordance with the results of experiments on plates. If (23.13) is applied to the results of Gebers' and Kempf's experiments, sufficient information is obtained to determine the numerical coefficient of (23.11). This becomes thus reduced to 0.455. When compared with the results of experiments however, the values obtained from (23.5) are systematically too low when the Reynolds numbers are large; this can be compensated to some extent by increasing the coefficient of that formula to 0.074.

Hence the following can be taken as practical formulae for calculating the resistance for a turbulent flow with friction

$$C_f = 0.074\,(R_x)^{-1/5} \tag{23.14}$$

and

$$C_f = 0.074\,(R_x)^{-1/5} - \frac{1700}{R_x} \tag{23.15}$$

respectively, for $R_x < 10{,}000{,}000$ or,

$$C_f = \frac{0.455}{(\log R_x)^{2.58}} \tag{23.16}$$

and

$$C_f = \frac{0.455}{(\log R_x)^{2.58}} - \frac{1700}{R_x} \tag{23.17}$$

for the entire domain. The number 1700 characterizes a state of affairs in which there is little or no disturbance in the fluid, as in Gebers' experiments, and it must be altered correspondingly for other cases. It is often necessary to use numbers rather smaller than 1700. It will be useful to insert here the formula for the laminar flow, to be applied for values of R_x which are less than 500,000

$$C_f = 1.328\,R_x^{-1/2} \tag{23.18}$$

(see **14**). Clark B. Millikan[1] has extended the above theory to the case of a long body with a form of revolution (airship form).

With the help of the results at the end of **22**, calculations corresponding exactly to the above can also be performed for the case of rough plates. All that is necessary is to replace the dimensionless quantity $\eta = v_* y/\nu$ by another dimensionless quantity $\theta = y/k$. The calculations proceed in the same fashion as in the case of smooth plates though the integrals which occur in place of (23.8) and (23.9) are somewhat less simple. Similar calculations can also be performed for the transition from the completely rough plate to one which though rough is still hydraulically smooth. Finally, to complete the discussion for the entire domain these various solutions must be suitably combined. These calculations will be omitted here, but Fig. 67 shows the relation between the shearing stress, the Reynolds number and the roughness of the wall, as obtained from these calculations. The diagram shows the relation between $\psi = \tau / 1/2\,\varrho u_1^2$ and R_x; the curves of constant relative roughness k/x and the curves of equal Reynolds numbers for the roughness $u_1 k/\nu$ are also shown.

These curves have the following obvious interpretations: if the velocity is changed at some definite position on the plate, k/x remains constant at that point; if, on the other hand, x is changed, the plate being

[1] Millikan, Clark B., The Boundary Layer and Skin Friction for Figure of Revolution. Transactions American Society of Mechanical Engineers, Applied Mechanics Section, Vol. 54, No. 2, p. 29, January 30, 1932.

uniformly rough and the velocity constant, $u_1 k/\nu$ is constant for all points on the plate. The coefficient of friction C_f is obtained by integrating ψ

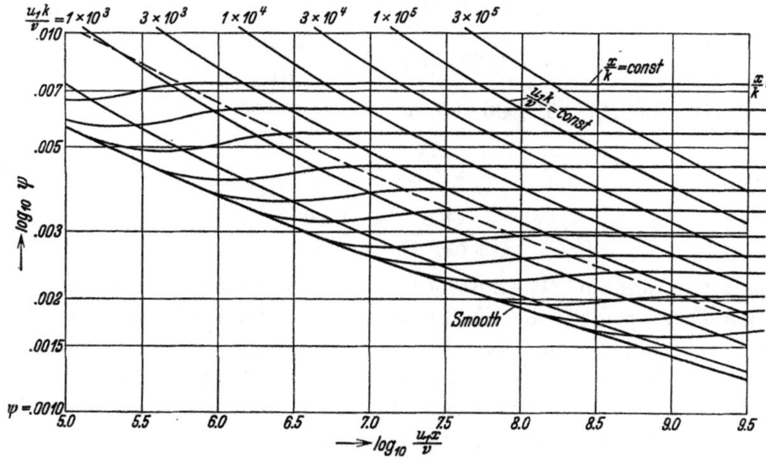

Fig. 67. Coefficient of shearing stress of rough plates.

along the plate and subsequently dividing by the length, *i. e.* by finding the mean value of ψ along the plate. This is represented in similar

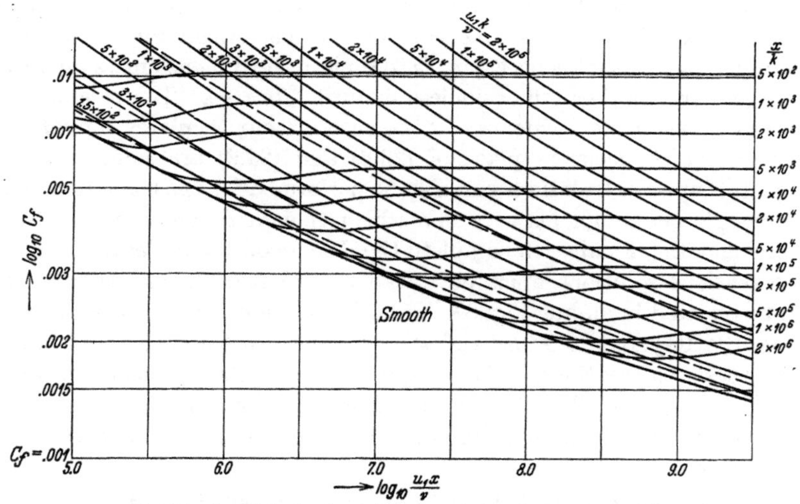

Fig. 68. Coefficient of frictional resistance of rough plates.

fashion in Fig. 68, on the assumption that the flow is turbulent from the leading edge onwards[1].

[1] The necessary calculations were made by H. Schlichting.

24. Turbulent Friction Layers in Accelerated and Retarded Flows. In addition to the comparatively simple cases already discussed the study of turbulent friction layers in more complicated flows, subject to considerable accelerations and retardations is of great practical importance. The necessary numerical calculations and values for such conditions not only furnish important information in such cases as the frictional drag of an airfoil or the frictional losses in the vanes of a turbine, but also make it possible to determine whether or not a given velocity and pressure distribution is possible without separation of the boundary layer. Certain very promising attempts, now to be briefly described, have already been made on these lines.

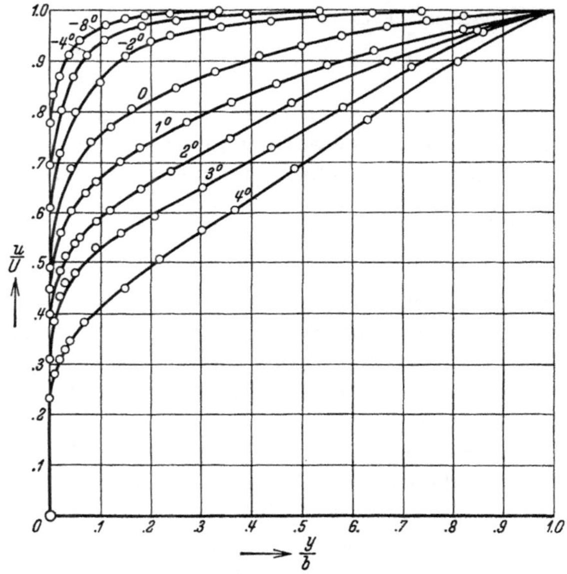

Fig. 69. Velocity distributions in accelerated and retarded turbulent flow, after Nikuradse.

It will be convenient first to note the facts which have been brought to light by experimental investigations. Experiments on gently converging or diverging canals carried out by Dönch and Nikuradse[1] using, air and water respectively, have shown that the velocity profiles depend, to a very marked degree, on the extent to which accelerations or retardations occur. Some of the velocity profiles from Nikuradse's paper are reproduced in Fig. 69. The Reynolds number also appears to have some influence and for the shearing force at the wall a formula along the lines of (22.6) is indicated. In the converging canal the r of this formula must be replaced by the thickness δ of the frictional layer; in the diverging canal, the frictional layer reaches to the middle, so that r can be replaced by half the width of the canal. If the Reynolds number $u_1 \delta/\nu$ is kept constant the form of the relation between the velocity profile and the pressure gradient along the wall, and hence also

[1] Dönch, F., Divergente und konvergente turbulente Strömungen mit kleinen Öffnungswinkeln. Göttingen Dissertation 1925, Forschungsarbeiten des V.D.I., Heft 292, 1926. — Nikuradse, J., Untersuchungen über die Strömungen des Wassers in konvergenten und divergenten Kanälen. Forschungsarbeiten des V.D.I., Heft 289, 1929.

between the velocity profile and the angle of convergence or divergence, is found to be conveniently expressed in terms of a parameter defined by

$$\omega = -\frac{\delta}{\varrho u_1^2} \cdot \frac{dp}{dx}$$

It represents the fall in pressure along a length equal to the thickness of the frictional layer, reduced to zero dimensions by division by ϱu_1^2.

Since
$$\frac{dp}{dx} = -\varrho u_1 \frac{du_1}{dx}$$

the parameter ω can also be written in the form

$$\omega = \frac{\delta}{u_1} \cdot \frac{du_1}{dx}$$

Comparisons between experiments performed by Dönch and Nikuradse involving various Reynolds numbers appear to show that it is advisable to replace ω by the following parameter in the region for which the power formula holds
$$\Gamma = \omega \left(\frac{u_1 \delta}{\nu}\right)^{1/4} = \frac{\delta}{u_1} \cdot \frac{du_1}{dx} \left(\frac{u_1 \delta}{\nu}\right)^{1/4}$$

Although the experiments mentioned have not provided a complete basis for the introduction of this parameter Γ, it will nevertheless be used in the present discussion, since it materially simplifies the calculations which follow. The formula for Γ was based on the considerations that the shear at the wall, τ_0, is proportional to $(u_1 \delta/\nu)^{-1/4}$ and that multiplication by $(u_1 \delta/\nu)^{1/4}$ eliminates the dependence on the Reynolds number. Analysis of the experiments by Dönch and Nikuradse did, in fact, show that Γ, as above, was a suitable "form parameter"; that is, a constant value of Γ always produced the same type of velocity profile.

We return now to the particular problem of connecting with (17.11) which was derived from the theorem of momentum. This formula is,

$$\tau_0 = \varrho u_1 \frac{du_1}{dx}\left(\delta^* + 2\delta^{**}\right) + \varrho u_1^2 \frac{d\delta^{**}}{dx} \tag{24.1}$$

where δ^* denotes the measure of the displacement and δ^{**} the measure of the momentum. This equation can be used in the following manner in order to obtain approximate values of the desired quantities[1]. Using the power formula and replacing δ by δ^{**} for reasons which will become apparent later, the shear τ_0 can be expressed as

$$\tau_0 = \zeta \varrho u_1^2 \left(\frac{u_1 \delta^{**}}{\nu}\right)^{-1/4} \tag{24.2}$$

On account of the use of δ^{**} in this equation, the numerical value ζ will not be the same as that of (22.6) but will in this case depend on the shape of the velocity profile. Hence if the results obtained from the experiments with converging and diverging canals can be applied to

[1] These calculations were made by Herr A. Buri at the suggestion of the present writer. See his Zürich dissertation „Eine Berechnungsgrundlage für die turbulente Grenzschicht bei beschleunigter und verzögerter Grundströmung". Zürich, 1931.

the present case, ζ will be a function of Γ, the parameter of form. The experiments of Dönch and Nikuradse mentioned above show that ζ does not vary very much. Substitution of the value of τ_0 from (24.2) in (24.1) leads to the equation

$$\zeta \left(\frac{u_1 \delta^{**}}{\nu}\right)^{-1/4} = \frac{1}{u_1} \frac{d u_1}{d x} (\delta^* + 2 \delta^{**}) + \frac{d \delta^{**}}{d x}$$

On multiplying by $(u_1 \delta^{**}/\nu)^{1/4}$, adding $(1/5)\,(\delta^{**}/u_1)\,(d u_1/d x)\,(u_1 \delta^{**}/\nu)$ to both sides and rearranging, we obtain

$$\left.\begin{aligned} \zeta - \frac{\delta^{**}}{u_1} \cdot \frac{d u_1}{d x} \left(\frac{u_1 \delta^{**}}{\nu}\right)^{1/4} \left(\frac{9}{5} + \frac{\delta^*}{\delta^{**}}\right) \\ = \frac{d \delta^{**}}{d x} \left(\frac{u_1 \delta^{**}}{\nu}\right)^{1/4} + \frac{1}{5} \frac{\delta^{**}}{u_1} \frac{d u_1}{d x} \left(\frac{u_1 \delta^{**}}{\nu}\right)^{1/4} \\ = \frac{4}{5} \frac{d}{d x} \left[\delta^{**} \left(\frac{u_1 \delta^{**}}{\nu}\right)^{1/4}\right] \end{aligned}\right\} \qquad (24.3)$$

One of the expressions on the left hand side of the equation can be recognized as the form parameter Γ, altered by the substitution of δ^{**}, the measure of momentum for δ. This parameter will therefore be called Γ_1. If Γ_1 is a form parameter as in the previous and simpler case of convergent and divergent canals, the ratio δ^*/δ^{**} must be a function of Γ_1. Hence the whole of the left hand side of (24.3) is a function of Γ_1, the form of which must now be determined from experimental results. Herr Buri was able to show that a graph of this function based on the Nikuradse results is nearly a straight line. Hence, if this function is denoted by $\psi(\Gamma_1)$, we may write

$$\psi(\Gamma_1) = a - b\, \Gamma_1$$

On substituting the abbreviation

$$z = \delta^{**} \left(\frac{u_1 \delta^{**}}{\nu}\right)^{1/4} \qquad (24.4)$$

in (24.3), it follows that

$$a - \frac{b}{u_1} \cdot \frac{d u_1}{d x} z = \frac{d z}{d x} \qquad (24.5)$$

This is a differential equation of the first order in z whose solution can easily be obtained, and is,

$$z = \frac{a}{u_1 b} \left(\int u_1^b \, d x + c \right) \qquad (24.6)$$

Hence if u_1 is given as a function of x, z and hence δ^{**} can both be easily evaluated. Finally from Γ_1 the form of the velocity profile can be obtained. From Nikuradse's experiments it is found that $a = 0.017$; the value of b can be taken as 4.75 for accelerated flows (Γ_1 positive) and 5.15 for retarded flows (Γ_1 negative). Since only approximate values are desired the more convenient value $b = 5$ for both cases is sufficiently accurate[1]. The important question of when separation may be expected reduces simply to the determination of the corresponding values of Γ_1. According to Nikuradse's experiments this value is found about -0.07. It should be mentioned however that determinations of the separation based on various other series of experiments lie within a rather large range of variations bounded approximately by -0.05 and -0.09.

Buri's calculations, it has been seen, are based on the assumption that the same profile occurs for the same value of Γ, *i. e.* for the same relative increase of velocity along a length in the direction of flow equal to the thickness of the frictional layer. When interpreted quite strictly this assumption is not valid, for each velocity profile develops in some regular manner from the velocity profiles situated farther upstream, *i. e.* the shape of the velocity profile depends absolutely on the previous history of the portion of fluid considered. No account of this factor is taken in Buri's calculations. If however, the transition of the velocity profiles proceeds very slowly and over a long distance, this factor has little influence and it is permissible to use the results of experiments such as those of Dönch and Nikuradse when the form of the velocity profiles at all parts of the flow varies but little. The errors which may be produced by neglecting the variation of the velocity profiles may produce uncertainty as to the values of Γ_1 which correspond to some definite profile. Nevertheless, Buri's system of equations are valuable, especially for the consideration of the qualitative features of a flow.

More accurate calculations have since been performed by Dr. E. Gruschwitz[2]. In addition to an equation similar to (24.5) he uses a second differential equation in terms of another form parameter, thus forming a system of two simultaneous equations of the first order. The empirical functions which occur in these equations are determined by a series of experiments in which widely differing states of acceleration and retardation are produced. On calculating the thickness of the frictional layer with the help of these equations, very satisfactory agreement was found both with the varying thickness of the frictional layer

[1] Buri finds as the result of his own experiments, which were however confined to the case of accelerated motion, that $a = 0.015$ and $b = 4$. The difference may possibly arise from the fact that in Buri's experiments the flow started from rest and possibly did not have completely developed turbulence in the stretch under observation. Nikuradse's experiments involved flow emerging from a pipe which was certainly turbulent from the very beginning. Where states of affairs similar to that of Buri's experiments are under consideration, his numerical values will naturally be preferred to those of Nikuradse.

[2] Gruschwitz, E., Die turbulente Reibungsschicht in ebener Strömung bei Druckabfall und Druckanstieg. Ingenieur-Archiv, Vol. II, p. 321, 1931.

A shorter account appeared under the title of "Über den Ablösungsvorgang in der turbulenten Reibungsschicht" in the Zeitschrift für Flugtechnik und Motorluftschiffahrt, Vol. XXIII, p. 308, 1932.

and with the series of values of the form parameter which determines the velocity profile. For this the calculations are however rather more involved than those given above. Buri's calculations may serve here as an example and the reader may be referred to the original papers for further details.

A very important relation can be discovered from the condition for separation obtained in Buri's approximations, *i. e.* from the equation

$$\Gamma_1 = \frac{\delta^{**}}{u_1}\frac{d u_1}{d x}\left(\frac{u_1 \delta^{**}}{\nu}\right)^{1/4} = \frac{z}{u_1}\frac{d u_1}{d x} = -0.07 \qquad (24.7)$$

For this equation shows that the point at which separation occurs depends chiefly on the percentage change in velocity over a stretch of length δ^{**} measured along the wall. And it may be noted that the reintroduction of δ, the thickness of the frictional layer, in place of δ^{**}, the measure of momentum, will only affect the factor of proportionality; for δ is proportional to δ^{**} for any definite shape of profile and will therefore be proportional to δ^{**} for the separation profile. Hence, if separation is to be avoided, the smaller the thickness δ of the frictional layer the greater the percentage decrease in velocity, and hence also increase in pressure for a given distance along the wall. This shows clearly *inter alia* that roughness of the surface, which naturally increases the thickness of the frictional layer, increases the danger of separation, whereas diminution of the thickness by suction of part of the frictional layer into the interior of the body diminishes the danger of separation, and hence permits of greater increases of pressure. In accordance with a previous remark[1] the Reynolds number has no effect on the appearance of separation. This can now be verified by calculating the value of z for the case of flow past a straight plate as previously discussed. Here u_1 is a constant and $\Gamma_1 = 0$ and hence in accordance with (24.6)

$$z = a\, x$$

This equation can also be applied as an approximation for other frictional layers, different numbers being substituted in place of a in accordance with the different requirements of each case. If such a value is substituted for z in (24.7) it is seen that

$$\Gamma_1 \sim \frac{x}{u_1}\frac{d u_1}{d x}$$

i. e. the form chosen for the initial assumption does in fact eliminate the influence of the Reynolds number, and the point where separation occurs is independent of the Reynolds number in the region where the power formula is valid. For this to be the case however, the same condition as for the validity of (23.4) is necessary, *viz.* that turbulence begins directly at the front of the body. Very often, however, the flow commences with a laminar portion in that region and in such cases, since the frictional drag is smaller in the laminar case than in the case

[1] When Γ was introduced on p. 156.

of turbulence, there will be a thinner frictional layer at the front of the body than in completely turbulent motion. Hence in accordance with (24.7) the thinner frictional layer produces less tendency towards separation. Various observations on the drag of spheres, cylinders, etc., having turbulent frictional layers, confirm the existence of such phenomena. If the place where turbulence begins shifts farther forward as the Reynolds number increases, then, on account of the thickening of the frictional layer, the point of separation also moves forward with consequent increase of drag. This increase of resistance or drag as the Reynolds number increases has been observed in experiments on spheres. Similar features are present in flows past thick airfoils with well-rounded leading edges. In the neighborhood of the stagnation point there is a laminar portion of the flow of an extent which diminishes as the Reynolds number increases and, as in the other cases mentioned, thickens the frictional layer. The effect of this is that the maximum angle of incidence for which the stream is still continuous along the airfoil, and hence also the maximum lift, diminishes somewhat as the Reynolds number increases. This phenomenon also has been observed experimentally now for some considerable time. The preceding explanations are due to Gruschwitz. Compare also the discussion in **26** on the question of the commencement of turbulence.

The above account of flows along walls, accelerated in the direction of motion, may be supplemented by a few remarks on flows with accelerations transverse to the direction of motion. Two distinct cases must be considered: transverse acceleration parallel to and perpendicular to the wall respectively. In the first case there are curved stream-lines lying in surfaces of flow which may or may not be plane, but which will lie nearly parallel to the wall. The flow in such a path experiences a fall of pressure toward the center of curvature of magnitude $\varrho u^2/r$ where u is the velocity of flow and r the radius of curvature of the path. Since the frictional layer by the wall has smaller velocities than the layers farther from the wall but receives its pressure from them, it has the same pressure gradient, and hence its stream-lines have smaller radii of curvature if they commence with the same initial directions as the remaining stream-lines. Hence the frictional layer is diverted away from the main direction of flow in the direction of decreased pressure and there is a difference of velocity between the main flow and the diverted portion. This calls frictional forces into play between the diverted portion and the main body of the fluid, the magnitude of which depends on the difference in velocity between the two portions. These and the corresponding component of the friction at the wall compensate the fall in pressure to some extent so that the deflection of the fluid does not increase beyond a certain amount. The whole of this phenomenon is termed "secondary flow".

Secondary flow can be observed, for example, in a flow through a curved pipe; it has the effect of superimposing upon the velocities in the direction of the curved axis of the pipe, additional velocities at the walls which are directed toward the center of curvature, and which in turn compel the rest of the fluid, by continuity, to stream slowly toward the outer wall (see Fig. 70). Another example is given by the flow around a curved airfoil to which is attached a surface perpendicular to its axis. Secondary flow is produced on the suction side of the attached surface; it is directed toward the foil, collects in the angle between surface and foil, and thereby produces a considerable increase of drag and tendency toward separation. The secondary flow on the other side of the attached surface moves away from the foil and has no detrimental action.

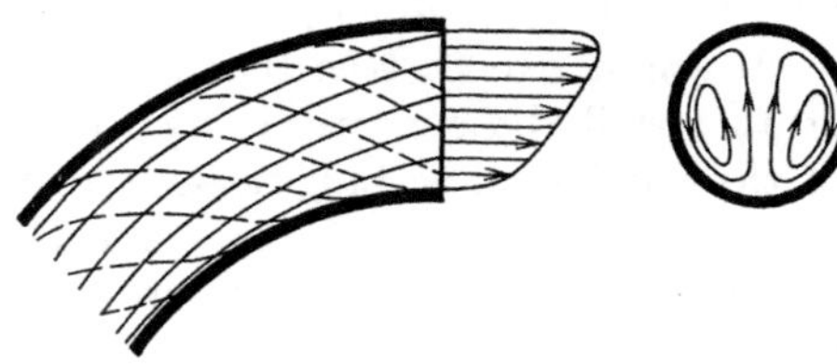

Fig. 70. Flow in curved pipes, with secondary flow.

The second of the two cases referred to above, acceleration normal to the wall, occurs when the wall is curved in the direction of flow. The resulting state of affairs depends on whether the wall is convex or concave. The curvature of the path taken by particles in layers near to the wall is practically determined by the curvature of the wall. Let the radius of curvature of the wall be r: the centrifugal force induced in particles of fluid has then the value $\varrho u^2/r$ per unit volume and is therefore greater for quickly moving particles than for relatively slow particles. Hence for a flow past a wall with concave curvature, the fast moving particles of the unretarded fluid press toward the wall more strongly than those particles which are near to the wall and are already retarded by friction. The consequence is that the faster portion of the fluid moves toward the wall while the more slowly moving portion is thrust into the interior of the fluid. These phenomena have some similarity to the state of affairs in the flow over a horizontal warm plate in which the particles near to the plate are retarded by friction and are warmed by the plate, with resultant decrease of density. The particles rise and are replaced by others moving with the unretarded velocity. In both cases the friction at the wall is considerably increased.

At a convex wall, the particles retarded by friction at the wall will have relatively smaller centrifugal forces acting upon them and will therefore show less tendency to leave the wall than the more swiftly moving flow in the interior of the fluid. This produces a stabilizing effect on the flow similar to the effect of the streaming of air over a cooled horizontal plate. In this latter case, the particles already near the plate tend to remain in position on account of their greater density.

This stabilizing effect considerably reduces the tendency toward intermixture of the fluid, and a very marked decrease in friction at the wall is to be expected.

The results of experiments recently performed, or about to be performed, should, after some time, permit the definite quantitative determination of the magnitudes involved in the phenomena just described. For the theory of these phenomena the reader may be referred to a lecture delivered by the present writer at the Aerodynamical Conference at Aachen[1]. Some experimental results on the same topic may also be found in a lecture by A. Betz on the same occasion[2].

It appears to be true that in a flow over the upper (or suction) side of an airfoil the curvature is sufficient to produce a slight diminution of the friction in certain cases, but the observations which have been made up to the present time are not quite sufficient to settle this point. In curved channels, the increase in friction on the wall concave to the flow more than balances the diminution of friction on the convex wall so that frictional losses are on the whole considerably greater than in a straight channel. The secondary flow also increases the amount of loss by removing fluid from the already thin frictional layer at the concave wall and transferring it to the passive layers on the other side.

25. Spread of Turbulence. In addition to turbulent flows past fixed walls, it is now necessary to consider cases involving turbulence without fixed boundaries. These include cases of gradual breaking up of free fluid jets through the formation of vortices resulting from instability of flow; and likewise the phenomena in the wake of a moving body. For such phenomena it has likewise proved useful to assume the existence of an apparent shearing force of magnitude

$$\tau = \varrho\, l^2 \left|\frac{\partial u}{\partial y}\right| \frac{\partial u}{\partial y} \tag{25.1}$$

which simultaneously reduces the mean velocity and increases the width of the jet as the flow proceeds. The increasing width is produced by the mixture of the fluid composing the jet and the surrounding fluid as a result of which more and more of the latter is gradually entrained into the motion. G. I. Taylor[3] has pointed out that this type of turbulence is different in character from the turbulence at boundary walls. This

[1] Prandtl, L., Einfluß stabilisierender Kräfte auf die Turbulenz. Gilles-Hopf-v. Kármán: Vorträge aus dem Gebiete der Aerodynamik und verwandter Gebiete, p. 1, Aachen, 1929, Berlin, 1930.

[2] Betz, A., Über turbulente Reibungsschichten an gekrümmten Wänden. *loc. cit.* p. 10, see also H. Wilcken, Turbulente Grenzschichten an gewölbten Flächen. Ing.-Arch., Vol. I, p. 357, 1930.

[3] Taylor, G. I., The Transport of Vorticity and Heat Through Fluids in Turbulent Motion. With an appendix by A. Fage and V. M. Falkner. Proc. Roy. Soc. A. 135, p. 685 and p. 702, 1932.

results from the fact that the propagation of temperature differences in a flow along a boundary follows the same law as for a difference of velocities, while with the propagations here considered, the spread of temperature differences follows a law different from that of the velocities, and according to which the turbulent exchange of difference of the mean rotation in the stream corresponds to that of the temperature differences. This different behavior can be given a theoretical explanation by assuming that vortices with axes parallel to the direction of flow predominate in flows past boundary walls, while the predominating direction of vortex axes in the spread of fluid jets is perpendicular both to the direction of flow and to the direction of the greatest change of velocity.

If *all* the vortices in the turbulent motion are arranged in the latter fashion, a two-dimensional flow results. If then viscosity is neglected, Helmholtz's vortex theorem shows that the angular velocity of every particle of fluid in this case remains constant with respect to time. It follows that where mixing occurs the mean value of the angular velocity of the particles follows the laws of mixture in somewhat the same way as in the case of heat-content, that is, proportional to the temperature. As the result of considerations which are reproduced in an appendix to this paragraph, G. I. Taylor obtains a relation which, in the symbolism of this Division, can be written as follows

$$\frac{\partial \tau}{\partial y} = \varrho l^2 \left| \frac{\partial u}{\partial y} \right| \frac{\partial^2 u}{\partial y^2}$$

It can be assumed in general for all such investigations of spread that the mixing length l in a direction perpendicular to the direction of flow is constant; the relation then integrates into

$$\tau = \frac{1}{2} \varrho l^2 \left| \frac{\partial u}{\partial y} \right| \frac{\partial u}{\partial y} \tag{25.2}$$

This formula only differs from the previous one in having $1/2\ l^2$ in place of l^2. This simply means that the mixing length according to Taylor is $\sqrt{2}$ times as great as that assumed in (25.1). The heat convection $c \varrho l^2 |\partial u/\partial y| \partial \theta/\partial y$ (c specific heat, θ temperature), which has the same formula according to both theories is, for equal amounts of spread of momentum in both cases, twice as large in the case of the spread of a jet as in the case of spread in the flow past a wall.

It is now necessary to consider the actual details of the flow. The first important question concerns the magnitude of the mixing length. The following rule has been verified in most cases: the rate of increase of the width of the mixing zone b with respect to time is comparable with v', the disturbance velocity across the flow. On writing $v' = l\ (\partial u/\partial y)$ as before, the mean value of $\partial u/\partial y$, as a first approximation, can be replaced by

$$(\text{Number}) \times \frac{u_{max}}{b}$$

Hence the alteration Db/dt in width of a given portion of fluid (differentiation following the motion of the fluid) is such that

$$\frac{Db}{dt} \sim \frac{l}{b} u_{max} \tag{25.3}$$

It has also proved permissible to take the ratio l/b as constant in any given case, but it is found that the numerical value of this ratio is different in various cases; thus, for example approximately 0.15 for the spread of a jet and 0.3 for that of a wake, respectively, b being half the diameter of the mixing zone. This may be connected with the fact that in the latter case the moving body produces, from the very beginning, more powerful vorticity than the uniformly spreading jet.

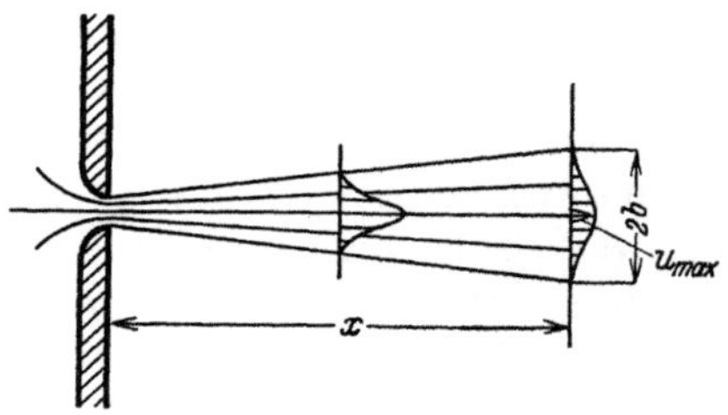

Fig. 71.

A number of important conclusions can be drawn from (25.3). Consider for example a jet emerging from a small opening (see Fig. 71). The relation between the velocity u_{max} and x, the distance of the cross-section from the opening, is not yet known, but the momentum of the jet (*i. e.* $\int \varrho u^2 dS$ over the cross-section) must have the same value for every cross-section. On writing the momentum M in the form

$$M = (\text{Number}) \times \varrho u_{max}^2 b^2$$

it follows that
$$u_{max} = \frac{(\text{Number})}{b} \cdot \sqrt{\frac{M}{\varrho}}$$

On the other hand the following substitution can be made in (25.3)

$$\frac{Db}{dt} = (\text{Number}) \cdot u_{max} \cdot \frac{db}{dx}$$

whence
$$\frac{db}{dx} = (\text{Number}),$$

i. e.
$$b = (\text{Number}) \cdot x + \text{const.}$$

This result could also have been obtained by recognizing that in this type of motion the initial diameter and velocity of the jet are unimportant, and that the fundamental quantity is the amount of momentum contained in the jet. If this is assumed to be true, the only length which is available to determine b is the distance from the opening (or, to be more exact, the distance from the point at which the jet, in accordance with the law of spreading which holds for considerable distances from the mouth, would have zero cross-section).

The relations in the wake of a moving body are somewhat different since the momentum must be calculated rather differently. In accordance with the above considerations it would be necessary to form the integral

$\varrho \iint u\,(u_1 - u)\,dS$ where u_1 is the velocity of the undisturbed flow, u the velocity in the wake, both referred to a stationary body. At considerable distances from the body however, the difference of velocity $u' = (u_1 - u)$ is very small so that as a first approximation for either of the expressions

$$u\,(u_1 - u) = (u_1 - u')\,u'$$

we can write $u_1 u'$. The calculations differ somewhat according as to whether the body has a form of revolution (sphere, airship-hull, etc.) or otherwise that of a long body lying across the flow (airfoil, strut, etc.). In the first case, the wake can be assumed to be circular in cross-section and therefore of area πb^2 where $2\,b$ is the diameter of the wake at distance x. In the second case, the cross-section of the wake is $2b\,l$ for a body of length l across the direction of flow. On writing the momentum of the wake as equal to the drag $[(1/2)\,\varrho\,u_1^2\,C_D\,S$ where $C_D =$ coefficient of resistance and $S =$ cross-section of the body] the following formula is obtained in the case of symmetry about the axis of flow

$$u_1\,u' \cdot \pi\,b^2 \sim u_1^2 \cdot C_D\,S$$

i. e.
$$\frac{u'}{u_1} \sim \frac{C_D\,S}{\pi\,b^2} \tag{25.4}$$

Since
$$\frac{D\,b}{d\,t} = u_1 \frac{d\,b}{d\,x}$$

(25.3) now gives
$$u_1 \frac{d\,b}{d\,x} \sim \left(\frac{l}{b}\right) u'$$

which in conjunction with (25.4) and $l/b = \beta$ leads to

$$b^2 \frac{d\,b}{d\,x} \sim \frac{\beta}{\pi}\,C_D\,S$$

or, after integration
$$b \sim \sqrt[3]{\beta\,C_D\,S\,x} \tag{25.5}$$

In the case of a body lying across the direction of flow, *i. e.* for a plane flow, the initial relation is

$$u_1\,u'\,b\,l \sim u_1^2\,C_D\,l\,d$$

i. e.
$$\frac{u'}{u_1} \sim \frac{C_D d}{2\,b} \tag{25.6}$$

and hence
$$2\,b \frac{d\,b}{d\,x} \sim \beta\,C_D\,d$$

and therefore
$$b \sim \sqrt{\beta\,C_D d\,x} \tag{25.7}$$

If (25.4) and (25.6) are borne in mind the following relations now hold for the velocity in the wake

$$u' \sim \frac{u_1}{\pi} \left(\frac{C_D\,S}{\beta^2\,x^2}\right)^{1/3} \tag{25.8}$$

and
$$u' \sim \frac{u_1}{2} \left(\frac{C_D\,d}{\beta\,x}\right)^{1/2} \tag{25.9}$$

It has therefore been shown that in the case of symmetry about the axis the diameter of the wake varies directly as the cube root of the distance x and the velocity inversely as the cube root of x^2; while in the case of two-dimensional flow the width varies directly as the square root of x and the velocity inversely as the square root of x.

The characterizing features of the motions considered are indicated very well by the above approximate relationships; it is however possible to go further and to obtain the velocity profile by integration of the hydrodynamic differential equations using the general friction theorem stated above. The flows which can be discussed most simply are those which depend on time and in which the field velocity of flow in the x direction at any given instant is a function of y only, and there are no velocities in other directions. For such flows therefore

$$u = u\,(y, t); \qquad v = w = 0$$

The "field" velocity denotes the velocity obtained by subtracting the turbulent fluctuations from the actual velocity; it can be obtained in this case by suitable methods of forming mean values. The following special assumption may be here made

$$u = u_0\,(t)\, f\left(\frac{y}{b\,(t)}\right) \tag{25.10}$$

This means that we assume the velocity profile to remain "affine"[1] for all time, and consequently the ratio of the velocities at corresponding points to vary as $u_0\,(t)$ and the width of the mixing region at any time to be $2b\,(t)$. On making the further assumption for the mixing length,

$$l = \beta b$$

and observing that Du/dt reduces to $\partial u/\partial t$ on account of the simplicity of the case considered, the following equation is obtained, according to Taylor's assumption which is valid for this kind of fluid motion.

$$\frac{\partial u}{\partial t} = \frac{1}{\varrho}\frac{\partial \tau}{\partial y} = l^2 \left|\frac{\partial u}{\partial y}\right| \frac{\partial^2 u}{\partial y^2} \tag{25.11}$$

In order to specialize the problem we now put

$$u_0 = A\,t^p \tag{25.12}$$

$$b = B\,t^q \tag{25.13}$$

and

$$\frac{y}{b} = \eta$$

[1] An "affine" transformation of a geometrical figure is one in which all parallel straight lines are transformed into parallel straight lines, whence it follows that all dimensions in a given direction remain proportional as between the original and the transformed figure. Thus a circle may transform into an ellipse wherein two diameters at right angles will transform into conjugate diameters of the ellipse.

On substituting in (25.11) and performing the differentiations the equation obtained is

$$\left.\begin{aligned} A\,t^{p-1}\,(p\,f(\eta) - q\,\eta\,f') &= \pm\,\beta^2\,B^2\,t^{2q}\frac{A^2}{B^3}\,t^{2p-3q}\,f'\,f'' \\ &= \pm\,\frac{\beta^2\,A^2}{B}\,t^{2p-q}\,f'\,f'' \end{aligned}\right\} \qquad (25.14)$$

The power of t must be a factor throughout this equation so that $(p - 1)$ must be equal to $(2p - q)$ *i. e.* $q = p + 1$.

Two cases will now be considered:

1) Let there be a given surface of separation, and let the initial velocities be constant and equal to U_1 on the one side, and constant and equal to U_2 on the other. It is required to find the details of the turbulent mixing phenomena which are characterized by the gradual transition between the velocities in a zone of increasing width. Since the velocities at the boundaries of the mixing zone remain constant with respect to time the index p must be equated to zero and hence from the above equation q must equal 1. On writing u in the following form

$$u = \frac{1}{2}\,(U_1 + U_2) + \frac{1}{2}\,(U_1 - U_2)\,f(\eta) \qquad (25.15)$$

the condition that u should have the values U_1 and U_2 respectively at the two edges of the mixing zone is satisfied if $f(\eta)$ has the values $+1$ and -1 respectively at these two boundaries. Hence, in (25.12), $A = (1/2)\,(U_1 - U_2)$. It follows that the differential equation for this case is

$$\eta\,f' \mp \frac{\beta^2\,A}{B}\,f'\,f'' = 0 \qquad (25.16)$$

One solution of this equation is $f' = 0$, *i. e.* $f =$ constant. If however f' is not zero, dividing the equation by f' leads to

$$f'' = \mp\,\frac{B}{\beta^2\,A}\,\eta$$

and hence

$$f(\eta) = \mp\,\frac{B}{\beta^2\,A}\,\frac{\eta^3}{6} + C_1\,\eta + C_2$$

If it is postulated that $f(\eta) = 0$ when $\eta = 0$, the constant C_2 vanishes and the solution is

$$f(\eta) = C_0\,\eta^3 + C_1\,\eta$$

where

$$C_0 = \mp\,\frac{B}{6\,\beta^2\,A}$$

Further, it has already been stated that $f(\eta) = 1$ and $f'(\eta) = 0$ when $y = b$, *i. e.* when $\eta = 1$. This condition yields two equations for C_0 and C_1

$$\left.\begin{aligned} C_0 + C_1 &= 1 \\ 3\,C_0 + C_1 &= 0 \end{aligned}\right\}$$

from which it follows that

$$C_1 = \frac{3}{2}\,, \qquad C_0 = -\,\frac{1}{2}$$

Hence $B = 3A\,\beta^2$ and, finally substituting the known value for A, the width of the mixing zone has the value

$$b = \frac{3}{2}\,(U_1 - U_2)\,\beta^2\,t \tag{25.17}$$

while the distribution of velocity is given by

$$u = \frac{1}{2}\,(U_1 + U_2) + \frac{1}{2}\,(U_1 - U_2)\left[\frac{3}{2}\,\frac{y}{b} - \frac{1}{2}\left(\frac{y}{b}\right)^3\right] \tag{25.18}$$

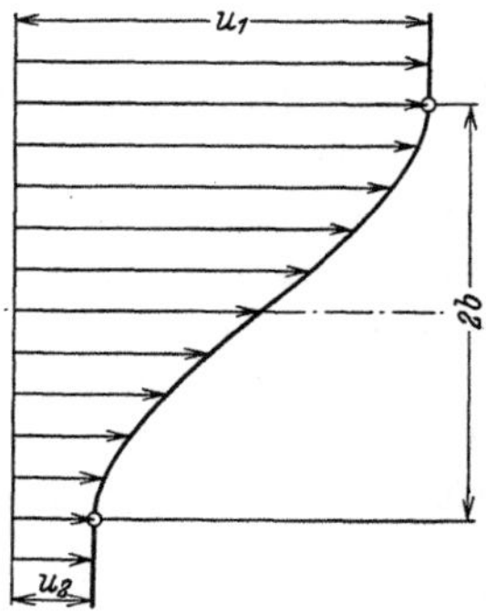

Fig. 72.

This function is represented in Fig. 72. It is a striking fact that the mixing zone ends with a discontinuity of $\partial^2 u/\partial y^2$ and then acquires the undisturbed velocity, but this is a general feature of solutions involving a turbulent friction term.

2) The second case to be considered is that of a mass of fluid in which a thin layer is set into motion parallel to its own plane while the rest of the fluid remains at rest. The momentum of the layer remains constant with respect to time since there are no frictional forces outside the layer, and no differences of pressure arise. If u_0 is the velocity in the middle of the zone where mixing occurs and b the half width of the zone (*i. e.* from the middle to the edge) it follows that

$$\int_0^b u\,d\,y = \text{constant}$$

and hence in accordance with the postulate expressed by (25.10)

$$u_0\,b = \text{constant}$$

On referring to (25.12) and (25.13) where the indices p and q show the variation of u_0 and b with respect to time, it is seen that in the present case, $p + q = 0$. In conjunction with the equation $q = p + 1$ this entails $p = -1/2$, $q = +1/2$. The general differential equation (25.14) now takes the form $\quad f + \eta\,f' = \mp 2\,\frac{\beta^2 A}{B}\,f'\,f''$

which can be integrated at once into

$$\eta\,f = \mp\,\frac{\beta^2 A}{B}\,f'^2 + \text{constant} \tag{25.19}$$

Since the maximum velocity occurs at $\eta = 0$, $f' = 0$ at $\eta = 0$ and obviously the constant of integration is zero. For real solutions, the plus sign must be taken when η is positive and the minus sign when η is negative. The square root of both sides of the equation can now be taken. The result is $\quad \frac{f'}{\sqrt{f}} = \pm\,K\,\sqrt{\eta}$

where $$\frac{\beta^2 A}{B} = \frac{1}{K^2}$$

and, on integration, $$2\sqrt{f} = \pm \frac{2}{3} K\eta^{3/2} + \text{constant}$$

The condition $f = 0$ for $\eta = 1$ necessitates the value $\mp (2/3)\, K$ for the constant and hence $$f = \mp \frac{K^2}{9} (1 - \eta^{3/2})^2$$

The following conditions still remain to be satisfied: $f' = 0$ when $\eta = 1$, $f = 0$ when $\eta > 1$ and $f = 1$ when $\eta = 0$. The first two are already satisfied by the differential equation, the last requires $K^2 = B/\beta^2 A = 9$, *i. e.* $B = 9\, A\beta^2$. In order to make the problem completely determinate some postulate is needed as to the total momentum contained in a portion of fluid of unit length in the direction of motion and unit width perpendicular to the plane of the figure. If this quantity is denoted by M, then

$$M = \varrho \int_{-b}^{+b} u\, d\, y$$

On writing $$u = A t^{-1/2} f(\eta)$$

and $$d\, y = b\, d\, \eta = B\, t^{1/2}\, d\, \eta$$

we have $$M = \varrho\, A\, B \int_{-1}^{+1} f(\eta)\, d\, \eta$$

The integration is easy to perform and gives the numerical value 0.9. Hence, using the above value for B

$$M = 8.1\, \varrho A^2 \beta^2$$

A can be found from this equation and so finally

$$u = \frac{0.351}{\beta} \sqrt{\frac{M}{\varrho\, t}} \left[1 - \left(\frac{y}{b}\right)^{3/2}\right]^2 \tag{25.20}$$

and $$b = 3.16\, \beta \sqrt{\frac{M\, t}{\varrho}} \tag{25.21}$$

The problem is therefore solved except for the value of β which can only be determined empirically. The velocity distribution corresponding to (25.20) is represented in Fig. 73. The above calculations also give a first approximation for the problem of the plane wake behind a moving body. It is only necessary to replace t in the equations used above by x/U where x denotes the distance of the point under consideration behind the body, and U is the velocity of the body. Experimental information is available as to the magnitude of β in the case of the wake produced by cylindrical rods. Schlichting[1] found $\beta = 0.293$. His experimental

[1] SCHLICHTING, H., Über das ebene Windschattenproblem. Ingenieur-Archiv. Vol. I, p. 533, 1930.

results agree very well with the relations, b proportional to $\sqrt{x}$ and u inversely proportional to $\sqrt{x}$, deduced from theoretical considerations, provided that x is measured not from the center of the body but from another point whose position depends on the shape of the cross-section. This is easily accounted for by the fact that the suppositions made in obtaining these formulae do not hold for the immediate neighborhood of the body.

In addition to the cases just discussed, certain steady flows have yielded exact solutions. Several of these have been found by W. Tollmien[1] and one of the simplest of his calculations will now be reproduced in some detail. The case investigated is that of the emergence from a nozzle

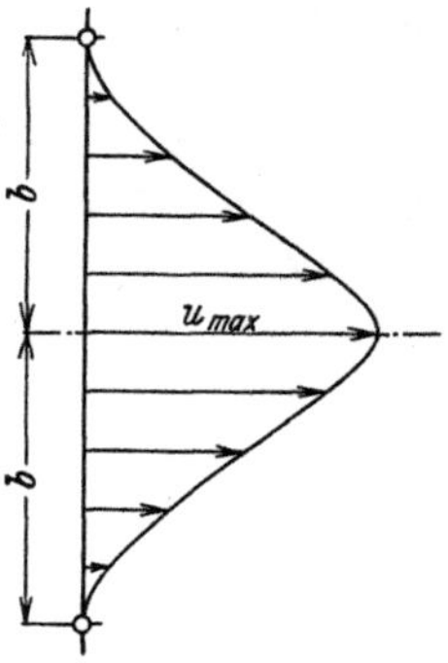

Fig. 73. Velocity distribution in a wake.

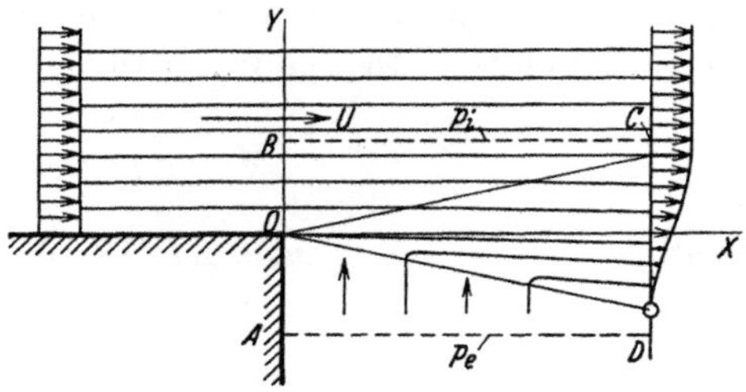

Fig. 74.

of a parallel stream of air of constant velocity U (see Fig. 74), which then proceeds to mix with the surrounding undisturbed air. The mixing begins at the very edge of the nozzle and is sufficiently marked to deserve attention in wind channels of the open jet type. The first step is to use (25.3) in order to obtain approximate information concerning the (varying) width. Hence $\frac{Db}{dt} \sim U \frac{db}{dx} \sim \frac{l}{b} U$

If the ratio l/b is assumed to be constant and equal to β as before, then $db/dx \sim \beta$ and hence $b = (\text{const.}) \times x$

Here again a quantity η is introduced which is to be proportional to y/b but since b is not yet known the definition of η is written

$$\eta = \frac{y}{x}$$

And since the mixing length is to be proportional to b it is written $l = cx$. The premise of our calculations can therefore be stated as

$$u = U f(\eta)$$

with the reservation that the values of $f(\eta)$ range from 0 in the undisturbed air to 1 in the undisturbed parallel stream.

[1] Tollmien, W., Berechnung turbulenter Ausbreitungsvorgänge. Zeitschrift für angewandte Mathematik und Mechanik, Vol. IV, p. 468, 1926.

Besides u there is a velocity component v connected with u by continuity. The conditions of continuity can be conveniently satisfied by introducing a stream function

$$\Psi = \int u\,dy$$

Then with $dy = x\,d\eta$, Ψ can be written in the form

$$\Psi = U x \int f(\eta)\,d\eta = U x F(\eta)$$

Hence
$$u = \frac{\partial \Psi}{\partial y} = U F'(\eta),$$
$$v = -\frac{\partial \Psi}{\partial x} = U\left[-F(\eta) + x\frac{y}{x^2}F'(\eta)\right]$$
$$= U[\eta F' - F]$$

Experimental determinations show that the pressure is practically constant in the mixing zone. This can also be seen by observing that the velocity v is everywhere small compared with U so that it is possible to neglect $\partial p/\partial y$, the fall of pressure in the y direction, the value of which depends on that of Dv/dt. Since $p =$ const., the first of Euler's equations is

$$u\frac{\partial u}{\partial x} + v\frac{\partial u}{\partial y} = \frac{1}{\varrho}\frac{\partial \tau}{\partial y}$$

where
$$\tau = \frac{1}{2}\varrho\, l^2\left(\frac{\partial u}{\partial y}\right)^2 = \frac{1}{2}\varrho\, c^2 x^2\left(\frac{\partial u}{\partial y}\right)^2$$

But
$$\frac{\partial u}{\partial x} = -\frac{U}{x}\eta F'' \quad\text{and}\quad \frac{\partial u}{\partial y} = \frac{U}{x}F''$$

so that
$$\tau = \frac{1}{2}\varrho\, c^2 U^2 F''^2$$

and
$$\frac{\partial \tau}{\partial y} = \frac{1}{x}\varrho\, c^2 U^2 F'' F'''$$

Moreover
$$u\frac{\partial u}{\partial x} + v\frac{\partial u}{\partial y} = \frac{U^2}{x}[-F'\eta F'' + (\eta F' - F)F'']$$
$$= -\frac{U^2}{x}F F''$$

This gives the differential equation

$$F F'' + c^2 F'' F''' = 0 \tag{25.22}$$

which again splits into two factors so that either $F'' = 0$ or $F + c^2 F''' = 0$. The first relation means that $F' =$ const., *i. e.* $u =$ const., while the second gives rise to a function which occurs in the theory of damped oscillations. The equation is of the third order; the solution contains three constants of integration and is

$$F = C_1 e^{-\alpha\eta} + C_2 e^{\alpha\eta/2}\cos\frac{\sqrt{3}}{2}\alpha\eta + C_3 e^{\alpha\eta/2}\sin\frac{\sqrt{3}}{2}\alpha\eta \tag{25.23}$$

where
$$\alpha = c^{-2/3}$$

The following conditions must now be satisfied by F:

1) At the boundary between the mixing zone and the flow from the nozzle, $u = U$ and $v = 0$, and also $\partial u/\partial y = 0$, on account of the continuous transition of the velocity profiles. Denoting this boundary by η_1, these conditions can be expressed as

$$F'(\eta)_1 = 1, \quad F(\eta_1) = \eta_1, \quad \text{and} \quad F''(\eta_1) = 0.$$

2) At the other boundary of the mixing zone, *i. e.* between it and the still air it is necessary that $u = 0$ and $\partial u/\partial y = 0$. It may however be noticed that v is not zero at this boundary, and indeed there is a weak flow in the v direction. Denoting this second boundary of the mixing zone by η_2 the conditions are

$$F'(\eta_2) = 0 \quad \text{and} \quad F''(\eta_2) = 0$$

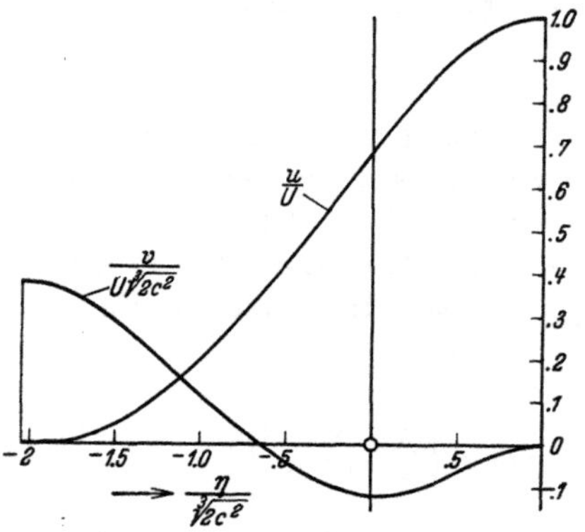

Fig. 75.

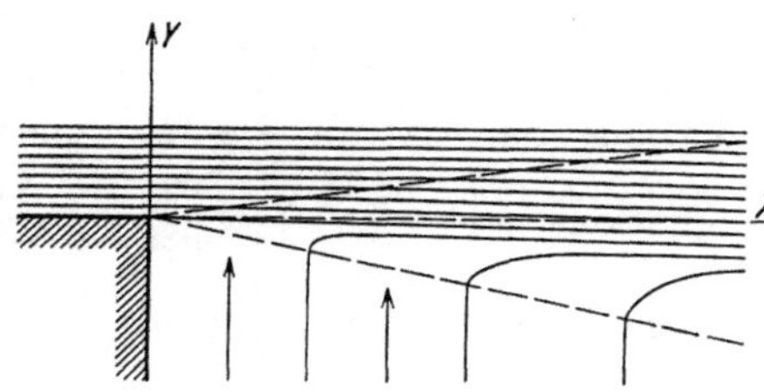

Fig. 76. Mixing of a parallel stream with the adjacent air.

Altogether five conditions have now been obtained and are sufficient to determine the five unknowns C_1, C_2, C_3, η_1 and η_2. The necessary numerical calculations, which will not be reproduced in this account, are somewhat complicated and involve the solution of a transcendental equation for $(\eta_1 - \eta_2)$. The values obtained are

$$\alpha\,\eta_1 = 0.981; \quad \alpha\,\eta_2 = -2.04; \quad C_1 = -0.0062;$$
$$C_2 = 0.987; \quad C_3 = 0.577.$$

The shapes of the graphs for $F'(\eta) = u/U$ and $\alpha(\eta F' - F) = \alpha\, v/U$ in the mixing region are shown in Fig. 75. Fig. 76 is a diagram of the calculated mean stream-lines. The degree to which these calculations agree with the observed values is shown in Fig. 77 by comparing the values of the calculated stagnation or total head pressures with the corresponding values observed at the jet of the large Göttingen wind tunnel. The value $c = 0.0246$, corresponding to $\alpha = 11.8$ has been found from this and other experiments. The total width of the mixing zone now becomes

$$b = \frac{x}{\alpha}(\eta_1 - \eta_2) = 0.0845\,x \cdot (3.02) = 0.255\,x$$

Also b', the width of that part of the emergent stream which is dragged into the zone of turbulent mixing is given by

$$b' = \frac{x}{\alpha}\,\eta_1 = 0.083\,x$$

For comparison with previous results it may be added that the value of l/b in this case $= 0.0246/0.255 = 0.0963$. It is remarkable that this numerical value is so much smaller than that obtained in the case of the wake produced by a moving body. In searching for a possible explanation it must be remembered that the numerical values in the two cases are, of course, not strictly comparable. For whereas in both cases the distance is measured from where $u = u_{max}$ to the point where

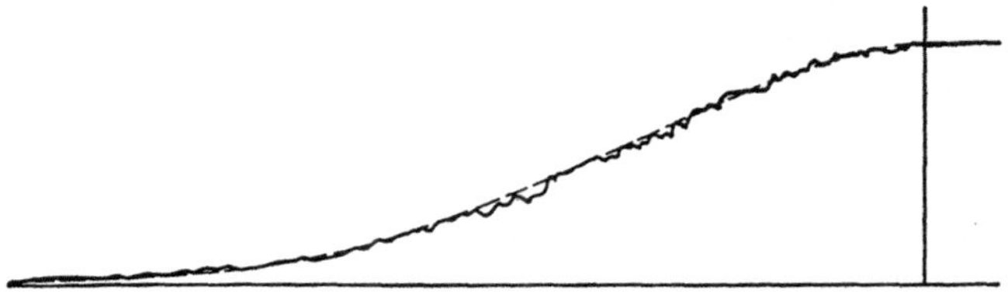

Fig. 77.

$u = 0$, b denotes different magnitudes—the whole width of the mixing zone in the present case, half the width in the previous case. And apart from this it is very possible that in the large Göttingen wind tunnel, l/b is unusually small on account of the exceptional degree of freedom from turbulence in that tunnel. If this is so, larger values of c and of l/b might be expected to occur in other wind tunnels and the flows produced by effluxes from jets in general.

With the help of the above solution it is also possible to calculate the differences of pressure in the mixing zone and hence for example, to obtain the magnitude of such difference between the interior of a jet of air and the undisturbed air surrounding it—a determination of importance in connection with experiments on such flows. The present discussion will be limited to some consideration of this pressure difference. Purely qualitative considerations indicate that the pressure inside the jet must be somewhat greater than the pressure outside, for the air which is sucked into the jet arrives with a velocity directed toward it and, as the stream-line diagram in Fig. 76 shows, the air is then carried forward and outward at a small inclination. This means that there is a retardation of air particles moving toward the interior, and their subsequent acceleration toward the exterior is naturally only possible with an increase of pressure in the mixing zone. The magnitude of this increase of pressure can be conveniently found by using the theorem of momentum in the region bounded by the rectangle $ABCD$ shown in Fig. 74. The procedure is to compare the pressure forces in the

y direction with the frictional forces and the component of momentum in the same direction. It is only necessary to consider the pressures across BC and AD since the pressures on the other two sides are parallel to the x axis; frictional forces occur only on the side CD. There is no flow between A and O; the direction of flow between O and B is exactly horizontal and hence gives no contribution to the momentum, while the flow is also horizontal along BC. Hence it is only necessary to take into account the momentum values at the sides AD and CD. If all quantities are calculated for unit thickness of the layer perpendicular to the plane of the rectangle, the area of the surface AD is equal to x and therefore the amount of momentum transported through AD is $\varrho v_0^2 x$, and its sense such as to be equivalent to a positive force, *i. e.* a force in the upward direction. Let the pressure on AD be p_e; the force due to pressure is then $p_e x$. If the pressure on BC is p_i, the corresponding force is in the negative direction of the y axis and of magnitude $-p_i x$. There remain the momenta and frictional forces on CD. In a layer of height dy a mass of $\varrho u\, dy$ per second is transported across the surface bringing with it a velocity v. The amount of Y momentum transported across CD per second is therefore $\varrho \int_D^C u v\, dy$, and since outflowing across the boundary, the corresponding force must be considered as negative. There is also the apparent friction $\int \tau\, dy$ directed upward and therefore positive. On equating these various terms with due regard to signs, we have

$$p_i - p_e = \varrho v_0^2 - \frac{\varrho}{x} \int_D^C u v\, d y + \frac{1}{x} \int_D^C \tau\, d y$$

The calculation to be performed involves quantities which are determined by the previous solutions. Omitting the details of this calculation, the result may be briefly stated as,

$$p_i - p_e = 0.090 \frac{\varrho U^2}{2} \left(\frac{b}{x}\right)^2 \tag{25.24}$$

which becomes, on substituting the value of b in the Göttingen wind tunnel $(b = 0.255\, x)$ $\qquad p_i - p_e = 0.00585 \frac{\varrho U^2}{2}$

The smallness of this value gives good justification of the assumption $p = \text{const.}$ made in the previous approximation and the calculation of velocity corrections becomes superfluous. At a great distance from the jet which emerges from the nozzle the pressure is greater than p_e by the stagnation pressure of the suction velocity v_0. Denoting the pressure of the air which is at rest by p_0 it follows that

$$p_i - p_0 = p_i - p_e + \frac{1}{2} \varrho v_0^2 = 0.074 \frac{\varrho U^2}{2} \left(\frac{b}{x}\right)^2 \tag{25.25}$$

or, for the special value $b = 0.255\,x$,

$$p_i - p_0 = 0.0048\,\frac{\varrho\,U^2}{2}$$

This formula is very valuable in studying instruments for measuring the static pressure in air-flows. Such an instrument can be tested by placing it in the central part of a jet of air emerging from a large nozzle, *i. e.* in air which is not yet part of the mixing zone. If the instrument measures correctly, then on comparing the reading with that of a Pitot-tube facing the flow v_0 of the surrounding air toward the jet, there should be a surplus of pressure of the magnitude given above. Since such instruments are the very basis of all aerodynamic measurements the importance of the above relation in the technique of such measurements scarcely needs further emphasis.

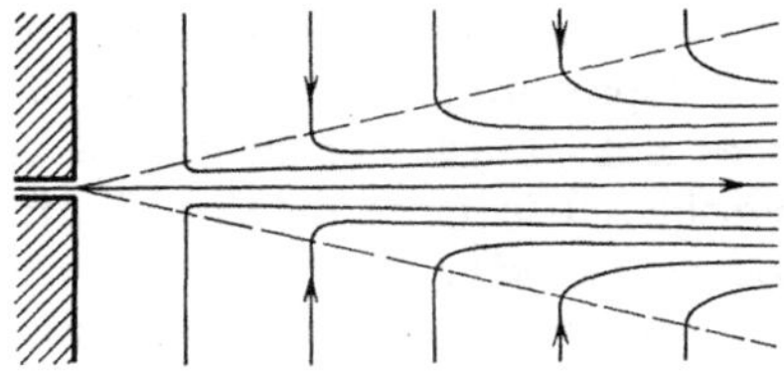

Fig. 78.
Mixing of a jet with the surrounding air.

In addition to this problem, Tollmien's paper considers the flow of a jet emerging from a circular aperture and calculates the corresponding hydrodynamical quantities at distances which are great compared with the diameter of the aperture. It discusses also the similar problem in the case of a jet emerging from an infinitely long narrow slit, that is, two-dimensional flow.

Fundamentally, both calculations proceed in the same way as those which have just been described but the functions involved are more complicated and must be expanded in series. The velocity profiles are very similar to those in Fig. 73 and the stream-lines for the case of symmetry about an axis are reproduced in Fig. 78. Experiments show that b, the semi-width of the jet is, in this case, equal to $0.214\,x$. No measurements are as yet known for the case of the two-dimensional flow, whose main features are, however, quite similar to those described above.

Appendix to 25. The following account will serve as a useful supplement to the exposition given at the commencement of 25. This theory is due to G. I. Taylor and is founded on the supposition that in the turbulent flow considered, not only the main flow but also the turbulent disturbing motion in all its details may be considered as two dimensional flows. Let the instantaneous values of the velocities be u and v, the corresponding value of the pressure, p, and let

$$\frac{1}{2}\left(\frac{\partial u}{\partial y} - \frac{\partial v}{\partial x}\right) = \zeta$$

be the instantaneous rotation. The flow is assumed to be frictionless. From Euler's equation we have

$$-\frac{1}{\varrho}\frac{\partial p}{\partial x} = \frac{\partial u}{\partial t} + u\frac{\partial u}{\partial x} + v\frac{\partial u}{\partial y}$$

If $v\,\partial v/\partial x$ is added and then subtracted from the right-hand-side the equation can be written as

$$-\frac{1}{\varrho}\frac{\partial p}{\partial x} = \frac{\partial u}{\partial t} + \frac{\partial}{\partial x}\left(\frac{u^2+v^2}{2}\right) + 2\,v\,\zeta \qquad (1)$$

In order to simplify the theory it is now assumed that a turbulent flow is being considered with mean velocities

$$U = f(y), \qquad V = 0$$

The mean values with respect to time of the quantities in (1) are now to be constructed. Since the velocity does not increase permanently with time and since it experiences no changes in the x direction except a fluctuation about a mean value, it is obvious that the mean values of $\partial u/\partial t$ and of $\partial/\partial x\,(u^2 + v^2)$ are both zero. If, however, there is a fall of pressure, as for example, in a tube, the mean value of $1/\varrho\,(\partial p/\partial x)$ does not vanish nor does that of $2v\,\zeta$. This last statement must be considered in more detail. The mean value of $2\,\zeta$ at the point y is obviously dU/dy. The fluctuation which is superimposed upon this can be expressed in the manner already used in the previous exposition in terms of the "path-length" L, *i. e.* the distance which a particle having initially the mean rotation of its neighborhood must be displaced so that it comes to the place actually occupied at the moment considered. Using only the first two terms of Taylor's series, $2\,\zeta$ can be expressed as follows

$$2\,\zeta = \frac{d\,U}{d\,y} + L\,\frac{d^2\,U}{d\,y^2}$$

In forming mean values (denoted by bars) we obtain from (1) since the mean value of v is zero, Taylor's equation

$$-\frac{1}{\varrho}\frac{\overline{\partial p}}{\partial x} = \overline{v\,L}\,\frac{d^2\,U}{d\,y^2}$$

The L of this equation has not the same meaning as l in previous equations and is to be regarded as a length whose value varies from particle to particle and can be both positive and negative. The mean value of $v\,L$ is not zero, for the particles which come from below have positive v and negative L, those which come from above have negative v and positive L, so that the product is predominantly negative. For reasons similar to those given in **21** it is now possible to write

$$v = L'\,\frac{d\,U}{d\,y}$$

where L' is a length somewhat similar to L and hence

$$\overline{v\,L} = -\left|\overline{L\,L'}\,\frac{d\,U}{d\,y}\right|$$

The absolute value is here introduced on account of the essentially negative character of $\overline{v\,L}$. The mean value $\overline{L\,L'}$ can be written as l^2 in the previous notation so that the equation now becomes

$$\frac{1}{\varrho}\frac{\partial p}{\partial x} = l^2\left|\frac{d\,U}{d\,y}\right|\frac{d^2\,U}{d\,y^2}$$

If we introduce again the apparent friction τ, the right-hand-side has the significance of $(1/\varrho)\,(\partial\tau/\partial y)$, because if the case is regarded as the equilibrium of a deformable body, we must have

$$-\frac{\partial p}{\partial x}+\frac{\partial \tau}{\partial y}=0$$

Hence the equation
$$\frac{\partial \tau}{\partial y}=\varrho\, l^2\left|\frac{dU}{dy}\right|\frac{d^2 U}{d y^2} \tag{2}$$

This equation should be compared with that which follows from (25.1) *viz.* $\tau=\varrho l^2\,|\,dU/dy\,|\,dU/dy$ and hence

$$\frac{\partial \tau}{\partial y}=\varrho\frac{\partial}{\partial y}\left(l^2\left|\frac{dU}{dy}\right|\frac{dU}{dy}\right) \tag{3}$$

If l varies with y, the significance of (2) is quite different from that of (3). If however $l=$ constant, (2) is integrable and gives the following equation for positive dU/dy

$$\tau=\frac{1}{2}\varrho\, l^2\left(\frac{dU}{dy}\right)^2$$

which differs from the previous equation only in having the factor 1/2. If l varies with y however, the formula obtained by Taylor has the inconvenience of not being integrable so that the developments of **22** and **23** could not be carried out with it. Fortunately, however, as Taylor stresses in his paper, experiments made by Éliás[1] on the propagation of heat prove that in the case of turbulent flow along a wall the form of propagation of heat coincides with that for velocity. This shows that our previous results hold in this case, and it must be assumed that this "wall turbulence" is an essentially three-dimensional phenomenon to which Taylor's equation as constructed in this appendix does not apply. Taylor himself shows that in the case where

$$U=f(y), \qquad V=0$$

results of the kind just described would be obtained by assuming perturbing velocities which lie entirely in the yz plane and are independent of x so that their rotation vector is parallel to the x axis. This can be seen immediately by observing that in this fashion fluid elements, linear in form and extended along the x direction, are shifted as a whole parallel to themselves in the y and z directions. If mean values are taken after a turbulent shift of this character, whether the shifted particles are thought of as mixed with their new neighbors or not, it is obvious that the same mixing law holds for differences of temperature as for differences of velocity. In practice of course, on account of the instability of such an arrangement, threads of fluid with different velocities could not exist side by side. With such an assemblage, essentially unstable, and with the development of vortices with axes in various directions, the

[1] Éliás, F., Der Wärmeübergang von einer geheizten Platte an strömende Luft. Abh. d. Aerodyn. Inst. Aachen, 9, 1930.

threads of fluid would become twisted together in a complicated manner, and further under the influence of viscosity and heat conduction, the velocities and temperatures of the fluid particles would become thoroughly mixed together.

26. The Development of Turbulence. Experiment shows that the value of the Reynolds number for which transition from laminar to turbulent motion occurs, is influenced by the magnitude of the initial disturbances; the smaller these are the greater in general is the value of the Reynolds number. The boundary value is called the upper critical Reynolds number; its exact value depends on the particular circumstances of the motion. There is also a lower boundary value below which all turbulence is extinguished however large the disturbances which have produced it. This value is called the lower critical Reynolds number and for flows in straight pipes its value, according to L. Schiller[1] is reached when $\bar{u}d/\nu = 2320$, while the upper critical number may have any greater value. By taking special precautions, it is even possible to reach $\bar{u}d/\nu = 40{,}000$. Since the work of Reynolds became known, it has always been assumed that if the retarding effect of viscosity is not powerful enough to damp out all casual disturbances, some influence is present which produces a weak instability in rectilinear flow through pipes. This conception is still valid but must be modified by the subsequent recognition that the presence of slight viscosity is responsible for a destabilizing influence which competes with the previous stabilizing effect of viscosity; for it has been found that viscosity produces an obviously important type of instability characterized by the growth of all disturbances however small initially. More will be said concerning this phenomenon in the following pages.

The theoretical discussion of this question starts by considering a weak disturbing motion superimposed upon a given basic flow, in order to investigate the conditions under which the disturbing motion diminishes or increases with time. The method used in the earliest investigations, of which the first are due to Reynolds[2] himself, was to assume the disturbance to be of some convenient type and then to discover whether the energy of the flow increased immediately afterward as a result of the combination of the basic flow and the disturbance. It may be objected however that in this method of approach the increase of energy thus demonstrated is only momentary so that the configuration may change as the flow proceeds, with subsequent decrease of energy. If however the result of such calculations shows that for some definite Reynolds number every arbitrary disturbance produces a decrease of energy, this

[1] Forschungsarb. d. Ver. Deutscher Ing., Heft 248, 1922.

[2] Reynolds, O., On the Dynamical Theory of Incompressible Viscous Fluids and the Determination of the Criterion. Philosophical Transactions of the Royal Society, 1895 or Scientific Papers, Vol. II, p. 535.

method will supply a valid proof of the stability of the flow under discussion. The method will in fact yield a lower limit for the critical Reynolds number.

In order to obtain better results it is necessary to require that the disturbing motion shall obey the hydrodynamic differential equations; the question then to be answered is whether there are disturbances satisfying the equations of motion which increase with time.

We will, therefore, start with a small disturbing motion superimposed upon a given basic flow and we will formulate the relations which can be deduced from the hydrodynamic differential equations. In order to avoid undue difficulties, the discussion will be restricted to a basic flow of the simplest possible type, one in which the velocity is parallel to the x axis at all points of the flow and is a function only of the distance from the x axis and, in some cases, of the time also. That is to say

$$U = f(y, t); \qquad V = 0$$

For simplicity, the disturbing motion is also assumed to be two-dimensional, so that

$$u' = u'(x, y, t); \qquad v' = v'(x, y, t)$$

The components of the total velocity are therefore

$$u = U + u'; \qquad v = v'$$

On making the assumption that the disturbing velocities are so small that expressions involving squares or products of these velocities or of their differential coefficients are small compared with expressions in which these terms occur linearly, the following equations are obtained,

$$\frac{\partial u'}{\partial x} + \frac{\partial v'}{\partial y} = 0 \tag{26.1}$$

$$\frac{\partial U}{\partial t} + \frac{\partial u'}{\partial t} + U\frac{\partial u'}{\partial x} + v'\frac{\partial U}{\partial y} + \frac{1}{\varrho}\frac{\partial p}{\partial x} = \nu\left(\frac{\partial^2 U}{\partial y^2} + \nabla^2 u'\right) \tag{26.2}$$

$$\frac{\partial v'}{\partial t} + U\frac{\partial v'}{\partial x} \qquad + \frac{1}{\varrho}\frac{\partial p}{\partial y} = \nu\,\nabla^2 v' \tag{26.3}$$

where

$$\nabla^2 = \frac{\partial^2}{\partial x^2} + \frac{\partial^2}{\partial y^2}$$

Moreover, the pressure p also consists of two portions; P produced by the basic flow alone, and p' produced by the disturbing flow. The first is obtained by putting $u' = v' = 0$ in (26.2) and (26.3) which then become

$$\frac{\partial U}{\partial t} + \frac{1}{\varrho}\frac{\partial P}{\partial x} = \nu\frac{\partial^2 U}{\partial y^2} \tag{26.2a}$$

and

$$\frac{1}{\varrho}\frac{\partial P}{\partial y} = 0 \tag{26.3a}$$

On subtracting (26.2a) from (26.2), (26.3a) from (26.3), multiplying the two equations thus obtained by u', v' respectively and adding, the following equation is obtained:

$$\left.\begin{aligned}\left(\frac{\partial}{\partial t}+U\frac{\partial}{\partial x}\right)\left(\frac{u'^2+v'^2}{2}\right)+u'v'\frac{\partial U}{\partial y}+\frac{1}{\varrho}u'\frac{\partial p'}{\partial x}+v'\frac{\partial p'}{\partial y}\\ =\nu\left(u'\nabla^2 u'+v'\nabla^2 v'\right)\end{aligned}\right\}\quad(26.4)$$

The first term is obviously the change of kinetic energy of the disturbing motion of a fluid particle of unit mass moving forward with the basic flow. This equation therefore constitutes a suitable starting point for the energy considerations mentioned above, which occur in earlier papers on the subject. The detailed form of theory chosen for exposition in this report is due to H. A. Lorentz[1]. In order to obtain the total change in energy of the disturbing motion in a definite region, it is necessary to integrate (26.4) throughout that region, which is assumed for simplicity to be a rectangle bounded on two sides by the lines $x = a$ and $x = b$, a and b being such that the disturbances are identical at both boundaries *i. e.* u', v' and p' have the same values at corresponding points in the two lines. This assumption simplifies the calculations which follow but is not absolutely necessary to insure the validity of the proof. The other two boundaries of the domain are taken to be the lines $y = c$ and $y = d$ at which

$$u' = v' = 0$$

It is not necessary to specify whether the walls at $y = c$ and $y = d$ are fixed, in which case the vanishing of the disturbances follows from the boundary conditions, or whether these lines are situated in the fluid outside the region of disturbed flow. The integration of (26.4) will now be performed term by term.

It has already been mentioned that on multiplying the first term by the density, we obtain the rate of increase of the kinetic energy of the disturbing motion. The terms involving p' can be transformed by partial integration, for

$$\int_a^b\int_c^d u'\frac{\partial p'}{\partial x}\,d x\,d y=\int_c^d\Big[u' p'\Big]_a^b d y-\int_a^b\int_c^d p'\frac{\partial u'}{\partial x}\,d x\,d y$$

In the first expression on the right hand side we have to obtain the difference in value of the product $(u' p')$ on the lines $x = a$ and $x = b$ at points which have the same y ordinate. In accordance with the previous assumptions this difference is zero and the first integral on the right hand side therefore vanishes. Similarly

$$\int_a^b\int_c^d v'\frac{\partial p'}{\partial y}\,d x\,d y=\int_a^b\Big[v' p'\Big]_c^d d x-\int_a^b\int_c^d p'\frac{\partial v'}{\partial y}\,d x\,d y$$

[1] LORENTZ, H. A., Abhandlungen über theoretische Physik, Vol. I, Leipzig, 1907. Über die Entstehung turbulenter Flüssigkeitsbewegungen und über den Einfluß dieser Bewegungen bei der Strömung durch Röhren, p. 43. See also KÁRMÁN, TH. v., Proceedings of the First Int. Congress of Applied Mechanics, p. 97, Delft, 1924.

Here again the first integral on the right hand side vanishes since it was assumed that v' was zero at the boundaries $y = c$ and $y = d$. The sum of the two double integrals is therefore

$$-\int_a^b\int_c^d p'\left(\frac{\partial u'}{\partial x}+\frac{\partial v'}{\partial y}\right)dx\,dy$$

Equation (26.1) shows that this integral is zero so that no net contribution is made by the terms involving p'.

The expression $u'\nabla^2 u' + v'\nabla^2 v'$ must now be integrated. If ζ' is the rotation velocity of the disturbing motion about the z axis, so that

$$\frac{\partial u'}{\partial y}-\frac{\partial v'}{\partial x}=2\zeta' \tag{26.5}$$

it is easily found, by evaluating the partial derivatives and using (26.1) that

$$2\left(u'\frac{\partial\zeta'}{\partial y}-v'\frac{\partial\zeta'}{\partial x}\right)=u'\nabla^2 u'+v'\nabla^2 v'$$

Hence, on integrating by parts, it follows that

$$\int_a^b\int_c^d u'\frac{\partial\zeta'}{\partial y}=\int_a^b\Big[u'\zeta'\Big]_c^d dx-\int_a^b\int_c^d\frac{\partial u'}{\partial y}\zeta'\,dx\,dy,$$

and

$$\int_a^b\int_c^d v'\frac{\partial\zeta'}{\partial x}=\int_c^d\Big[v'\zeta'\Big]_a^b dy-\int_a^b\int_c^d\frac{\partial v'}{\partial x}\zeta'\,dx\,dy$$

The first terms on the right hand side of these equations vanish for the same reasons as before. By combining the two terms which remain, the expression $\partial u'/\partial y - \partial v'/\partial x$, whose value is $2\zeta'$ [see (26.5)] is once again obtained. Finally therefore, the value of the integrated term is

$$\iint (u'\nabla^2 u' + v'\nabla^2 v')\,dx\,dy = -4\iint \zeta'^2\,dx\,dy$$

Hence, writing $\mu = \nu\varrho$, we arrive at the equation

$$\frac{DE}{Dt}=-\varrho\iint u'v'\frac{dU}{dy}dx\,dy-4\mu\iint\zeta'^2\,dx\,dy \tag{26.6}$$

Let us denote the expressions which are multiplied by ϱ and μ by M and $-N$ respectively; M may be either positive or negative according to the nature of the disturbing motion, but $-N$ must always be negative. With these abbreviations (26.6) becomes

$$\frac{DE}{Dt}=\varrho M-\mu N \tag{26.6a}$$

Both M and N can be given physical interpretations. The discussions of **20** have shown that the apparent mean shearing force produced by interchange of momentum has the mean value $-\varrho\overline{u'v'}$ [(see (20.2)]. Hence $(-\varrho u'v')\,dU/dy$ can be regarded as the work done by the momentary apparent shearing force due to the shearing movement connected with the basic flow. In terms of energy, therefore, ϱM represents a quantity of work produced by the basic flow and obviously contributed to the disturbing motion. The expression μN represents the "dissipation" of the disturbing motion, *i. e.* the amount of work lost by the fluid in the disturbing motion as a result of viscosity and transformed into heat. It may be observed that $4\mu\int\int\zeta'^2\,dx\,dy$ can be transformed by integration by parts and with the help of (26.1), into the sum of two terms

$$2\mu\int\int\left[\left(\frac{\partial u'}{\partial x}\right)^2+\left(\frac{\partial v'}{\partial y}\right)^2\right]d\,x\,d\,y+\mu\int\int\left(\frac{\partial u'}{\partial y}+\frac{\partial v'}{\partial x}\right)^2 d\,x\,d\,y$$

each of which can be given a quite definite interpretation. The first is the work done by the normal stress due to the disturbing motion acting through the viscosity, and the second the work of the corresponding tangential shear.

It is now necessary to examine the sign of DE/Dt. It is always possible to arrange that M shall be positive by choosing a suitable form for the disturbances; it is sufficient, in fact, to make the predominating sign of the products $(u'\,v')$ in the integral, opposite in sign to that of $d\,U/d\,y$. The first fact to be noticed is that the sign of DE/Dt is unaffected by the absolute magnitude of the disturbances, for when the magnitude of the disturbing velocities are allowed to vary, the values of M and N are altered in the same ratio, namely, in proportion to the square of the disturbing velocity. On the other hand the value of the ratio (density/viscosity) is of extreme importance.

Given any basic flow, it is obviously possible to find a value of ν, *i. e.* of μ/ϱ, such that the dissipation term μN outweighs the other terms for all conceivable forms of disturbance and the energy decreases in consequence. Hence, the flow will always be stable for sufficiently large values of the viscosity. Moreover, (26.6) can be manipulated in such a way as to obtain a definite Reynolds number. For this purpose it will be supposed that the width of the domain already considered, *i. e.* the distance from $y = c$ to $y = d$ is h and two new non-dimensional variables will be used, defined by the equations $\xi = x/h$ and $\eta = y/h$ respectively. Let U_0 and u_0 be two velocities, respectively characteristic of the basic flow and of the disturbing velocities, and by which we shall divide the corresponding velocities in the two integrals which follow. M and N can now be expressed as,

$$M = -h\,u_0^2\,U_0\left(\int\int\frac{u'\,v'}{u_0^2\,U_0}\,\frac{\partial U}{\partial\eta}\,d\,\xi\,d\,\eta\right)$$

and

$$N = u_0^2\int\int\frac{1}{u_0^2}\left(\frac{\partial u'}{\partial\eta}-\frac{\partial v'}{\partial\xi}\right)^2 d\,\xi\,d\,\eta$$

in which the integrands have become completely independent of the units of measurement. If these new integrals are denoted by M' and N'

respectively the following equation is obtained in place of (26.6a)

$$\frac{DE}{Dt} = (\varrho\, h\, U_0 M' - \mu\, N')\, u_0^2 \qquad (26.6\text{b})$$

The condition of stability can now be written in such a fashion that the only question is whether $\varrho h\, U_0/\mu$ is smaller or greater than N'/M'. H. A. Lorentz has investigated the so-called Couette case for which $U = U_0\; y/h$ *i. e.* the case of a flow between a fixed wall at $y = 0$ and a wall moving with velocity U_0 at $y = h$. He assumed the existence of an elliptic vortex and found that the least value of N'/M' was 288. Orr[1] and Hamel[2] have investigated the variational problem and found 177 to be the absolutely smallest lower limit for stability. The critical value found experimentally by Couette[3] is considerably higher. His value, 1900, can be accounted for by remembering that the disturbances which occur in practice must be hydrodynamically possible and contribute only a small surplus toward the growth of the disturbances. The theory sketched above does not consider the question whether the disturbances are hydrodynamically possible and only investigates how the velocities must be distributed in order that the disturbances may increase as much as possible during the next instant.

It is to be expected that a more exact theory would be obtained by taking into account only those disturbances which are hydrodynamically possible. In looking for such disturbances it is possible to start with (26.1)—(26.3). The first equation can be satisfied by introducing a stream function Ψ such that

$$u' = \frac{\partial \Psi}{\partial y}; \qquad v' = -\frac{\partial \Psi}{\partial x} \qquad (26.7)$$

The disturbance will in general have along the axis, a wavelike form. Any such form can be separated by Fourier analysis into a set of sine waves and, since (26.1) — (26.3) are linear in u' and v', each sine wave can be investigated separately. Accordingly Ψ will be of the form

$$\Psi = f(y)\, e^{i(\alpha x - \beta t)} \qquad (26.8)$$

where $f(y)$ is the amplitude which in general will be a complex function of y. The wave-length λ is given by $\lambda = 2\pi/\alpha$ and the velocity of propagation $c = \beta/\alpha$. The pressure p can be eliminated from (26.2) and (26.3) by differentiating the first with respect to y, the second with respect to x and subtracting. This gives the relation

$$(U - c)(f'' - \alpha^2 f) - \frac{d^2 U}{d y^2} f = \frac{\nu}{i\alpha}(f'''' - 2\alpha^2 f'' + \alpha^4 f) \qquad (26.9)$$

The problem is now to solve (26.9) given both the basic velocity $U(y)$ and boundary conditions for the disturbing motion. Here again zero-

---

[1] Orr, W. M. F., Proc. Royal. Irish Academy, 17 (A) p. 124, 1907.
[2] Hamel, G., Göttinger Nachrichten, 1911.
[3] Couette, M., Ann. chim. phys. (6) 21, p. 433, 1890.

dimensional quantities can be introduced by using the variable $\eta = y/h$ and dividing the velocities by U_0. Let us put

$$f(y) = A\,\varphi(\eta),$$

and let a be defined by $\quad a = \alpha h = \dfrac{2\pi h}{\lambda}$

then, if the primes mean now differentiation with respect to η, (26.9) becomes

$$\frac{U-c}{U_0}(\varphi'' - a^2\varphi) - \frac{U''}{U_0}\varphi = \frac{\nu}{i\,a\,U_0 h}(\varphi'''' - 2a^2\varphi'' + a^4\varphi) \qquad (26.9\,\mathrm{a})$$

$U_0 h/\nu$ is the Reynolds number and may be denoted by R. Since R is a large number it can be seen that it is sufficient to take only a first approximation to the value of the right hand side of this equation. Important information can, in fact, be obtained even by putting the right hand side equal to zero, an approximation which is equivalent to neglecting the viscosity of the fluid. Lord Rayleigh[1] made investigations on these lines. The most important result of Lord Rayleigh's researches was the fact that a convex profile corresponding to that of Fig. 79 can always execute stable oscillations whereas certain profiles, in which the angular discontinuity is both positive and negative are unstable. This is shown in the calculations by the fact that a complex value is obtained for c, the velocity of propagation. O. Tietjens[2] has applied Rayleigh's theory to profiles intended to reproduce laminar flow along a wall, and found also stability for convex profiles but instability for profiles with a re-entrant angle (see Fig. 80). This is a most important result. It is to be expected that similar marked instability is present when the profile is continuously curved and has a point of inflection. From information in a letter received by the present author, it appears that W. Tollmien in an as yet unpublished paper, has confirmed this conjecture. No critical Reynolds number for such profiles with point of inflection is as yet known. It is, however, to be expected that the growth of disturbance depends upon the counter-curvature beyond the point of inflection and that the greater this curvature the more quickly will disturbances increase.

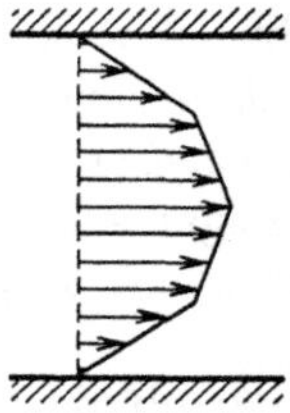

Fig. 79.

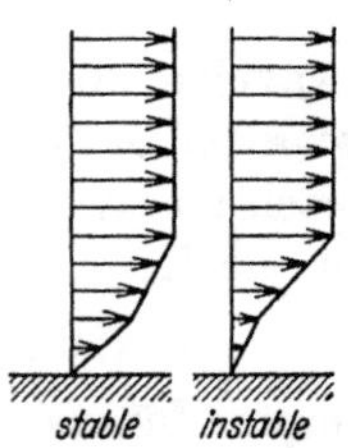

Fig. 80.

[1] Lord Rayleigh, On the Instability of Certain Fluid Motions. Proc. London Math. Soc. 11, p. 57, 1880 and 19, p. 67, 1887 (Scientific Papers Vol. I, p. 474 and Vol. III, p. 17).

[2] Tietjens, O., Beiträge über die Entstehung der Turbulenz. Göttingen Dissertation, 1922. Extract in Zeitschr. f. angew. Math. u. Mech. 5, p. 200, 1925.

A theory in which the frictional terms were not neglected was first developed by A. Sommerfeld[1] and carried further by L. Hopf[2], W. M. F. Orr and others[3]. All these writers have restricted their attention to Couette's case ($U'' = 0$) for in the general case the term $(U''/U_0)\varphi$ has awkward properties where $U - c = 0$ and where, therefore, if the friction is neglected $\varphi'' - a^2\varphi$ becomes infinite. Experiment has shown that in Couette's case as well as in the cases previously considered a critical Reynolds number does exist. All the calculations mentioned above, however, agree on the point that small oscillations are stable for all Reynolds numbers. In order to obtain a different result it was necessary therefore to investigate cases for which U'' is not equal to zero. A complete solution of this problem is not yet available. An approximation can however be obtained in the following manner. There are two places where the friction is effective, even when the Reynolds number is large. It should next be noted that frictionless oscillations involve a sliding of the fluid along the wall. In order that the flow should adhere to the wall it is therefore necessary to add a compensating solution of (26.9a) to reduce the velocity at the wall to zero; but the additional velocity is of such a nature that its value decreases very rapidly at a small distance from the boundary. On this account it is sufficient to retain only the term $(1/ia\,R)\,\varphi''''$ on the right hand side of (26.9a) for in comparison with it the other terms are very small. Similarly, the other side of the equation can be restricted to $(U - c)\,\varphi''/U_0$, with a linear expression for U. The second place where friction must be taken into consideration is where U becomes equal to c and where, without the effect of friction, as already noted, φ'' would become infinite and φ' logarithmically infinite. In his paper, O. Tietjens investigated the effect of the thin frictional layer at the wall when the profile was of the stable polygonal type (see Fig. 81). In this case because of vanishing U'', only the first of the two frictional layers comes into consideration. It appeared that there really was a weak instability due to the fact that the compression and extension of the frictional layer in the direction of x produced small velocities in the y direction and hence an impulse on the oscillations which otherwise are assumed to be frictionless. The effect of this boundary layer may be estimated in the following manner.

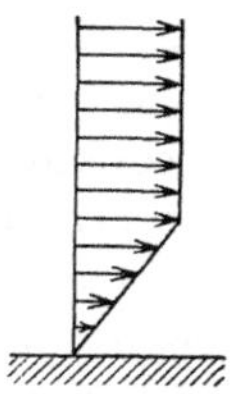

Fig. 81.

The energy considerations which led to (26.6a) furnish a starting point. A more detailed investigation shows that the dissipation μN is of subordinate importance and that it is the term ϱM which has the

[1] Sommerfeld, A., Atti. del congr. intern. dei Mat. Rome, 1908.

[2] Hopf, L., Annalen der Physik, Vol. 43, p. 1, 1914.

[3] A comprehensive report by F. Noether appeared in Zeitschr. f. angew. Math. u. Mech. 1, p. 125, 1921.

preponderating effect in determining whether an oscillation increases or decreases. This question is therefore reduced to determining the phases in which u' and v' cooperate, that is, to determining whether M becomes positive or negative. This is equivalent to finding out whether energy passes to or from the basic flow, the energy of which is very large compared with the energy of the disturbing motions with which the interchange takes place.

The differential equation when simplified in the above manner shows that the thickness of the layer whose oscillations suffer a displacement of phase due to friction is of the order of magnitude $\delta \sim \sqrt{\nu/\alpha c}$ (in accordance with **10**, $\delta \sim \sqrt{\nu t}$ and t is inversely proportional to β which equals αc). An expression is now required for the work output ϱM. The reader will observe that in the usual wave motion the phases of u' and v' differ by 90^0 and that the integral of $(u' v')$ vanishes in consequence. A non-vanishing contribution is obtained, however, from the product of a velocity additional to the frictionless v' (produced by the continuity in the boundary layer, the phase of which is shifted against v'), and the velocity u in the remainder of the fluid. In order to estimate the magnitude of this supplementary v component, v_1', let the tangential velocity of the frictionless oscillations at the wall be u_0. Then by continuity the associated v component at a distance δ from the boundary is

$$v'(\delta) = -\int_0^\delta \frac{\partial u'}{\partial x} d y \sim \alpha u_0 \delta = u_0 \sqrt{\frac{\nu \alpha}{c}}$$

It is easily seen that the phase-displaced velocity v_1' is of this same order of magnitude. The wave-lengths which actually occur in the more exact theory of stability now under consideration are sufficiently large to insure that v_1' is almost undiminished in strength throughout the layer of height h. Hence the following estimate is obtained for the work ϱM

per unit length $$\varrho M \sim \int_\delta^h \varrho u' v' \frac{d U}{d y} d y \sim \varrho u_0 v_1' U' h$$

or $$\varrho M \sim \varrho u_0^2 U' h \sqrt{\frac{\nu \alpha}{c}}$$

The increase of the disturbing motion with respect to time can be obtained by dividing the value of ϱM by the disturbance energy E. An approximate value for E in the case of long wave disturbance is $E \sim 1/2 (\varrho u_0^2/h)$. Hence the order of magnitude of the increase of

the oscillations is $$\gamma_1 = U' \sqrt{\frac{\nu \alpha}{c}} \qquad (26.10)$$

The calculations of Tietjens showed that in the case of the polygonal profile, damping occurred for very long waves but magnification was present when dealing with shorter wave-lengths. The maximum increase observed was of the order of magnitude found in (26.10) above. No critical Reynolds number could however be found; all that could be shown was that on account of the increased effect of the dissipation the arguments became invalid for Reynolds numbers below some boundary value. This value was situated in the region of the limit obtained above from energy considerations.

A more satisfactory solution was found later by W. Tollmien[1] by taking into account the effect of friction in the critical layer $U(y) = c$. It was already known that c, the velocity of propagation of the wave motion, is always smaller than the maximum velocity of the basic flow, so that the place where U becomes equal to c is always inside the profile. Closer investigation showed that the peculiarities which arise depend on the curvature of the profile of the given basic flow, *i. e.* on U''. The exact calculations proved to be difficult and are too complicated to be reproduced or even sketched in this report. The conclusions obtained were that the thickness of this layer which is under the influence of viscosity is of an order of magnitude $\sim (\nu/\alpha U)^{1/3}$ and that there is a discontinuity of phase of the wave motion in this layer proportional to $(U''/U')\, y_{cr}$. (U' and U'' are the values in the critical layer $y = y_{cr}$ where $U = c$.) In the same manner as above, an estimation of the value of the v component of the disturbing velocity in the critical layer can be found, as proportional to $\alpha\, u_0\, y_{cr}$. The portion of this component which differs in phase from the remaining wave motion is obtained approximately by multiplying the value of this supplementary component by the phase difference. Hence

$$v_2' \sim \alpha\, y_{cr}^2 \cdot \frac{U''}{U'}\, u_0$$

The considerations applied to the boundary layer can also be used for estimating ϱM since the critical layer is, in most cases, rather close to the wall; the order of magnitude thus obtained for the damping is

$$-\gamma_2 \sim \alpha\, y_{cr}^2\, U'' \qquad (26.11)$$

This entire treatment gives only a somewhat crude estimate of the diminution, for actually both the critical distance from the wall y_{cr}, and the velocity of wave propagation c are as yet unknown and must be deduced as special values from the solution of the problem. For further details the reader must be referred to the original paper. The considerations set forth above are not intended to replace the actual solution but rather to give a physical interpretation of the various factors which affect stability. It has been seen that U'', the curvature of the profile at the critical point, is of extreme importance in influencing

[1] Tollmien, W., Über die Entstehung der Turbulenz. Nachrichten der Gesellschaft der Wissenschaften, p. 21, Göttingen, 1929.

the extinction of oscillations. The more exact calculations performed by Tollmien in the paper already mentioned shows, in accordance with this result, that the critical Reynolds number varies considerably with the shape of the profile. He thus found a particularly small critical value for the flow along a flat plate, *viz.* $U_0\,\delta^*/\nu = 420$ (where δ^* is the displacement thickness of the basic flow) corresponding to $U_0 x/\nu = 60{,}000$ where x is the length along the plate from the leading edge. In comparing this number with experimental results it is necessary to emphasize the fact that it characterizes the state of affairs in which oscillations can occur for the first time without being extinguished. This, however, is not sufficient to produce noticeable turbulence in practice. The condition for this is that some disturbance initially quite small should be able to grow to a strength sufficient to alter the character of the stream. The Reynolds numbers for which actual turbulence is observed vary from $U_0\,\delta^*/\nu = 950$ approximately, for slight initial disturbances, to 500 for large disturbances; these correspond to values of $U_0\,x/\nu$ from 300,000 down to 85,000.

The critical velocities calculated by H. Schlichting[1] for the initially developing flow in the Couette case are considerably higher. The profile curvature is somewhat greater in this case. The lowest critical value of $U_0\,\delta^*/\nu$ obtained was 1530, which value increased as δ^*/h increased, h being the width of the channel. The minimum value of $u_0\,h/\nu$ occurred when δ^* equalled 0.115 h, and was equal to $2220/0.115 = 19{,}300$, which is considerably larger than the critical number observed by Couette. The reader should remember, however, that up to the present time all theories have only considered two-dimensional disturbances; it is possible that three-dimensional disturbances would produce greater instability.

The wave-lengths of the disturbances for which there is increase of the disturbance are rather large. According to Tollmien the smallest value is 17.1 δ^*.

Undulatory movement of this kind is not turbulence in the proper sense of the word. It is more likely that the actual development of turbulence takes place somewhat as follows: as a result of the increase in the wave motion, the velocity profile develops points of inflection at those places where back flow occurs, and, as briefly sketched above, the velocity profile becomes unstable and soon dissolves into separate vortices which spread out very rapidly. This point of view is confirmed by experiment.

In general, it is possible to say that the origin of turbulence can always be found in the instability of a flow. Flows in the form of jets with free boundaries and likewise flows in the wake of a body are especially unstable and therefore yield particularly small critical Reynolds

[1] Schlichting, H., Über die Stabilität der Couette-Strömung. Annalen der Physik, Vol. V, p. 905, 14 (1932).

numbers. Experimental values are in the neighborhood of 100. Flows in boundary layers at walls are considerably stabilized by the wall if the velocity profile is convex everywhere. This is the case for flows with fall of pressure and although measurements have as yet not been taken, very high critical Reynolds numbers may be here expected. Flows with increase of pressure have a point of inflection in the velocity profile, and according to the remarks made above, are unstable if friction is absent. Hence, such flows, which can only be stabilized by friction, will exhibit turbulence for comparatively small Reynolds numbers and the turbulence will become more marked as the increase of pressure is greater. This phenomenon may be noted in further detail: an increase of pressure in a laminar flow usually produces separation of the flow from the wall and consequently back flow, but the free layer thus separated is unstable and splits up into vortices. Provided that the increase of pressure is not too great, the vortices will drive the back flow forward again, thus producing a strongly turbulent flow with a mean forward motion. Greater increase of pressure produces stronger turbulence for if weak turbulence does not prevent back flow, this immediately increases the instability and causes the formation of more powerful vortices. The flow in an expanding diffusor can be described in this way.

The beginning of turbulence in flows through pipes or along smooth walls can be often regarded as caused by three-dimensional disturbances of the following form. Let us assume the existence of feebly disturbing vortex motions with axes parallel to the main direction of flow. The effect of these vortices is to shift sideways the boundary layer which develops along the walls, to thicken it at some points and also occasionally to drag part of it into the unretarded flow. A band of retarded fluid brought by such means in between portions of fluid moving with greater velocities is however unstable and very soon breaks up into separate vortices. Apparently events of this kind are responsible in most cases for the production of turbulence in straight pipes with well rounded mouths. The initial vortices whose existence has been postulated owe their origin to disturbances which occur before the fluid enters the pipe.

In practice it is observed that vortices produced by the disintegration of an unstable flow multiply, as a rule, very quickly, so that once such vortices are present, turbulence soon spreads throughout the whole flow. The appearance of fully developed turbulence is properly attributed to this phenomenon of which, however, no theoretical explanation has hitherto been possible. The reason for this is that the disturbances are of finite magnitude whereas the calculations involved in the theory can only be successfully carried to a conclusion by assuming disturbances so small that terms of the second order in the additional velocities may be neglected. The wide differences between the calculated and observed values of the critical Reynolds number in the Couette flow may possibly be accounted for by the fact that finite disturbances unstable in character may very well arise, and of which the instability is of a different type from that of the very small disturbances which must be assumed for the theoretical discussion. A definite answer to the question as to whether or not, under varying circumstances, turbulence will arise, can only be furnished by experimental observation, and wherein naturally, the theory of small disturbances will often be found a serviceable guide. Much work remains to be done on this problem.

27. The Drag on Bodies Moving Through Fluids[1]. Attempts have often been made to analyze the drag experienced by bodies when moving through fluids or gases into the sum of a number of terms. A starting point for one form of such an analysis is obtained by applying the principle of conservation of energy and by examining the way in which the work done by the force which impels the body forward against the drag is dissipated. If v is the velocity of the body and D the drag, the amount of work done per unit time is obviously Dv. In accordance with the principle of conservation of energy, this amount of work remains as energy in the fluid. This energy takes three principal forms:

1) Energy transmitted to infinity by wave motion.

2) Kinetic energy of vortex motion, remaining practically localized in portions of fluid which have been close to the body at some time.

3) Heat, produced by the effect of friction.

The first component needs consideration only in cases where wave motion occurs, such as the movement of a ship on a free water-surface or the motion of bodies through air with velocities greater than the velocity of sound. This form of energy is of no importance for motion through air with velocities below that of sound. In such cases the energy is composed of the kinetic energy of vortices and of heat due to friction; but since friction disintegrates the vortices in the course of time, so that their energy is changed into heat, the division between the two forms of energy is not sharp. The proportion of heat to kinetic energy will decrease at points taken successively nearer to the body and increase at points taken successively further from it. In many cases well defined orderly movements of vortices are observed; these may be distinguished from the remaining disorganized motion of the fluid. Since heat is itself a disorganized motion of molecules it is sometimes convenient to combine the heat energy with the energy of the disorganized flow of the fluid. This division is convenient for the purpose of certain hydrodynamic theories based on the study of organized vortex motion.

[1] The term "drag" is used in the following two sections rather than "resistance" since attention is directed primarily to the influence of the viscous fluid on the body and since likewise the analysis involves terms which have come into use more recently in connection with aeronautic terminology.

Quite another form of analysis is obtained by considering how the drag comes into existence at the surface of the body itself. At each surface element of the body the force exerted on it by the fluid can be resolved into a normal and a tangential component. The first is obviously the pressure on the element and the second the force due to friction. It is now possible to compound the pressure components on all the elements into a single force and similarly the frictional components into a second force, thus resolving the total drag into a pressure drag and a friction drag. The separation of the drag into such components can also be performed experimentally. The distribution of pressure at the surface of the body can be observed by boring a sufficient number of apertures into the surface and connecting them to manometers by tubes in the interior of the body. The component of the pressure-drag D_p in any direction s is then given by the formula

$$D_{ps} = \iint p \cos(n, s)\, dA$$

The frictional drag too can be directly determined by using a procedure elaborated by Fage and Falkner[1]. The method is laborious and difficult to carry out. Measurements are taken of the increase of pressure produced in a tube, projecting only slightly and topped by a sharp edge as in Fig. 82. The values obtained are proportional to the shear τ in the neighborhood of the point where measurements are taken. The instrument is calibrated in a laminar flow through a narrow channel where the shear can be calculated from the observed decrease of pressure. If t is the tangential direction of the flow the component of frictional drag D_f in the direction s is

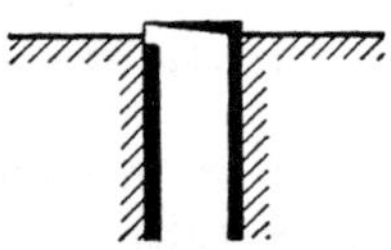

Fig. 82. Measurement of frictional drag, after Fage and Falkner.

$$D_{fs} = \iint \tau \cos(t, s)\, dA$$

It is however, simpler to measure the total drag of the body with the aid of a balance and to calculate the frictional drag D_f as the difference between the total drag D and the pressure drag D_p.

Writers have frequently spoken of "surface drag" and "form drag", identifying these terms with the frictional drag and pressure drag respectively. This terminology is, however, based on the theory that while the frictional drag is proportional only to the surface area of the body and while the coefficient of the frictional drag depends on the Reynolds number, the pressure drag is a function of the shape of the body alone; but closer examination shows that the frictional drag is also a function of the form of the body so that the separation of the drag into components due to the surface area and the shape must be rejected as unsound.

[1] Fage, A. and Falkner, V. M., An Experimental Determination of the Intensity of Friction on the Surface of an Aerofoil. Proc. Roy. Soc. A. Vol. 129, p. 378, 1930.

We now propose to investigate the drag of a body from the hydrodynamic standpoint especially, by applying the theory of ideal fluids.

In all cases of flows around bodies as considered in the older hydrodynamic theories, the drag in the direction of flow is found to be zero, a result which formerly appeared paradoxical but can now be seen to be an obvious consequence of the nature of the flows investigated. For these flows were the so-called "potential" flows and therefore consistently separated in front of the body to reunite behind it, leaving, of course, no disturbances in the flow. Hence there is no place for the energy, equivalent to the work done by the resistance, and therefore the drag *must* be zero. Furthermore, since forces perpendicular to the direction of flow do no work, the preceding argument is unaffected by the presence either of angular momenta or of forces perpendicular to the direction of motion. In fact, even in the early stages of the theories mentioned, it was found that potential flows might have angular momenta and in recent times other flows of this kind have been found to be associated with transverse forces (lift).

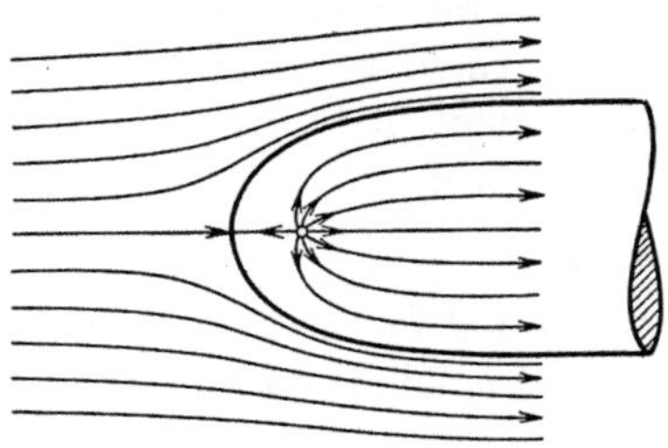

Fig. 83. Flow with a source.

It is possible to deduce from the behaviour of a potential flow that if any actual stream recombines behind the body in almost the same manner as in the ideal case, the drag will be very near to zero. This has been very well confirmed by experiment. It is found, for example, that the drag of a body whose shape is approximately that used for airships is not much greater than the frictional drag always to be expected with a body of the same surface and axial length. Although the flows in such cases are actually turbulent, it may be observed that the theory of the potential flow yields very useful approximations in several other respects, as for example the pressure distribution.

Very instructive results may be obtained by considering the behavior of the pressure drag at the head of a very elongated body. In order to fix definitely the ideas, it will be supposed that the body is in the shape of a cylinder extending to infinity at the rear and suitably rounded off at the head in somewhat the same fashion as the front part of an airship hull. The object of the investigation is to determine the drag experienced by such a "half-body" in an ideal fluid which approaches it with velocity u_0 parallel to its longitudinal axis. The stream function for a body of this kind can be generated by taking a suitable distribution of sources near the head of the body. The shape of body obtained by assuming the existence of a single concentrated source is shown in Fig. 83; the contour must always coincide with the stream-lines separating that part of the flow which approaches from infinity from that which proceeds from the source, and which we thus identify with the surface of the "half-body". If Q denotes the strength of the source, *i. e.* the volume of fluid emerging from the source per unit time, the cross-sectional area A of the cylinder at a great distance from the head will satisfy the simple equation $Q = A u_0$. Since the body is not closed at any finite distance, the pressure drag is not uniquely defined; but it is possible to obtain a unique definition of this quantity by postulating that the body is completely separated into two portions at a sufficient distance from the head (see Fig. 84) and that the pressure in the fissure of separation is that of the undisturbed fluid. By this means the first part of the body is completely enclosed by its surface and its pressure drag can therefore be clearly defined. Information concerning the value of the pressure drag can now be obtained by applying the momentum theorem. It will be convenient to commence by deducing a lemma concerning the momentum of a source which is situated in a flow of velocity u_0. Due to the source, additional velocities radiating from it, of magnitudes inversely proportional to the square of the distance from the source, must be superimposed upon the velocity u_0. If the velocity due to the source is w at a distance r from the source, Q must equal $4\pi r^2 w$, the total volume of fluid flowing in unit time across a sphere of radius r, so that $w = Q/4\pi r^2$. A cylindrical surface is now drawn around the source having its axis parallel to the direction of u_0 and closed by two transverse planes at distances from the source which are great in comparison with the radius of the cylinder. We have now to calculate the amount of momentum transported across this surface (Fig. 85) by the combined flow. The net contribution of the two ends of the cylinder can be neglected, for w decreases rapidly with distance from the source so that the difference between the amounts of momentum transported out of and into the cylinder at its two ends can be made arbitrarily small by prolonging the length of the cylinder. Again, in summing the contributions to the total transported momentum due to cylindrical surface elements, it is possible to combine the contributions of pairs of surface elements situated at equal distances in front of and behind the source respectively. If the x component of the velocity at one such element is $u_0 - (Q/4\pi r^2) \sin\varphi$, then the velocity at the other element of the pair is $u_0 + (Q/4\pi r^2) \sin\varphi$ so that the mean of the two values is exactly equal to u_0. By combining all the surface elements in this fashion and

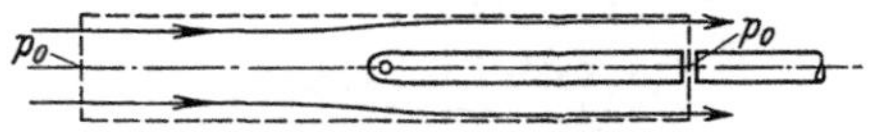

Fig. 84. Half-body.

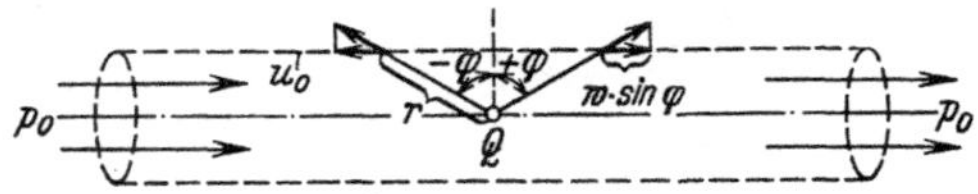

Fig. 85.

observing that the amount of fluid flowing out of the cylinder from its interior must equal the total strength Q of the source, it is found that the amount of momentum transported across the surface of the cylinder is $\varrho Q u_0$. In order to apply the theorem of momentum it is necessary also to find the value of the integral of the pressure in the flow taken over the chosen surface. In this integral however, the contributions from the two ends of the cylinder very nearly cancel out and may be neglected as before, while the pressure on the curved surface is perpendicular to the x direction at every point, and hence the value of the entire integral is zero. In the momentum theorem, we have therefore to deal simply with the momentum $\varrho Q u_0$ and the equivalent force, *i. e.* the reaction due to the flow *out* of the cylinder, which acts in a direction opposite to u_0 and is a force towards the head of the body, or in other words, a negative drag.

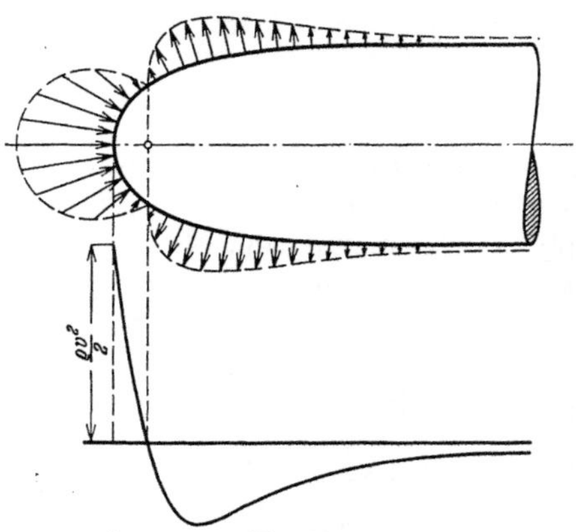

Fig. 86.
Pressure distribution on a half-body.

On applying this lemma to the case of the body of Fig. 84 it is easy to see from the diagram that on account of the presence of the body there is a deficit of $\varrho Q u_0$ in the amount of momentum leaving the downstream end as compared with the corresponding amount for the auxiliary cylindrical body used in establishing the lemma. This result, taken in conjunction with the lemma, means that the amount of momentum emerging from the auxiliary cylindrical surface in the case of flow around the infinitely long "half-body" exactly equals the deficit on the trailing surface, so that the total amount of momentum is unaltered. Also the pressure integral vanishes as before so that the resultant force exerted by the flow on the body is zero.

This result appears paradoxical at first sight but the state of affairs involved may be clarified by other methods, for example by choosing a reference system with respect to which the fluid is at rest. In such a system the velocities at any instant are those produced by the given distribution of sources, and noticeable velocities are found only near to the sources, *i. e.* near to the head of the body, while at large distances from the head the fluid is practically at rest (relative to the axes chosen). Hence the forward motion of the "half-body" does not impart a permanent velocity to the fluid in the direction of its own motion and the body needs no applied force to ensure its further movement. Consideration of the pressure distribution also verifies the correctness of this result and assists in its interpretation. In the neighborhood of the stagnation point, the pressure will be above the field pressure, while at other points it will be below, only regaining the normal field pressure as the transition to parallel flow gradually develops (see Fig. 86). On constructing the integral $\iint \varrho \cos(n, s)\, dA$ it will then appear in each individual case that the resultant of the super pressures in the front region exactly balances the resultant of the sub pressures elsewhere.

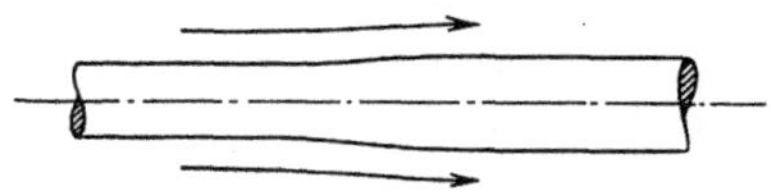

Fig. 87.

Naturally it is also possible to form a surface out of other stream-lines selected from the flow around a source and to regard it as the bounding surface of a body. It immediately follows that bodies of the form represented in Fig. 87 also experience no drag in the ideal fluid. Since this result, when true, applies equally for sinks as for sources, the flow of Fig. 87 can be regarded as proceeding either from the right or the left. This result is of some importance in applying the theorem of momentum to propellers, windmills and the like, for in such cases the exterior stream-lines run somewhat as in Fig. 87. The theorem which has just been established shows in fact in such cases, that contraction or expansion of the interior stream-lines does not evoke any reactive force on that part of the flow surrounding these lines.

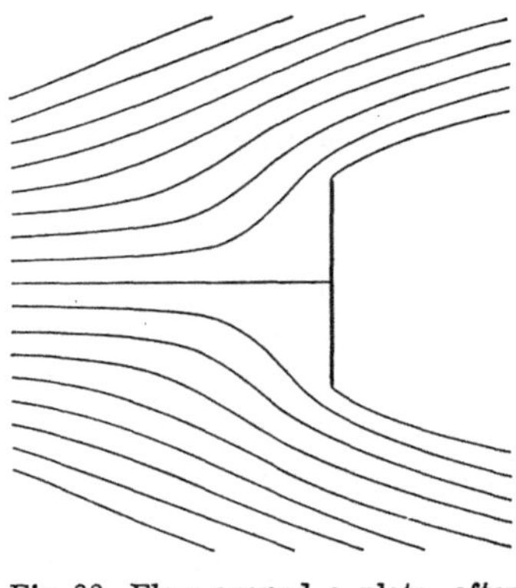

Fig. 88. Flow around a plate, after Kirchhoff.

It can also be deduced from the same theorem that a body with a "dead water" region behind it would experience no drag in an ideal fluid if the cross-section of the "dead water" asymptotically approached a fixed value; for the exterior flow in such conditions coincides with that due to a "half-body" similar to the one discussed. Flows with a "dead water" region in an ideal fluid, such as were studied by Kirchhoff, are however of quite a different nature, since the width of the wake continually increases with the distance from the body and to a first approximation varies as the square root of the distance (see Fig. 88). It can be shown that the drag is zero for flows around bodies whose width continually increases, provided that the order of the width remains less than the square root of the distance.

28. The Drag in Non-Perfect Fluids. Actual flows where drag is always present differ considerably from the Kirchhoff type for there is never a "dead water" region at rest relative to the body, and the vortex structures which occur instead are not stationary, but induce interchanges of momentum between wake and exterior flow. In other words the wake is active and no longer "dead"; it is more correct to say that turbulent mixing ensues, and certainly in regions which are at rather large distances from the body, takes forms which have already been described in **25**. Fig. 73 shows the distribution of velocities in flows with wakes of this character, referred to a fluid at rest. Since the width of the wake continually increases with time and since intermixture continues to take place, the velocity continually decreases, but in such a way that the momentum of the wake remains constant, as

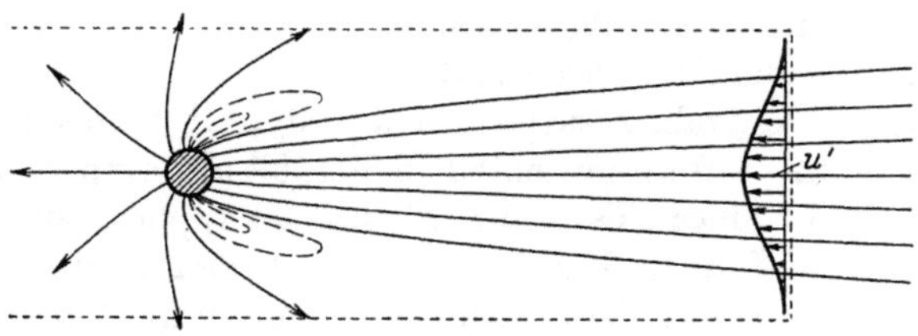

Fig. 89. Flow in the wake with the associate source.

already discussed in **25**. The quantity M in the last of these references is identical with $\varrho\, Q\, u_0$ in the remarks which follow.

The entire flow produced by the motion of a body through a fluid originally at rest consists partly of the wake which has just been described and which occupies a comparatively narrow region behind the body, and, for the rest, of a potential flow which, except at points in immediate proximity to the body may be considered as a flow due to a system of sources. If the stream-lines of the wake which run into the moving body are imagined to be prolonged through the body it follows from continuity that the strength of the equivalent "source" coincides with the strength of the wake, so that each stream-line of the wake is the continuation of some stream-line issuing from the source (see Fig. 89). In **25** it has been shown that the width of the wake increases as the cube root of the distance for a three-dimensional flow symmetrical about an axis, but is proportional to the square root of the distance in a two-dimensional flow. This case lends itself to considerations of momentum similar to those just applied to the "half-body" of the preceding lemma. If the cross-section of the auxiliary cylindrical surface is chosen sufficiently large so as to include the entire domain of the wake, then with a great length of the cylinder we can evidently realize a condition in which, so far as the sources are concerned, the difference in the momentum transferred through the two end planes compared with that for the cylindrical surface will vanish. This, however, will not be the case if the diameter of the end section is taken proportional to its length. Hence the flow generated by the equivalent fluid sources, according to the lemma, produces a negative drag of amount $\varrho Q u_0$ as before. Owing to the turbulent flow in the wake the velocity in a certain bounded domain is smaller at the back of the body than in front (see Fig. 90). If the effect of the source is left out of consideration the velocity of flow is u_0 at points outside the region of disturbance and hence, if the disturbing velocity is u' relative to the fluid, the velocity in the turbulent region of a steady flow, referred to the motionless body, will be $u_0 - u'$. The pressure at each end has the undisturbed value so that the amount of momentum transported across the two ends, each of cross-sectional area S_0, say, is

$$\varrho u_0^2 S_0 - \varrho \iint (u_0 - u')^2 \, dS$$

This expression can be simplified into

$$2 \varrho u_0 \iint u' \, dS - \varrho \iint u'^2 \, dS$$

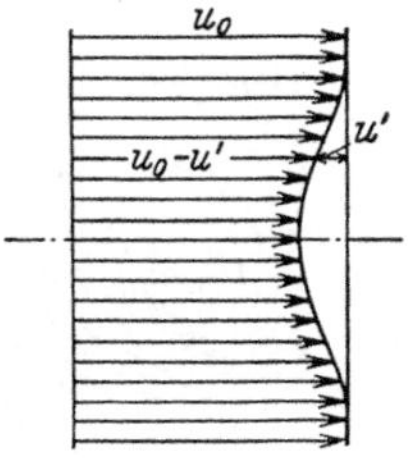

Fig. 90.

The integral $\iint u' \, dS$ is, however, equal to the strength of the wake which is identical with Q, the strength of the fluid source, and hence the first part of the above formula can be replaced by $2 \varrho Q u_0$. Since the velocity in the integrand is from the back of the body toward the front, the expression clearly denotes a positive drag. The remainder of the formula contains squares of the disturbing velocities and becomes proportionately smaller in comparison with the first term as the distance from the body is increased. At the limit, therefore, as the length of the cylinder increases indefinitely, the second term of the above expression vanishes. Hence, on combining the terms of the lowest order in the disturbing velocities, an amount $2 \varrho Q u_0$ is obtained in the direction of positive drag and $\varrho Q u_0$ in the opposite direction. Finally, therefore, the momentum equivalent to the drag[1] is $\varrho Q u_0$ and hence the strength of the wake is given by the equation

$$Q = \frac{D}{\varrho u_0}$$

If the drag is written in the form

$$D = C_D S \frac{1}{2} \varrho u_0^2$$

the wake strength can be expressed as

$$Q = \frac{1}{2} C_D S u_0$$

The last equation can easily be interpreted in physical terms. It can be seen *inter alia* that the wake-strength is directly proportional to

[1] The relation obtained at this stage also applies to the wake in very viscous fluids as calculated in **12**. The reader may convince himself of this by performing the necessary calculations. Since the flow surrounding the wake, at large distances from the body, reduces to the flow due to sources, exactly as in the cases already considered, no further forces need be included in the momentum considerations, and the identity of the results in the two cases is therefore comprehensible.

$C_D\, S$, the drag-area of the body. The fact that this has a constant value accords with the fact that all the stream-lines in the wake come from infinity into the vicinity of the body as shown[1] in Fig. 89.

In the theories described above, the flow in the wake has been treated as a steady motion and that portion of momentum due to turbulent oscillations of the velocity has been neglected. In order to take account of them it would be necessary to use a term proportional to $\varrho \iint u'^2\, dS$, since the velocities of turbulent oscillations are proportional to the velocity of the wake. It has therefore been correct to neglect the contribution of turbulence since the integral $\iint u'^2\, dS$ has already been neglected in comparison with larger terms. If however, the system investigated contains an established system of strong vortices as in the case of the so-called Kármán vortex street, deviations from the relations obtained above must be expected.

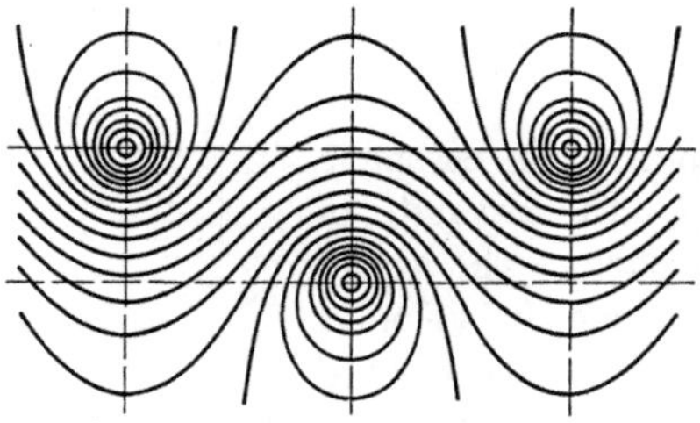
Fig. 91. Kármán vortex street.

v. Kármán[2] has shown in a very well-known paper that such series of vortices are in general unstable but that one arrangement can be found which is stable with respect to most disturbances (though it is merely neutral for *one* such disturbance).

The ratio of breadth to the distance between consecutive vortices in the same row in this arrangement (see Fig. 91) is given by the equation

$$\frac{h}{l} = \frac{1}{\pi} \cosh^{-1} \sqrt{2} = 0.2806$$

Kármán explained how considerations of momentum could be applied to a vortex street of this kind and proved it was possible to obtain a numerical value for the drag of the body by means of a photograph and by determining the undisturbed velocity and counting the number

---

[1] This is true however only to the extent that $\iint u'^2\, dA$ can be neglected, and the presence of this term produces somewhat greater wake strength near to the body. This requires an extra fluid source and a corresponding distribution of sinks in the wake itself in order to absorb the surplus, so that the same relations still hold at greater distances from the body. It is easily seen that this supplement to the potential flow will result in closed stream-lines (represented by dotted lines in Fig. 89). According to the shape of the body other sources and sinks situated in the body may also be required; blunt bodies also require a vortex-pair behind the body.

Kármán, Th. v., Nachrichten der Gesellschaft der Wissenschaften zu Göttingen, Math.-phys. Klasse, p. 509, 1911; p. 547, 1912.

[2] Kármán, Th. v. and Rubach, H., Über den Mechanismus des Flüssigkeits- und Luftwiderstandes. Phys. Zeitschr., p. 49, 1912.

Bénard, H., Comptes Rendus, Vol. 147, 1908; Vol. 156, 1913; Vol. 182, 1926; Vol. 183, 1926.

of vortices produced in unit time. The result obtained for the coefficient of drag was

$$C_D = 1.587\frac{u_1}{u_0} - 0.628\left(\frac{u_1}{u_0}\right)^2 \cdot \frac{l}{d}$$

In this equation u_0 denotes the velocity of the body, u_1 the velocity of the vortices, both referred to the fluid assumed motionless; l is the distance between consecutive vortices on the same line and d is a dimension of the body for which the coefficient C_D is being calculated. The length l can be measured directly from the photographs while the quantity u_1/u_0 is connected in a simple manner with the period of the vortices induced by the body. For if T denotes the time required for the formation of one right-handed and one left-handed vortex, then, on account of the fact that the system of vortices is advancing relative to the body with a velocity $(u_0 - u_1)$, the following relation holds[1].

$$l = (u_0 - u_1)\,T,$$

and therefore

$$\frac{u_1}{u_0} = 1 - \frac{l}{u_0 T}$$

When plates or aerofoils of finite span are set obliquely across a stream of fluid, flow takes place around the plate or airfoil ends in order to equalize the differences of pressure between the upper and lower sides, and remains in the form of vortices long after the body has passed through the fluid. In such cases considerable kinetic energy is left in the fluid, and work must be done to produce these vortices even if all frictional effects are left out of account. There will therefore be a drag due to this factor, whose magnitude can be investigated by applying the theory of ideal fluids. The relation between this part of the drag and the lift (generated at the same time) can be calculated, and is found to agree very well with experiment. (See further on this point, Divisions E and J.) Following Munk this kind of drag is termed "induced drag", for the formulae which arise in calculating it are very similar to those of electromagnetic induction. That part of the total drag due to turbulence of the flow is called profile drag. Its value can be obtained by subtracting the calculated value of the induced drag from the value of the entire drag as measured by a balance. Experiments of this kind have shown, with a marked degree of accuracy, that the profile drag does in fact depend only on the shape of the profile, the effective angle of incidence, and naturally, on the Reynolds number; whereas the total drag depends quite definitely on the aspect ratio, or, in more general terms, on the contour of the aerofoil.

29. Experimental Methods for the Determination of Resistance. The *results* of experimental determinations of the resistance in air experienced

[1] For a more recent account of the application of considerations of momentum to this case, see TIETJENS, Hydro- und Aeromechanik nach Vorlesungen von L. PRANDTL, Vol. 2, § 91.

by bodies of various shapes are discussed in Division J on the experimental characteristics of the airfoil and in Division K on the resistance of bodies without lift. (See Vol. IV.) It is, therefore, unnecessary here to enter into a detailed investigation of the results. It may, however, be of interest to discuss, in conjunction with the above sections, the *experimental methods* used in the determination of resistances in air and water. The evaluation of the pressure resistance by integration of the pressure distribution has been mentioned in 27 as also the evaluation of the frictional resistance, following Fage and Falkner. Both methods can be applied not only to models but to large bodies, although in actual practice the frictional resistance has not yet been determined in this manner.

The determination of the total resistance occupies a much more important place in experiment and in connection with this, three methods may be noted: The first uses the free motion of a body in a motionless medium. In the second, the body is dragged through the motionless medium and the propelling force is either so regulated that when the desired velocity is attained the motion continues without acceleration, or the force is recorded by a self-acting balance. In the third method the body remains at rest and the medium is made to flow past it with known velocity. The body is connected to one or more balances from which the existing forces can be read.

To the first group belong those experiments in which bodies are allowed to fall through some medium. It is applied particularly to small bodies, as for example when small spheres are allowed to fall through viscous fluids or where microscopic drops of oil fall through air, thus serving to determine the law of resistance at very small Reynolds numbers (Stokes' formula, etc.). The method is also applicable to larger bodies so long as they are light enough to attain their terminal velocities within a short distance. It is then only necessary to determine the velocity in the steady state and the net propulso-motive force (weight minus buoyancy). Since there is no acceleration the resistance is exactly equal to this net force. At larger Reynolds numbers this method is only applicable with some restriction, as many bodies, the sphere for example, tend in free motion, to take on a more or less irregular lateral motion, and naturally the resistance, when there is free lateral motion, is larger than when the body falls in a perfectly straight path.

Fundamentally, the determination of the form of the polar curve of an aeroplane (resistance coefficient against lift coefficient) in gliding is connected with this method. For this purpose experiments are made on a number of gliding flights at different angles of gliding and in them the velocity of flight, the vertical component of the velocity, and the inclination of a fixed line in the body to the vertical are measured. Since gravity is here the only driving force, these measurements, together with the weight of the airplane provide the necessary data for the evaluation of the lift, resistance and angle of inclination.

Another method belonging to the first group is one which is applied to airships and which, apparently, was first given by Pannell and Frazer[1]. The following considerations provide the foundation for this method. If we assume that the air resistance is proportional to the square of the velocity, we obtain the following differential equation

$$m \frac{d v}{d t} = - K v^2$$

[m is the mass of the airship, including the mass of the air which is dragged along with it, K is equal to $(1/2) \varrho S C_D$]. This equation can be integrated immediately and gives

$$v = \frac{m}{(K t + C)}$$

where C is a constant of integration to be determined from the initial velocity. If then the reciprocal of v is plotted as a function of t we get a straight line, and from the inclination of this line the quantity K/m can be evaluated. Figure 92 shows the results of an experiment of this nature taken from the work of Pannell and Frazer. The points on the graphs lie approximately on straight lines and we see therefore that the coefficient of resistance, within the limits of experimental error, is independent of the velocity.

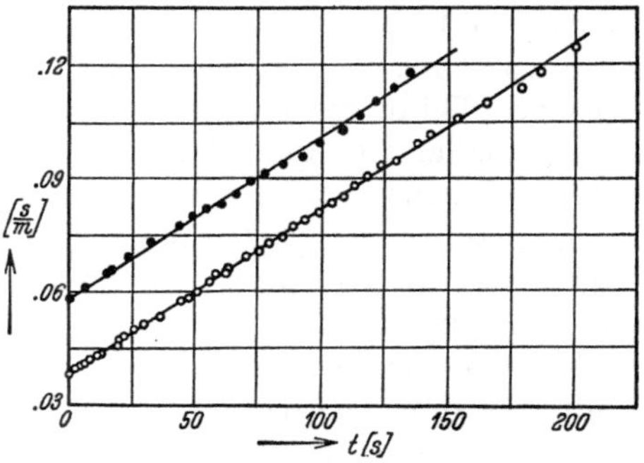

Fig. 92.

The methods of the second group are applied more particularly to the determination of the resistance of ship models and have been developed in great detail. In one of these methods the propulsive force is derived from falling weights the magnitudes of which are so arranged that the model, after having reached the desired velocity, experiences approximately no further acceleration. The recording of the velocity by the aid of some timing instrument shows also the small additional accelerations which must be taken into account. A more usual method is that in which the model is carried forward by a carriage running on rails, the carriage containing some apparatus which records the resistance experienced by the model. The methods of the second group cannot be applied successfully to experiments in air since the masses to be set in motion are usually very large in comparison with the mass of air dragged forward with the body, and on account of this, small accelerations produce comparatively large errors. Likewise the use of a

[1] "Experiments on Rigid Airship R 33", Br. A.R.C. R. and M. 668, 1919.

carriage in order to draw the models forward is objectionable since the carriage itself produces large disturbances in the air which are appreciable even on the model itself. In experiments on ship's models this state of affairs does not exist since the carriage does not move in the water but in the air above it.

The third method is rarely used to determine the resistance in water but is, on the contrary, commonly used to determine resistance in air. The models are placed in a stream of air which, as far as possible, is made uniform in space and in time. The models are kept in position by attachments which are very thin, and so disposed as to produce the smallest possible disturbances in the air, and the steady forces on the model are then measured by the aid of balances. There are three fundamental operations in the experiment: First the balances must be read in still air. Second, they must be read when the stream has a known velocity. Third, the apparatus must be calibrated in such a way that forces, similar to those of the air-resistance, are produced by weights and the balances must again be read. When several components are being measured it is preferable to reproduce in calibrating, all the rather large force components at the same time by weights acting on rollers and wires in order to eliminate the effect of the elastic deformation of the suspension of the model. In many cases where a body experiences a lift, the drag is very small in comparison with the lift and in such cases it is essential to verify that the undisturbed air stream is accurately horizontal, for if not, the above method of calibration only gives the horizontal and vertical components of the resulting system of forces which differ in this case from the components parallel and perpendicular to the stream, which by definition are drag and lift. If necessary, the model must be measured in two positions differing by 180^0.

Velocity determinations are best made by the use of Pitot-static-tubes or, from the Bernoulli equation, by measuring differences of pressure associated with changes of cross-section of the air stream.

More recently another method for determining resistance has been developed which is connected with the state of affairs in the wake behind the body. Following the discussion in the first section of 28 it can be seen that in this manner the resistance (drag) experienced by bodies of no lift, and what is very important, the profile drag as distinct from the induced drag, can be evaluated. In the practical application of this method, which is due to A Betz[1], it is necessary to refine our previous considerations of momentum as applied to the wake in such a way that they are applicable at a rather small distance from the body. Since this method is intimately connected with the discussions embodied in previous sections, it may be described here in somewhat greater detail.

---

[1] Betz, A., Ein Verfahren zur direkten Ermittlung des Profilwiderstandes. Z. Flugtechn. Vol. 16, p. 42, 1925.

In comparison with previous considerations, it must be realized that it is necessary, first, to take account of the term in the wake expression which is of the second power in the velocity, second, to remember that the pressure in the neighborhood of the body, where the wake-velocity is measured, does not coincide with the pressure at infinity. In Betz's procedure, two extended planes are taken, both at right angles to the direction of motion, one behind the body and one in front of it (see Fig. 93). Let the velocities and the pressure in the front plane be u_1, v_1, w_1, p_1 and the corresponding magnitudes in the other be u_2, v_2, w_2, p_2. At infinity let $u = u_0$, $v = w = 0$ and $p = p_0$. The theorem of momentum in its general form now gives the following equation

$$D = \iint (p_1 + \varrho u_1^2)\, dS - \iint (p_2 + \varrho u_2^2)\, dS \qquad (29.1)$$

where the domain of integration extends over both infinite planes. The problem is now to transform the integrals in such a way that it is sufficient to integrate over the domain in the wake. The following total head pressures are introduced

$$g_1 = p_1 + \frac{1}{2}\varrho\,(u_1^2 + v_1^2 + w_1^2)$$

and $$g_2 = p_2 + \frac{1}{2}\varrho\,(u_2^2 + v_2^2 + w_2^2)$$

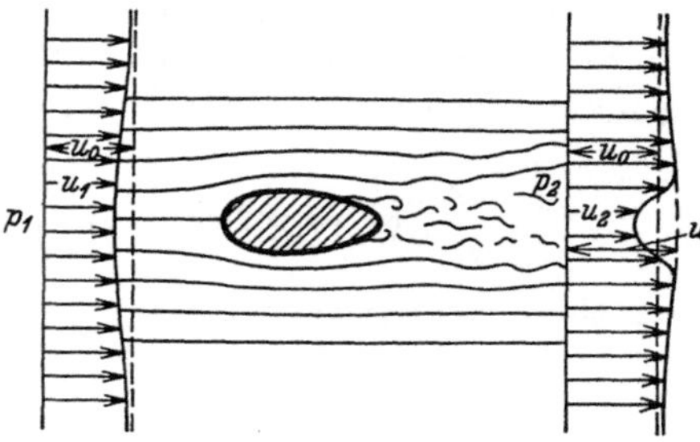

Fig. 93.

In accordance with Bernoulli's theorem, g is constant on every stream-line which is unaffected by friction or apparent turbulent friction. Hence $(g_1 - g_2)$ only differs from zero in the wake, and on introducing g_1 and g_2 into (29.1), we have,

$$\left.\begin{aligned} D = \underset{(I)}{\iint (g_1 - g_2)\, dS} + \frac{1}{2}\varrho \underset{(II)}{\iint (u_1^2 - u_2^2)\, dS} \\ + \frac{1}{2}\varrho \underset{(III)}{\iint \Big[(v_2^2 + w_2^2) - (v_1^2 + w_1^2)\Big] dS} \end{aligned}\right\} \qquad (29.2)$$

From what has been said it can be seen that expression I need only be integrated in the wake. In order to deal with expression II a hypothetical flow is introduced which coincides at every point outside the wake with the real flow at the back plane, but differs from the actual flow in the wake by having $g_2' = g_1$, throughout the wake, so that no energy is lost through friction or turbulence. This can be achieved by changing the x component of the velocity. Let the new velocity be u_2'. Since the actual flow is incompressible the same cannot be the case for the hypothetical auxiliary flow which must have a distribution of sources of total strength Q, say. Then clearly

$$Q = \iint^{W} (u_2' - u_2)\, dS$$

The symbol W over the sign of integration denotes that the domain of integration can be restricted to the wake, since outside, u_2' and u_2 are equal.

The theorem of momentum in the form of (29.2) will now be applied in two steps, first of all by calculating the differences between u_1, v_1, w_1, p_1 and u_2', v_2, w_2, p_2 and then those between u_2', v_2, w_2, p_2 and u_2, v_2, w_2, p_2; the sum of the two quantities so obtained will be the result required. In the first step expression I is zero since $g_2' = g$, but expression III presents certain difficulties. It becomes comparatively small when only flow due to sources occurs but may be considerable in the presence of steady vortices such as are produced by aerofoils. Its total value may be denoted by D_i and more detailed consideration will be deferred for the moment. Apart from the value of D_i, since we have the assumed distribution of sources between the auxiliary planes, and also no loss of energy in the first step, the lemma previously proved in **27** shows that expressions I and II together provide a forward force $\varrho Q u_0$. Since expression I is zero it follows that for expression II

$$\frac{1}{2}\varrho\iint (u_1^2 - u_2'^2)\, d\,S = -\varrho Q u_0 = -\varrho u_0 \overset{W}{\iint} (u_2' - u_2)\, d\,S$$

In the second step expression I becomes

$$\overset{W}{\iint} (g_1 - g_2)\, dS,$$

and expression II

$$\frac{1}{2}\varrho \overset{W}{\iint} (u_2'^2 - u_2^2)\, d\,S = \frac{1}{2}\varrho \overset{W}{\iint} (u_2' - u_2)(u_2' + u_2)\, d\,S$$

while expression III is zero.

Altogether, the following equation is now obtained:

$$D = \overset{W}{\iint} (g_1 - g_2)\, d\,S + \frac{1}{2}\varrho \overset{W}{\iint} (u_2' - u_2)(u_2' + u_2 - 2u_0)\, d\,S + D_i \quad (29.3)$$

These two integrals are now to be integrated over the wake. The following may be said with regard to the quantity D_i, where

$$D_i = \frac{1}{2}\varrho \iint \left[(v_2^2 + w_2^2) - (v_1^2 + w_1^2)\right] d\,S \quad (29.4)$$

If the y and z velocities involved are due to a source it would be possible to choose the auxiliary planes at equal distances from the source, in which case, both portions of D_i would be equal and would cancel each other.

The case is somewhat altered when considering the motion of aerofoils through fluids, for then steady vortices proceed from the trailing edge as explained at the end of (28). In this case, velocities are produced in the back plane with no equivalent velocities in the front plane, and D_i is then a non-vanishing drag which is easily recognizable as the "induced drag" obtained as a result of the energy considerations in 28. Since the right hand side of (29.3) denotes the total drag, *the sum of expressions I and II must denote the profile drag.*

In a practical case, numerical calculations can be undertaken by first measuring by Pitot tube, the distribution of values of g_2. Outside the wake g_2 coincides with g_1 which is the Bernoulli constant of the undisturbed flow and therefore has a fixed value (see Fig. 94). It is necessary also to determine the static pressure p_2 in order to deduce the value[1] of $u_2 = \sqrt{2\,(g_2 - p_2)/\varrho}$. The value of u'_2 can usually be found from $u'_2 = \sqrt{2\,(g_1 - p_2)/\varrho}$ for although this relation neglects $(v_2^2 + w_2^2)$ it is permissible to do so in most cases.

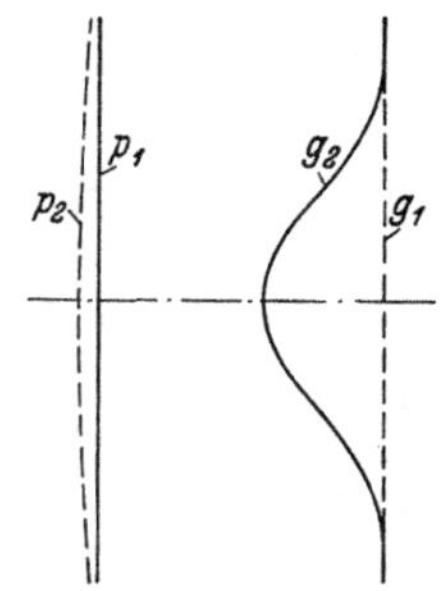

Fig. 94.

The magnitudes which occur in *I* and *II* now being known the numerical values of the integrals can be determined. Expression *I* yields the major contribution and gives the same result as the formula given in the previous section for large distances from the body (small wake velo-

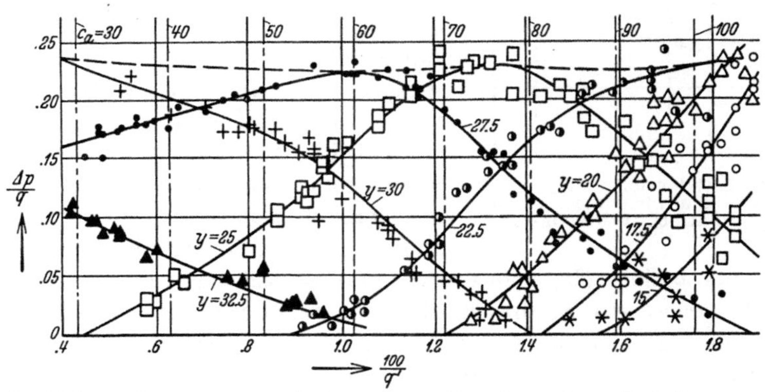

Fig. 95. Values of $g_1 - g_2$ on an abscissa representing the reciprocal of the uncorrected value of the stagnation pressure $g = 1/2\, \varrho u_0^2$: From M. Schrenk.

city, flattened wake). Expression *II* supplies a correction which becomes quite small at large distances but is important at points near to the body. Expression *III*, which is not restricted to the wake, cannot be calculated in this fashion. The method is primarily applicable when

[1] This is true on the assumption that in the wake the amounts $(1/2)\,\varrho\,(v_2^2 + w_2^2)$ can be neglected in comparison with $(g_2 - p_2)$. They are very small in all practical cases.

either the "profile drag" composed of I and II is alone required, or when expression III $(= D_i)$ is of no importance[1].

This method can be applied not only in the wind tunnel but also on airplanes in flight. The checking by the use of balances of the results thus obtained in wind tunnels has often revealed some small differences which are chiefly due to the fact that the spatial uniformity of the flow in wind tunnels is not complete, so that in consequence the wake is wider at some places than at others. At points very close to the body difficulties arise through the intrusion of the strongly disturbed pressure field into the wake. At some distance from the body the method is however quite reliable and has proved to be especially instructive for comparing the effects of varying roughnesses or various methods of treatment of the airfoil surface. Measurements on airplanes themselves[2] can be made by having, first, a Pitot-static tube in some place which, as far as possible, is free from disturbance, so as to measure the value of g_1 and the stagnation pressure $q = 1/2\ (\varrho u_0^2)$, and second, at some distance behind the trailing edge of the wing, a Pitot-tube and a static pressure tube near together, or again a Pitot-static tube. If gliding flights are made at different angles of incidence the wake is displaced relative to the instruments and by plotting the readings of the instruments against C_L, points are obtained through which reasonably smooth curves can be drawn. The observations are repeated with the instruments located at various heights. In addition, corresponding to each value of C_L, $(g_1 - g_2)$ and $(g_2 - p_2)$ can be expressed as functions of the vertical position of the instruments. An example of this is

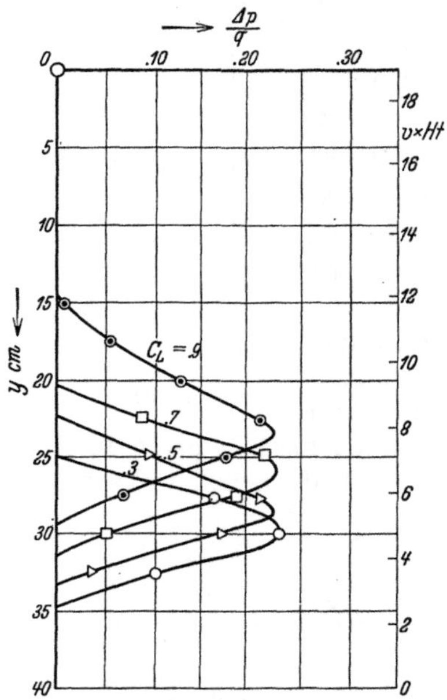

Fig. 96. Distribution of $(g_1 - g_2)$ in the wake of a wing under various angles of attack (values of C_L).

[1] In the procedure described above it is no longer true that the contribution to III of the flow due to sources can be neglected. In practice, no measurements are made in the front auxiliary plane which is, as it were, shifted to infinity where v_1 and w_1 are zero. The values of v_2 and w_2 due to sources will, however, be quite small in most practical cases and it is probable that no great mistake is made by neglecting their effect in III.

[2] Weidinger, H., Profilwiderstandsmessungen an einem Junkers-Tragflügel. Jahrb. Wiss. Ges. Luftfahrt, p. 112, Munich, 1926.

Schrenk, M., Über Profilwiderstandsmessung im Fluge nach dem Impulsverfahren. Luftfahrtforschung, Vol. II, Part I, Munich, 1928.

shown in Figs. 95 and 96. Practical measurements show considerable differences when the surface of the airfoil is made of fabric, is lacquered, or polished, or made of metal, etc., and emerging from the results is the fact that in all cases smoothness of surface is greatly to be desired.

Bibliography.

The following short bibliography will serve to furnish references to some of the more important contributions to the development of the subject of the present Division, as well as to certain treatises dealing with the same general subject. Additional references will be found at suitable points in the text.

1. Memoirs.

a) Historical.

NAVIER, C. L., Mémoire sur les Lois du Mouvement des Fluides. Mém. de l'Acad. des Sciences, Vol. 6, 1822.

STOKES, G. G., On the Theories of the Internal Friction of Fluids in Motion. Cambr. Trans. Vol. 8, 1845 or Coll. Papers Vol. I.

— On the Effect of the Internal Friction of Fluids on the Motion of Pendulums. Cambr. Trans, Vol. 9, 1851 or Coll. Papers Vol. III, p. 48.

REYNOLDS, O., An Experimental Investigation of the Circumstances which Determine whether the Motion of Water Shall Be Direct or Sinuous, and of the Law of Resistance in Parallel Channels. Phil. Trans. Roy. Soc. 1883 or Coll. Papers Vol. II, p. 51.

— On the Dynamical Theory of Incompressible Viscous Fluids and the Determination of the Criterion. Phil. Trans. Roy. Soc. 1895 or Coll. Papers Vol. II, p. 535.

b) Modern.

LORENTZ, H. A., Über die Entstehung turbulenter Flüssigkeitsbewegungen und über den Einfluß dieser Bewegungen bei der Strömung durch Röhren. Zittingsverslag Akad. van Wetensch. Amsterdam Vol. 6, p. 28, 1897 (Dutsch) and Abhandlungen über theoret. Physik, Vol. 1, p. 43, Leipzig 1907 (German).

BLASIUS, H., Grenzschichten in Flüssigkeiten mit kleiner Reibung. Zeitschr. für Math. u. Phys. Vol. 56, p. 1, 1908.

OSEEN, C. W., Über die Stokes'sche Formel und über eine verwandte Aufgabe in der Hydrodynamik. Arkiv för mathematik, astronomi och fysik, Vol. 6, 1910; Vol. 7, 1911.

LAMB, H., On the Uniform Motion of a Sphere through a Viscous Fluid. Phil. Mag. (6) 2, p. 112, 1911 or Hydrodynamics, 5th Edition § 340—343.

PRANDTL, L., Über den Luftwiderstand von Kugeln, Göttinger Nachrichten, math.-phys. Klasse, p. 177, 1914.

v. KÁRMÁN, TH., Über laminare und turbulente Reibung. Zeitschr. f. angew. Math. u. Mech., Vol. I, p. 244, 1921.

POHLHAUSEN, K., Zur näherungsweisen Integration der Differentialgleichung der laminaren Grenzschicht. Zeitschr. f. angew. Math. u. Mech., Vol. I, p. 252, 1921.

SCHMIDT, W., Der Massenaustausch in freier Luft und verwandte Erscheinungen, Hamburg 1925.

TOLLMIEN, W., Berechnung turbulenter Ausbreitungsvorgänge. Zeitschr. f. angew. Math. u. Mech. Vol. 4, p. 468, 1926.

GOLDSTEIN, S., Concerning Some Solutions of the Boundary Layer Equations in Hydrodynamics. Proc. Cambr. Phil. Soc. Vol. 26, 1930.

v. Kármán, Th., Mechanische Ähnlichkeit und Turbulenz. Göttinger Nachrichten, math.-phys. Klasse, p. 58, 1930.

Taylor, G. J., The Transport of Vorticity and Heat through Fluids in Turbulent Motion. Proc. Roy. Soc. A Vol. 135, p. 685, 1932.

2. *Treatises.*

Lamb, H., Hydrodynamics, 5th Edition, Cambridge.

Prandtl, L., The Flow of Liquids and Gases. Chapter V of "The Physics of Solids and Fluids", London.

Tollmien, W., Two Contributions on Boundary Layer and Turbulence in "Handbuch der Experimental-Physik", Vol. IV, Part 1, p. 241 and 291, Leipzig 1931.

Müller, W., Einführung in die Theorie der reibenden Flüssigkeiten, Leipzig 1932.

Tietjens, O., Applied Hydro- and Aeromechanics, based on Lectures of L. Prandtl. Translated by J. P. den Hartog. New York 1934.

PLATES I—V
(Division G)

PLATE VI
(Division H)

Figs. 26–31.
Separate photographs showing flow past a cylindrical body with reverse flow, separation, and formation of vortices (from PRANDTL-TIETJENS, Hydro- und Aeromechanik II).

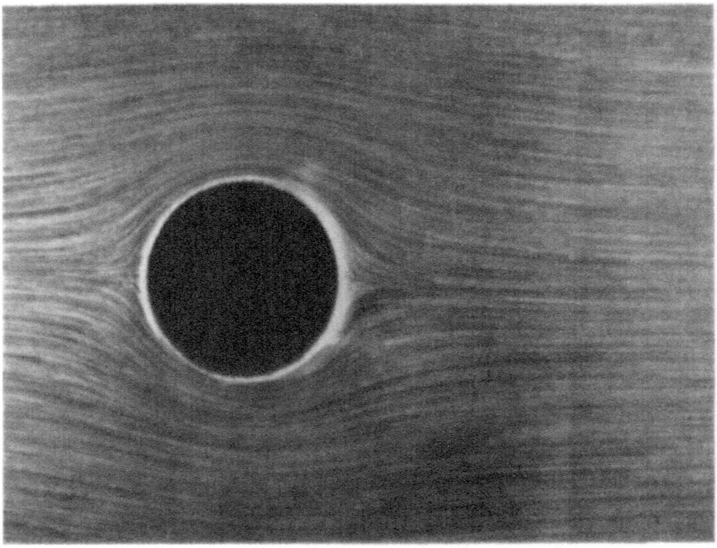

Fig. 26.

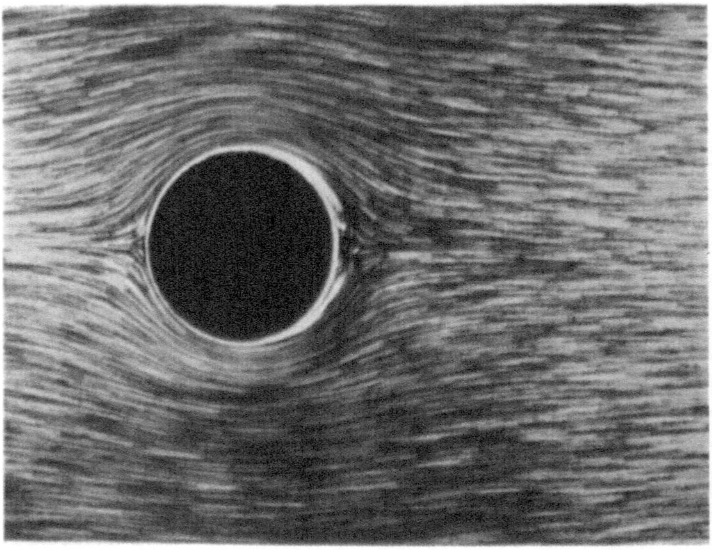

Fig. 27.

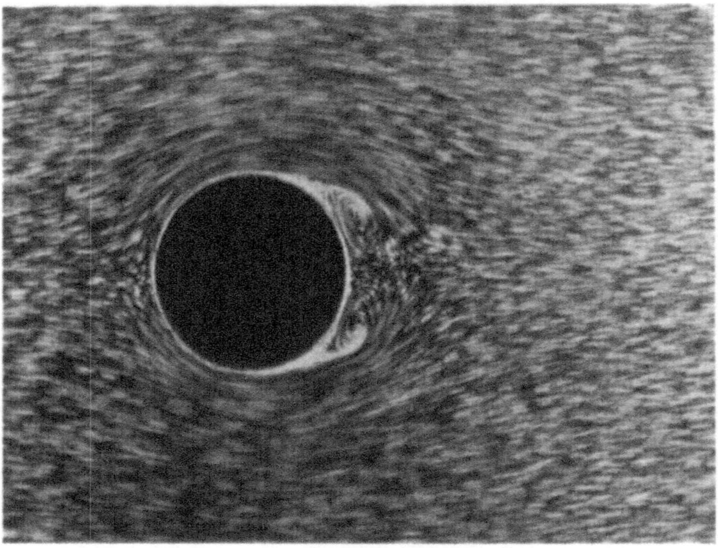

Fig. 28.

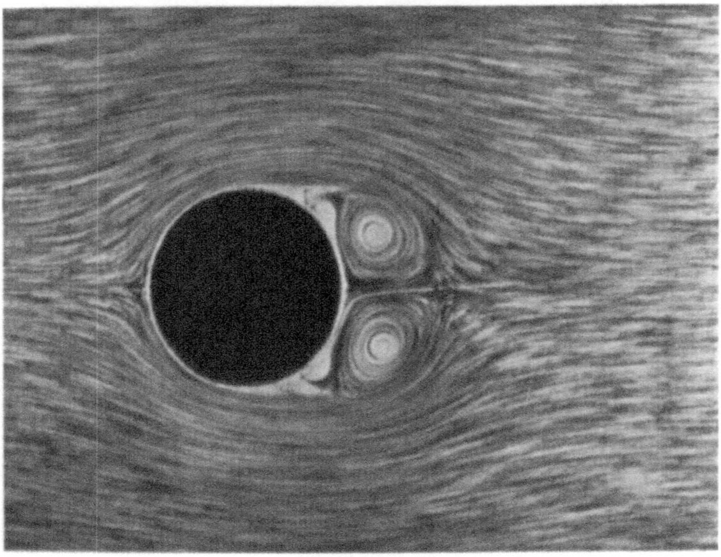

Fig. 29.

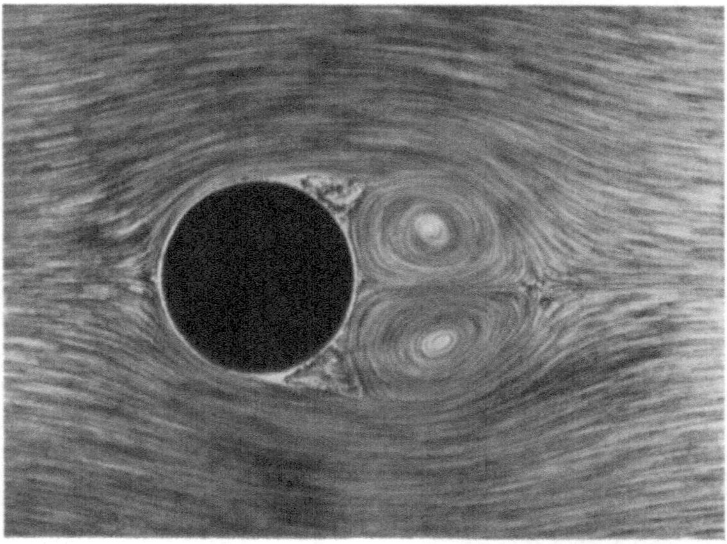

Fig. 30.

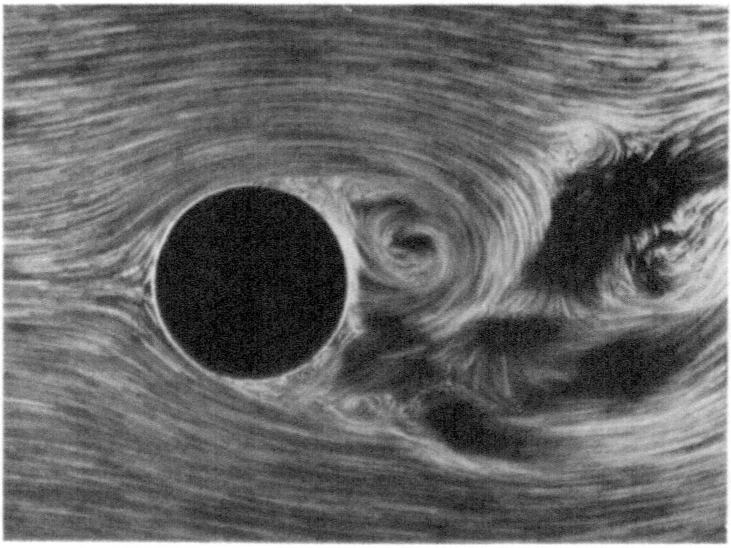

Fig. 31.

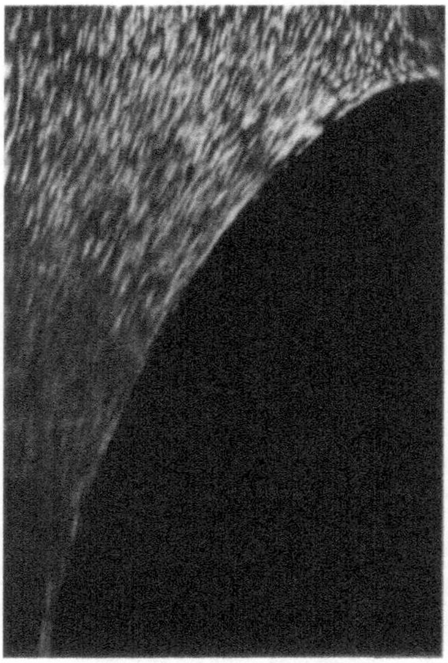

Fig. 32.

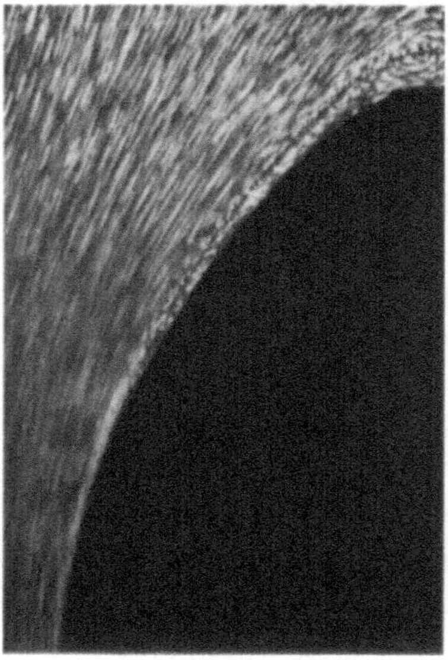

Fig. 33.

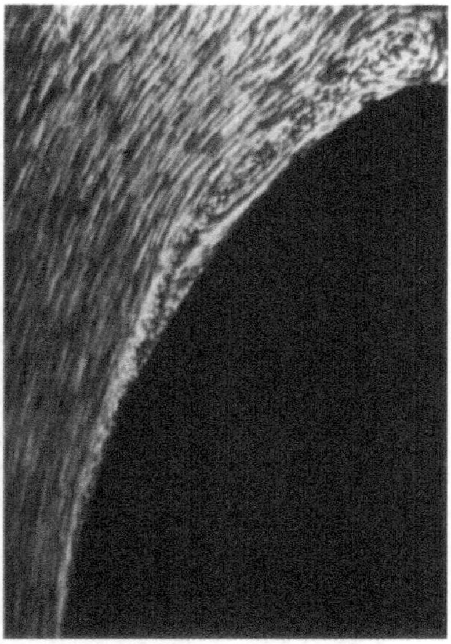

Fig. 34.

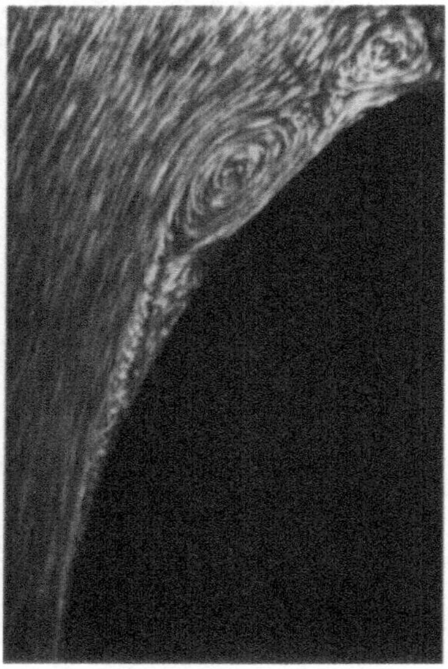

Fig. 35.

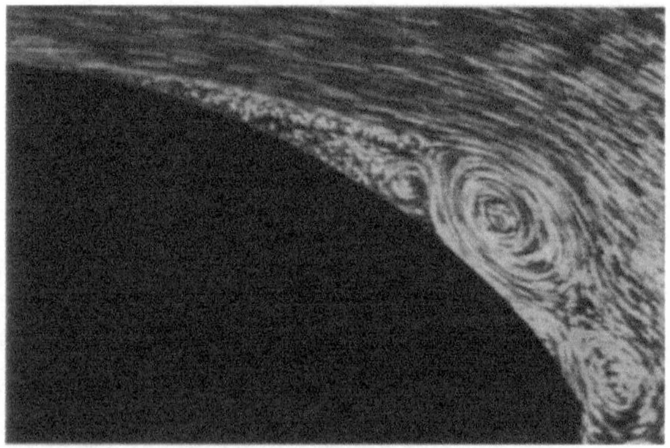

Fig. 36.

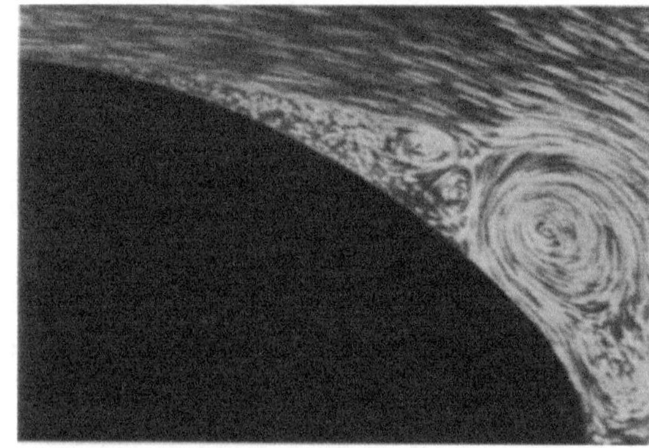

Fig. 37.

Figs. 32–37.
Successive prints from a cinematographic film showing flow past an elongated body with reverse flow, separation, and formation of vortices (from PRANDTL-TIETJENS, Hydro- und Aeromechanik II).

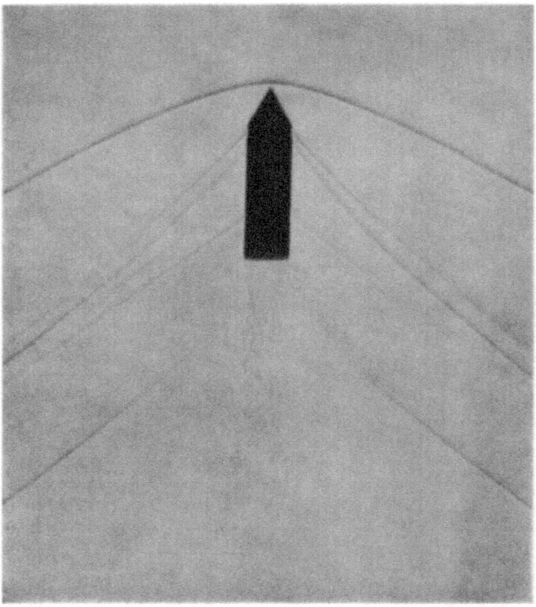

Fig. 1. Photograph of a Bullet in Flight.

Fig. 7. Shock Wave in an Expanding Channel (from PRANDTL's paper in Phys. Zeitschr. 8, 1907).

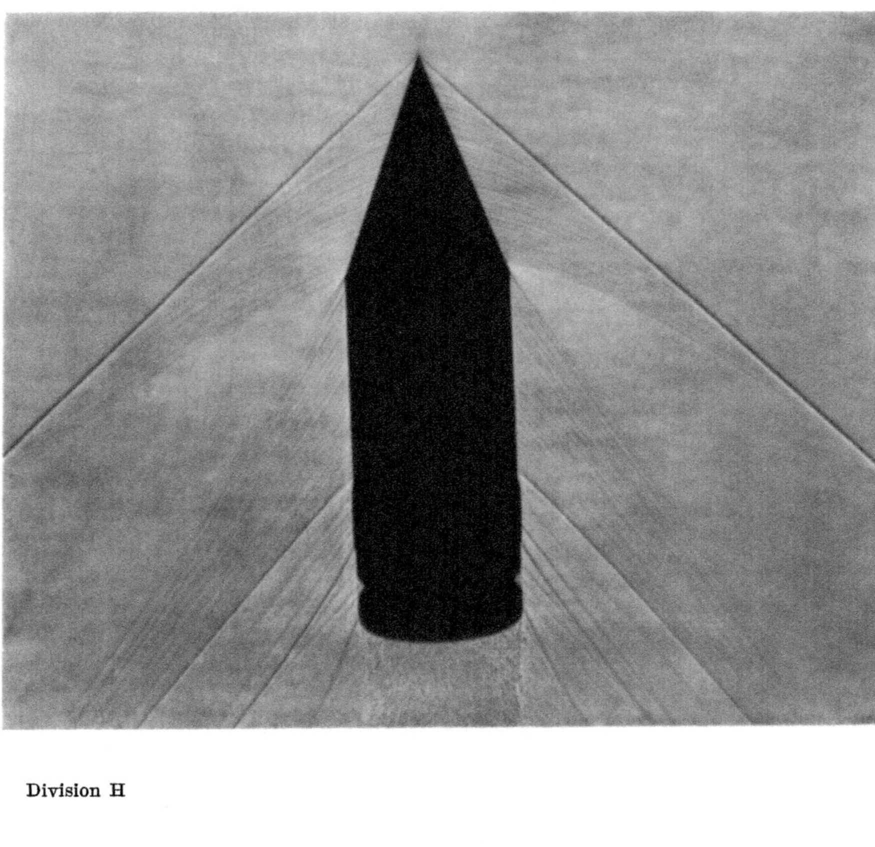

Fig. 18. Photograph of a Conical Headed Projectile in Flight.

DIVISION H
THE MECHANICS OF COMPRESSIBLE FLUIDS

By

G. I. Taylor and **J. W. Maccoll**
Cambridge, England Woolwich, England

EDITOR'S PREFACE

The ideal fluid medium of the mathematician permits of the development, by suitable mathematical methods, of a very considerable body of aerodynamic theory, as in Volume II of this series, Division E. With actual fluids, however, endowed with both viscosity and compressibility, this classical body of theory is very sharply limited, and in general the attack on problems where these characteristics of the fluid become of significance, is found to require supplementary and special methods.

In the present Division, the consequences of compressibility in a fluid medium are discussed and it is shown that by a suitable extension of classical mathematical methods, the conditions within a fluid in relative motion with a solid and undergoing changes of volume, may, for a considerable number of cases, be treated with satisfactory precision. In particular it is shown that the criterion for the significance of compressibility effects is furnished by the ratio between the velocity of flow to the velocity of sound in the medium, and that it is only when this ratio approaches unity that these consequences begin to assume serious importance.

The practical significance to aeronautics of these consequences of compressibility turns upon their influence on the force reactions between the fluid and the moving body at these high speeds, influencing as they must, the lift and drag of supporting surfaces, the effective operation of control surfaces, and the operation of the propeller as regards efficiency and capacity.

Another important practical application of the theoretical development is found in connection with the problem of the design and operation of wind tunnels for airspeeds near and above the velocity of sound. The general advance in speed in all matters relating to airplane design and operation (especially in the matter of propeller tip speeds) renders imperative more definite experimental knowledge regarding these compressibility effects. A discussion of these important problems will be found in Chapter II.

Chapter III summarises the methods for dealing with high velocity flow at speeds less than that of sound.

The Division closes in Chapter IV with some discussion of the flow at supersonic velocities past thin airfoils, wedge and conical forms, around a corner and over curved surfaces.

W. F. Durand.

CHAPTER I

THE PROPAGATION OF PLANE DISTURBANCES

1. Introduction. When an ideal incompressible non-viscous fluid moves in irrotational motion without circulation, the velocity at every point of the fluid depends only on the normal component of the velocity of the boundaries. For this reason an instantaneous or impulsive change in velocity of any bounding surface produces an impulsive change in velocity at all points in the fluid. This instantaneous change in velocity is brought about by an impulsive pressure which may be visualized as a very large pressure acting during a very short time. If this short interval of time is denoted by τ the impulsive pressure $\int_0^\tau p\, d\, t$, is equal to $\varrho\,(\varphi_2 - \varphi_1)$ where φ_1 and φ_2 are the velocity potentials before and after the action of the impulse[1]. With an ideal compressible non-viscous fluid, irrotational motion without circulation is still possible, and indeed all motions started from rest by movement of the boundaries are of this type, for Helmholtz's circulation theorem still applies; but since the density now depends on the pressure, the motion set up by movement of the boundaries will depend not only on their velocities at the instant under consideration, but also on the acceleration with which these velocities have been set up, and in fact on the complete history of their movements. Impulsive pressure cannot exist in a compressible fluid but the boundaries could theoretically be made to move impulsively by giving impulses to the solids of which they are the surfaces. The effect of such an impulsive movement of a boundary could not be felt instantaneously throughout the fluid, for this would necessitate the existence of an infinitely large pressure acting for an infinitely short time, a condition which is incompatible with compressibility, but it would ultimately produce some effect at all points in the fluid. This idea may be expressed by the statement that the effects of an impulsive movement are propagated outward at some finite speed. Thus we may foresee that a study of some simple cases of propagation of disturbances in a compressible fluid will be a useful preliminary to theoretical investigations of the effect of compressibility on fluid motion.

2. Propagation of Small Plane Disturbances. Even in the simplest case when the motion is limited to one dimension and ϱ, the density,

[1] Division B I 8.

is a function of p, the pressure, only, the solution of the problem is not simple unless the motion be assumed small so that the term $u \cdot \partial u/\partial x$ in the equation of motion may be neglected compared with the term $\partial u/\partial t$. Under these circumstances the equation of motion is

$$\frac{\partial u}{\partial t} = -\frac{1}{\varrho} \cdot \frac{\partial p}{\partial x} = -\frac{1}{\varrho} \cdot \frac{dp}{d\varrho} \cdot \frac{\partial \varrho}{\partial x} = -\frac{dp}{d\varrho} \cdot \frac{\partial \log \varrho}{\partial x} \tag{2.1}$$

and the equation of continuity is

$$\frac{1}{\varrho}\frac{\partial \varrho}{\partial t} + \frac{\partial u}{\partial x} + \frac{u}{\varrho}\frac{\partial \varrho}{\partial x} = \frac{\partial \log \varrho}{\partial t} + \frac{\partial u}{\partial x} + u\frac{\partial \log \varrho}{\partial x} = 0 \tag{2.2}$$

and if the variation in ϱ is negligible compared with ϱ, the term $u \frac{\partial}{\partial x}(\log \varrho)$ may be neglected compared with $\partial u/\partial x$. In the circumstances of small motion, $dp/d\varrho$ may be regarded as a constant which we may denote by a^2 so that $\log \varrho$ can be eliminated between (2.1) and (2.2), the resulting equation for u being

$$\frac{\partial^2 u}{\partial t^2} = a^2 \frac{\partial^2 u}{\partial x^2} \tag{2.3}$$

The complete solution of (2.3) is

$$u = f_1(x - at) + f_2(x + at) \tag{2.4}$$

as can be seen at once by substitution for u in (2.3).

The functions f_1 and f_2 are quite arbitrary and represent two disturbances, one, f_1, travelling forward along the axis of x with speed $a = (dp/d\varrho)^{1/2}$, and the other, f_2, travelling backward with the same speed. Fixing our attention on the forward-moving disturbance, it will be seen that a small disturbance represented by $u = f(x - at)$ can be propagated forward without change in form, at a constant speed a. This speed is called the speed of sound. In the case of gases for which the relationship between p and ϱ in adiabatic changes is $p\,\varrho^{-\gamma} =$ constant,

$$\frac{dp}{d\varrho} = \gamma \frac{p}{\varrho} = \gamma R T \tag{2.5}$$

where T is the absolute temperature (*i. e.* 273 + temperature in centigrade degrees) and R is a constant for each gas. For air $\gamma = 1.405$ and at normal atmospheric pressure of 760 mm. of mercury and a temperature of 0^0 C, $p = 76 \times 13.6 \times 981$ dynes per sq. cm. and $\varrho = 0.001293$ so that

$$a = \left(\frac{1.405 \times 76 \times 13.6 \times 981}{0.001293}\right)^{1/2} = 33193 \text{ cm. per sec.} = 1089 \text{ ft. per sec.}$$

For water at 10^0 C, $a = 4370$ ft. per sec. .

To find the relationship between u and ϱ in a progressive wave, substitute $f_1(x - at)$ for u in (2.1) and integrate with respect to x. Thus

$$\log \frac{\varrho}{\varrho_0} = \frac{1}{a} f_1(x - at) = \frac{u}{a} \tag{2.6}$$

where ϱ_0 is the density at points where $u = 0$. Since the motion is assumed small, ϱ is nearly equal to ϱ_0 so that approximately $\log(\varrho/\varrho_0) =$

$(\varrho - \varrho_0)/\varrho_0$. The ratio $(\varrho - \varrho_0)/\varrho_0$ is sometimes called the condensation and is denoted by the symbol s so that (2.6) becomes

$$s = \frac{u}{a} \tag{2.7}$$

In a progressive wave therefore, the condensation at any point is equal to the velocity of the fluid at that point, divided by the velocity of sound. Provided therefore the velocities of the fluid are everywhere small compared with the velocity of sound, the changes in density are everywhere small compared with the density. Though this has been proved here only in the case of plane disturbances, it is equally true in the more general case of three-dimensional motion. Taking the case of the steady flow past an obstacle for instance, if the variation in density is small the condensation s can be calculated from Bernoulli's equation. If p, q are the pressure and velocity at any point and p_1, U those in the stream at points far from the obstacle,

$$p - p_1 = \frac{1}{2}\varrho\,(U^2 - q^2) \tag{2.8}$$

The condensations s due to the pressure change $p - p_1$ is $s = (p - p_1)\left(\varrho \frac{d\,p}{d\,\varrho}\right)^{-1}$ or $s = (p - p_1)/\varrho a^2$. Hence from (2.8), $s = (U^2 - q^2)/2a^2$. It will be seen therefore that s is small when U/a and q/a are small. Steady flow at speeds which are small compared with the speed of sound is therefore nearly identical with that of an incompressible fluid which suffers no change in density owing to changes in pressure.

3. Dynamic Similarity. The significance of the ratio q/a in problems of steady motion may be extended to cases where q/a is not small. In such cases the motion differs from that of an incompressible fluid but when any one flow system of this type has been determined, a complete set of dynamically similar systems is also known. From the equations of motion of a non-viscous compressible fluid, it can be seen at once that if the dynamical equations are satisfied for one system they are also satisfied for any other geometrically similar system for which $q^2\varrho/p$ has the same value at similar points. If the two systems are denoted by suffixes 1 and 2, and $q_2 = m q_1$, it will be seen from the equation of continuity that $\varrho_2/\varrho_1 = \text{constant} = n$. Hence the dynamical equations of the second system will be satisfied if

$$\frac{p_2}{p_1} = m^2\,n \tag{3.1}$$

In the case of two gases $p_1 = c_1\,\varrho_1^{\gamma_1}$, $p_2 = c_2\,\varrho_2^{\gamma_2}$ where c_1 and c_2 are constants. Hence (3.1) cannot be satisfied except where $\gamma_1 = \gamma_2$. In this case, however, the two flow patterns are similar provided

$$\frac{q_1^2\,\varrho_1}{p_1} = \frac{q_2^2\,\varrho_2}{p_2} \quad \text{or} \quad \frac{q_1}{q_2} = \frac{a_1}{a_2}$$

where $$a_1 = \left(\frac{\gamma p_1}{\varrho_1}\right)^{1/2} \quad \text{and} \quad a_2 = \left(\frac{\gamma p_2}{\varrho_2}\right)^{1/2}$$

are the "local" velocities of sound at corresponding points of the two systems.

The practical importance of this similarity theorem will appear later when the results of experiments with high speed wind tunnels are discussed; it seems worth while, however, at this stage to mention that when air is released from a high pressure container into a vacuum, the maximum possible speed which it can attain is only 2.2 times the velocity of sound in the container; the temperature of the air in the discharging stream is, however, lower than that of the container owing to adiabatic expansion, so that the velocity of sound is reduced. Hence it comes about that it is possible to obtain for experimental purposes a steady stream of air flowing many times as fast as the velocity of sound in the air at the point where observations are made. The late Sir Thomas Stanton, for instance, used a jet in which he was able to obtain a speed of 3.3 times the velocity of sound. Though the actual speed of this jet was only 1920 ft. per sec., the flow around models tested in it was dynamically similar to that around bodies of similar shape moving at 3420 ft. per sec. through air at 15° C.

4. Propagation of Large Disturbances and Shock Waves. The connection between the velocity of sound and the effect of compressibility on air flow has so far appeared only as a mathematical conception. The flow patterns around a body moving steadily in still air at speeds which are small compared with that of sound are similar to those around a body moving through an incompressible fluid, the effect of compressibility being merely to distort the stream-lines slightly. One might imagine the steady state of motion to be acquired by giving the body a series of small separate impulses. Each of these impulses would produce a small disturbance in the surrounding air which would spread out from it with the velocity of sound and so long as the speed of the body was less then that of sound, the flow pattern would extend to an infinite distance from the body. On the other hand when the body attains a speed greater than that of sound, the flow pattern cannot extend to great distances in front of it, because by the time any disturbance has spread to a large distance from the body it must necessarily have become a small one and must therefore be propagated with the speed of sound, so that the body itself would tend to overtake the disturbance produced by it. Close in front of the body there must be some disturbance, for the stream-lines of the motion relative to the body must spread out so as to enclose it. Hence we are led to the conclusion that a disturbance due to the body must extend to a finite distance in front of it and must be propagated into still air with a speed greater than that of sound. Disturbances of this kind are visible in the well-known photographs of bullets in flight, a specimen of which is given in Fig. 1 (on Plate VI at the beginning of this Division). All photographs of blunt-nosed bodies travelling at speeds higher than that of sound have the same general features. Immediately in front of the body a black line which at some point is perpendicular to the direction of motion shows that a disturbance confined to a thin sheet is propagated into still air at a speed greater than that of sound. At points some distance from the body, this line, or in some cases several such lines, become straight and inclined to the direction of flight at an angle which we may denote by M. If U is the velocity of the projectile and a_1 that of sound in still air, it is found that $\sin M = a_1/U$ so that, as might be expected in the case of a narrow disturbance of small intensity confined to a conical sheet, the speed of propagation perpendicular to the sheet is that of a small plane disturbance, namely that of sound.

The fact that the thickness of the lines in bullet photographs indicating air disturbance is always small compared with their radii of curvature, indicates that the dynamics of such disturbances can be studied mathematically as a problem in plane or one-dimensional air flow, and two outstanding questions at once suggest themselves for investigation. First, why are the disturbances confined to thin sheets, and second, how great must be the condensation or rarefaction in such a disturbance in order that it may be propagated into still air at any given speed greater than that of sound?

The equations of motion of a gas in one-dimension have been solved completely by Riemann. His analysis involves necessarily disturbances of finite magnitude propagated in both directions; but, in the case when the disturbance is propagated in one direction only, the work can be much simplified. The equation of motion of a fluid in one-dimension is

$$\frac{\partial u}{\partial t} + u\frac{\partial u}{\partial x} + \frac{1}{\varrho}\frac{dp}{d\varrho}\frac{\partial \varrho}{\partial x} = 0 \tag{4.1}$$

while the equation of continuity is

$$\frac{\partial \varrho}{\partial t} + u\frac{\partial \varrho}{\partial x} + \varrho\frac{\partial u}{\partial x} = 0 \tag{4.2}$$

p is a function of ϱ, so that $dp/d\varrho$ is also a function of ϱ. In the study of the propagation of small disturbances given in **2** it was found that in a forward moving wave the condensation is $u/(dp/d\varrho)^{1/2}$ while in a backward moving wave it is $-u/(dp/d\varrho)^{1/2}$, so that when all disturbances move in one direction only, the condensation is simply proportional to u. Using this as an indication, we may search for modes of motion for which u is a function of ϱ.

In that case (4.1) and (4.2) become

$$\frac{\partial u}{\partial t} + u\frac{\partial u}{\partial x} + \frac{1}{\varrho}\frac{dp}{d\varrho}\frac{d\varrho}{du}\frac{\partial u}{\partial x} = 0 \tag{4.3}$$

$$\frac{d\varrho}{du}\frac{\partial u}{\partial t} + u\frac{d\varrho}{du}\frac{\partial u}{\partial x} + \varrho\frac{\partial u}{\partial x} = 0 \tag{4.4}$$

(4.3) and (4.4) are consistent if $\frac{1}{\varrho}\frac{dp}{d\varrho}\frac{d\varrho}{du} = \frac{\varrho}{d\varrho/du}$, or

$$\frac{du}{d\varrho} = \pm\frac{1}{\varrho}\left(\frac{dp}{d\varrho}\right)^{1/2} \tag{4.5}$$

so that

$$u = \pm\int_{\varrho_0}^{\varrho}\left(\frac{dp}{d\varrho}\right)^{1/2}\frac{d\varrho}{\varrho} \tag{4.6}$$

where ϱ_0 is the value of ϱ at points where $u = 0$. Taking the positive sign and substituting $\varrho\,(dp/d\varrho)^{-1/2}$ for $d\varrho/du$, (4.3) becomes

$$\frac{\partial u}{\partial t} + \frac{\partial u}{\partial x}\left[u + \left(\frac{dp}{d\varrho}\right)^{1/2}\right] = 0 \tag{4.7}$$

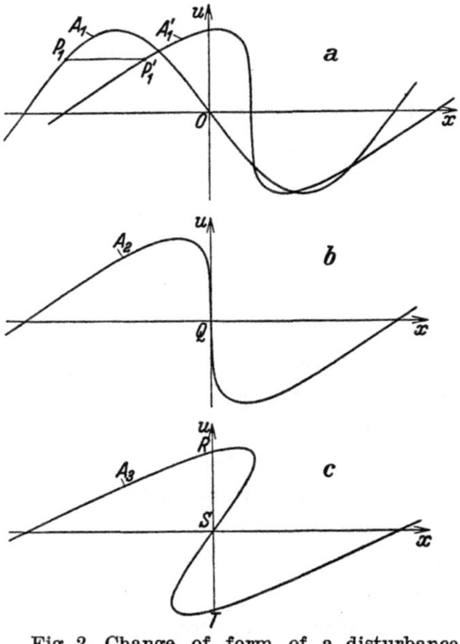

Fig. 2. Change of form of a disturbance of finite amplitude.

The meaning of (4.7) can be seen by considering the speed at which a point $\overline{x}$ must move in order that u, and consequently ϱ, may have a constant value. The rate of change in u at $\overline{x}$ is $\partial u/\partial t + (\partial u/\partial x)\,(d\overline{x}/dt)$ and this is zero if $d\overline{x}/dt = -\frac{\partial u/\partial t}{\partial u/\partial x}$. Hence from (4.7), $d\overline{x}/dt = u + (dp/d\varrho)^{1/2} = u + a$, where a is the local velocity of sound at any point. It has been shown therefore that disturbances for which $u = +\int_{\varrho_0}^{\varrho}\left(\frac{dp}{d\varrho}\right)^{1/2}\frac{d\varrho}{\varrho}$, are propagated forward in such a way that u and ϱ are constant at points which move with speed $u + a$. It is clear that in general this type of disturbance suffers a progressive change in form as it proceeds. A simple way in which this change can be appreciated is to make a diagram showing in the form of a curve some arbitrary distribution of velocity. Such a diagram is shown in Fig. 2, where the ordinates represent u and the abscissae x. The arbitrary initial form of the curve is shown as A_1. To find the form after time τ, take any point P_1 on A_1 and draw parallel to Ox the line $P_1\,P_1'$ to represent $(u_1 + a_1)\tau$, where a_1 represents the value of $(dp/d\varrho)^{1/2}$ when $u = u_1$. In this way the curve A_1' is formed. Following this construction it will be seen that it is only when $a + u$ or

$$\left(\frac{dp}{d\varrho}\right)^{1/2} + \int_{\varrho_0}^{\varrho}\left(\frac{dp}{d\varrho}\right)^{1/2}\frac{d\varrho}{\varrho} = \text{constant} \tag{4.8}$$

that the disturbance can progress without change of form. The condition (4.8) is satisfied if $\varrho^2 (dp/d\varrho)$ is constant, *i. e.* if $p = A + B/\varrho$. There is no known fluid for which the pressure-density relationship is of this form. In the case of gases, $a + u$ is found by integrating (4.6), to be $a_0 + \left(\frac{\gamma + 1}{2}\right) u$ where a_0 is the velocity of sound at points where $u = 0$.

If the actual position of the wave in space is not required, the curve A'_1 can therefore be derived from the curve A_1 by simply translating all points on A_1 to the right by an amount $\left(\frac{\gamma + 1}{2}\right) u\,\tau$, or in other words subjecting the whole plane on which the curve is drawn to a shearing strain parallel to the axis of x of amount $\left(\frac{\gamma + 1}{2}\right) \tau$. Successive positions are shown as A_1, A_2 and A_3.

It will readily be seen that such a process will make steeper the part of the curve where the density and velocity are decreasing in the direction of propagation, so that regions of condensation get narrower while conversely regions of expansion get wider. After a time the steepest part of the curve will become vertical as shown at the point Q in the curve A_2. If the greatest initial value of $-du/dx$ is represented by $-(du/dx)_M$, the time necessary for the curve to become vertical at one point is $\frac{2}{(\gamma + 1)(du/dx)_M}$. At later times the construction described above can still be carried out as is shown in curve A_3, but the curve so produced has no physical meaning, for within a certain region the density at every point would be 3-valued as is shown by the ordinate RST which cuts the curve A_3 in three points R, S, T.

It appears therefore, that all plane disturbances[1] of finite amplitude exist only for a finite time, at the expiration of which a discontinuity is formed where there is an infinitely rapid change in velocity and density.

The preceding analysis affords an explanation of the sheets observed in bullet photographs. They are regions of extremely rapid change in density formed in the manner illustrated in Fig. 2. They are known as waves of percussion or shock waves.

5. Propagation of Discontinuities. The analysis which explains the formation of these discontinuities does not give any information about their mode of propagation after they have been formed. For this purpose we must start again, assuming the discontinuity to exist.

Suppose that a plane discontinuity can exist such that at all points on one side the velocity, pressure and density, are u_1, p_1, ϱ_1, while on the other they are u_2, p_2, ϱ_2. If c is the velocity of the discontinuity the equation of continuity is

$$\varrho_1 (u_1 - c) = \varrho_2 (u_2 - c) = m \tag{5.1}$$

[1] Except those in which the velocity is everywhere opposite to the direction of propagation.

where m is the mass which crosses unit area of the plane of discontinuity per second. The equation of conservation of momentum is

$$p_2 - p_1 = m(u_1 - u_2) \tag{5.2}$$

These equations are not by themselves capable of determining c when, say, p_1, u_1, ϱ_1 and u_2 are given, for there are then three unknowns, namely p_2, c and ϱ_2. One more condition is necessary. This is supplied by the condition of conservation of energy. The work done per unit area per second by the pressure on one side of the gas is $p_1 u_1$ so that the rate at which work is done on a column of unit area of cross-section by pressure is $p_1 u_1 - p_2 u_2$. The rate of increase of kinetic energy per second is $(1/2)\ m\ (u_2^2 - u_1^2)$. The change in intrinsic energy of unit mass as it passes through the discontinuity is $\frac{1}{\gamma-1}\left(\frac{p_2}{\varrho_2} - \frac{p_1}{\varrho_1}\right)$ so that the equation for conservation of energy is

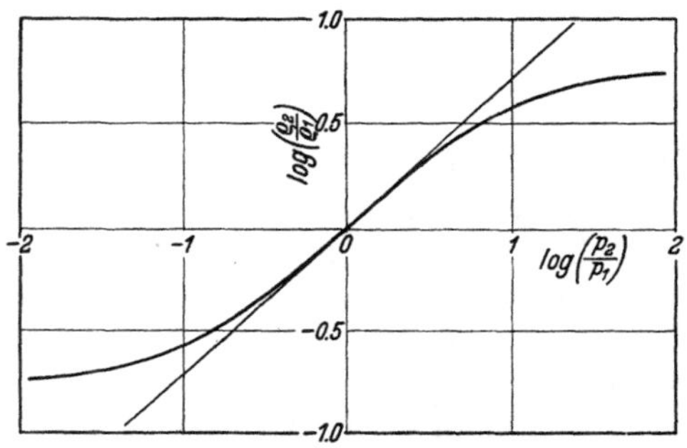

Fig. 3. Rankine-Hugoniot relationship between density and pressure ratios.

$$p_1 u_1 - p_2 u_2 - \frac{1}{2} m (u_2^2 - u_1^2) = \frac{m}{\gamma - 1}\left(\frac{p_2}{\varrho_2} - \frac{p_1}{\varrho_1}\right) \tag{5.3}$$

In view of (5.2), $(1/2)\ m\ (u_2^2 - u_1^2) = -(1/2)(p_2 - p_1)(u_2 + u_1)$ so that (5.3) may be written

$$\frac{1}{2}(p_1 + p_2)\left(\frac{u_1 - u_2}{m}\right) = \frac{1}{\gamma - 1}\left(\frac{p_2}{\varrho_2} - \frac{p_1}{\varrho_1}\right)$$

or

$$\frac{1}{2}(p_1 + p_2)\left(\frac{1}{\varrho_1} - \frac{1}{\varrho_2}\right) = \frac{1}{\gamma - 1}\left(\frac{p_2}{\varrho_2} - \frac{p_1}{\varrho_1}\right) \tag{5.4}$$

This relationship, due to Rankine[1] and rediscovered by Hugoniot[2], between the pressures and densities on the two sides of a discontinuity, differs from the adiabatic relationship. The adiabatic relationship is that for which there is no change in entropy so that according to the second law of thermodynamics the discontinuity represented by (5.1), (5.2) and (5.4), which must give rise to a change in entropy, cannot be reversible, though there is nothing in these equations to determine in which, if either, of the two possible directions the gas can flow past the discontinuity.

To settle this point it is necessary to find whether the gas of density ϱ_1 or that of density ϱ_2 has the greater entropy. For this purpose the diagram shown in Fig. 3 may be constructed in which the ordinates represent ϱ_2/ϱ_1 on a logarithmic scale while the abscissae, also on a

[1] Phil. Trans. Roy. Soc., Vol. 160, p. 277, 1870.

[2] Journal de l'Ecole Polytechnique, Paris, Cahiers 57—59, 1887—1889.

logarithmic scale, represent p_2/p_1. In these coordinates the Rankine-Hugoniot relationship (5.4) is

$$log \frac{\varrho_2}{\varrho_1} = log \frac{\frac{p_2}{p_1}\left(\frac{\gamma+1}{\gamma-1}\right)+1}{\frac{p_2}{p_1}+\left(\frac{\gamma+1}{\gamma-1}\right)}$$

This is shown in Fig. 3 for the case of air ($\gamma = 1.405$) as a curve, while the adiabatic relationship is shown as the straight line

$$log \frac{\varrho_2}{\varrho_1} = \frac{1}{1.405} \, log \frac{p_2}{p_1}$$

The curve represented in Fig. 3 touches the straight line at the origin but falls below it for values of $\varrho_2/\varrho_1 > 1$, while when $\varrho_2/\varrho_1 < 1$ the curve lies above the straight line. Since the adiabatic line is one of constant entropy and at a constant pressure the entropy of a gas increases as its density decreases, it will be seen that all points lying below the adiabatic line in Fig. 3 represent states of the gas for which entropy is greater than the initial state p_1, ϱ_1, while all points above it represent states of lower entropy.

In order that the second law of thermodynamics may be satisfied, the Rankine-Hugoniot relationship can only apply when the direction of flow through the discontinuous sheet involves an increase in entropy, *i. e.* the flow relative to the sheet must be from the smaller to the greater density. This conclusion is in agreement with what might have been expected from consideration of the manner in which the discontinuities might be formed. It was shown in 4 that a continuous disturbance may become discontinuous in a region where the density is decreasing in the direction of propagation, *i. e.* in a region where the air flow relative to the disturbance is from smaller to greater density. On the other hand, regions where the density is increasing in the direction of propagation show no tendency to form discontinuities.

6. Conditions Within a Shock Wave. In our theoretical discussion we have so far considered the shock wave as a sheet where there is a discontinuity of velocity, that is, a place where the rate of change in velocity and density become infinite. In an actual gas there can be no true discontinuities. When, owing to the change in type which occurs in a disturbance of finite magnitude, a region is formed where the rate of change in velocity and density in the direction of propagation is very large, the effect of viscosity and conductivity for heat can no longer be neglected. It is of interest, therefore, to inquire whether a disturbance of permanent type can be propagated in a viscous and conducting gas which, in the limiting case of very small viscosity and conductivity, can approximate to the ideal discontinuity of a non-viscous and non-conducting gas. First it may be pointed out that if such a disturbance can exist, equations (5.1), (5.2) and (5.4), which determine the conditions on the two sides of a discontinuity, apply equally well to any case where there is a transition layer of permanent type between two infinitely extended regimes defined by u_1, ϱ_1, p_1 and u_2, ϱ_2, p_2. The problem is to determine the distribution of u, ϱ and p in the transition layer so that it may remain permanent in type. The equations may be simplified by reducing the whole system to rest by imposing a velocity equal but opposite to the speed of propagation. In this case the equation of continuity is

$$\varrho u = \text{constant} = m \tag{6.1}$$

and the equation of steady motion of a viscous gas in one dimension is[1]

$$\varrho u \frac{d u}{d x} = -\frac{d p}{d x} + \frac{4}{3} \mu \frac{d^2 u}{d x^2} \tag{6.2}$$

where μ is the coefficient of viscosity.

Since ϱu is constant, (6.2) may be integrated, giving

$$p + m u - \frac{4}{3} \mu \frac{d u}{d x} = \text{constant} = p_1 + m u_1 = p_2 + m u_2 = m B \tag{6.3}$$

where B is a constant.

The intrinsic energy of the gas contained between a cylinder of unit area of cross-section and length δx is $\frac{p}{(\gamma - 1)\varrho} (\varrho\, \delta x)$. This increases at the rate $\frac{u \varrho\, \delta x}{\gamma - 1} \frac{d}{d x}\left(\frac{p}{\varrho}\right)$. The increase is due to the action of the stress[2] $p_{xx} = -p + \frac{4}{3} \mu \frac{d u}{d x}$ over the ends of the cylinder, and conductivity which causes heat energy to enter at the rate $\varkappa J \frac{d^2 \theta}{d x^2}\, \delta x$ ($\varkappa$ = coefficient of conductivity, J = mechanical equivalent of heat and θ = absolute temperature). The rate of increase in kinetic energy is $\varrho\, \delta x\, u \frac{d}{d x}\left(\frac{1}{2} u^2\right)$. The energy equation is therefore

$$\frac{d}{d x}\left(-p u + \frac{4}{3} \mu u \frac{d u}{d x}\right) + \varkappa J \frac{d^2 \theta}{d x^2} = \frac{m}{\gamma - 1} \frac{d}{d x}\left(\frac{p}{\varrho}\right) + \frac{m}{2} \frac{d}{d x} (u^2) \tag{6.4}$$

which may be integrated to

$$\left.\begin{aligned} p u - \frac{4}{3} \mu u \frac{\partial u}{\partial x} + \frac{1}{2} m u^2 + \frac{m}{\gamma - 1}\left(\frac{p}{\varrho}\right) - \varkappa J \frac{d \theta}{d x} &= \\ = p_1 u_1 + \frac{1}{2} m u_1^2 + \frac{m p_1}{(\gamma - 1)\varrho_1} &= \\ = p_2 u_2 + \frac{1}{2} m u_2^2 + \frac{m p_2}{(\gamma - 1)\varrho_2} &= m A \end{aligned}\right\} \tag{6.5}$$

where A is a constant.

The gas law gives the relationship $p/\varrho = R\theta$ and remembering that $p/\varrho = p u/m$ it will be seen that (6.5) is an equation between $p u$, u and x. From (6.3) we then have

[1] LAMB, H., Hydrodynamics, Sixth Edition, Chapter XI, § 328.
[2] LAMB, H., Hydrodynamics, Sixth Edition, Chapter XI, § 326.

$$p u = - m u^2 + \frac{4}{3} \mu u \frac{d u}{d x} + (p_1 + m u_1) u$$

and by substitution of this value, the resulting equation is of the form

$$\left.\begin{aligned} y \left[\left(\frac{2 \varkappa J}{R} + \frac{4 \mu}{3 (\gamma - 1)} \right) u - \frac{\varkappa J B}{R} \right] - \frac{4}{3} \frac{\mu \varkappa J}{m R} y \frac{d y}{d u} = \\ = \frac{\gamma + 1}{2 (\gamma - 1)} m u^3 - \frac{m \gamma}{\gamma - 1} B u^2 + m A u \end{aligned}\right\} \quad (6.6)$$

where $y = u\,(du/dx)$. This equation does not appear to be easily solved but it will be seen that if u_1 and u_2 are the two roots of

$$\frac{\gamma + 1}{2 (\gamma - 1)} u^2 - \frac{\gamma}{\gamma - 1} B u + A = 0 \quad (6.7)$$

then $y/u = du/dx$ must vanish when either $u = u_1$ or $u = u_2$.

The values u_1 and u_2 are evidently the two velocities determined by Rankine's equations, for neither $\varkappa$ nor μ occur in (6.7). The fact that du/dx vanishes when $u = u_1$ or $u = u_2$ indicates that the solution of (6.6) must represent a layer of transition between two regions where the velocities u_1 and u_2 do not vary with x. Taking u_1 as the larger root, the solution of (6.6) can be obtained, if either $\varkappa = 0$ or $\mu = 0$, in the form $x = C_1 \log (u_1 - u) + C_2 \log (u - u_2)$.

It appears therefore that under the action of viscosity or conductivity, or conductivity and viscosity combined, a disturbance of permanent type can be propagated, and from the form of (6.6) it is obvious that reducing both $\varkappa$ and μ in any given ratio merely reduces the dimensions of the system in the direction of x in the same ratio, *i. e.* it reduces the thickness of the layer of transition in the same ratio. The actual thickness of this layer of transition can be calculated if it is assumed that $\varkappa = 0$ or $\mu = 0$, or if neither $\varkappa$ nor μ is zero but u_1 nearly equal to u_2.

The last case is the most interesting because for air μ and $\varkappa$ are almost equally effective in controlling the transition layer. If u_1 is nearly equal to u_2, both are nearly equal to the velocity of sound, *i. e.* to $(\gamma p_1/\varrho_1)^{1/2}$ so that

$$m B = p_1 + m u_1 = m u_1 \left(\frac{1}{\gamma} + 1 \right)$$

The order of magnitude of the right hand side of (6.6) is $(u_2 - u_1)^2 \varrho_1 u_1^2$, so that y is of order $(u_2 - u_1)^2 \frac{\varrho_1 u_1^2}{\mu u_1}$, *i. e.* $(u_2 - u_1)^2 \frac{\varrho_1 u_1}{\mu}$. The expression $\frac{\mu \varkappa J}{m R} y \frac{d y}{d u}$ is therefore of order $\frac{y^2}{(u_2 - u_1)} \frac{\mu^2}{m}$, *i. e.* of order $\frac{(u_2 - u_1)^4 \varrho_1^2 u_1^2}{(u_2 - u_1) \varrho_1 u_1}$ or $\varrho_1 (u_2 - u_1)^3 u_1$ and this is small compared with $(u_2 - u_1)^2 \varrho_1 u_1^2$, so that the term in $y\,(dy/du)$ may be neglected and equation (6.6) then reduces to

$$\frac{d u}{d x} \left[\frac{\varkappa J}{R} \left(1 - \frac{1}{\gamma} \right) + \frac{4 \mu}{3 (\gamma - 1)} \right] u_1 = - \frac{(\gamma + 1) m}{2 (\gamma - 1)} (u_1 - u)(u - u_2) \quad (6.8)$$

The solution of (6.8) is

$$x = \frac{D}{u_1 - u_2} \log \frac{u_1 - u}{u - u_2} \tag{6.9}$$

where
$$D = \left[\frac{\varkappa J}{R}\left(1 - \frac{1}{\gamma}\right) + \frac{4}{3}\frac{\mu}{\gamma - 1}\right]\frac{2(\gamma - 1)}{\varrho_1(\gamma + 1)} \tag{6.10}$$

The regime in the layer of transition is shown in Fig. 4 where the curve represented by (6.9) is shown. The thickness of the layer of transition is, strictly speaking, infinite, but the greater part of it is confined between the points A and B, A being where $u = u_1 - (1/10)(u_1 - u_2)$ and B where $u = u_2 + (1/10)(u_1 - u_2)$. The thickness of this layer between A and B where 8/10 of the change in velocity from u_1 to u_2 occurs is

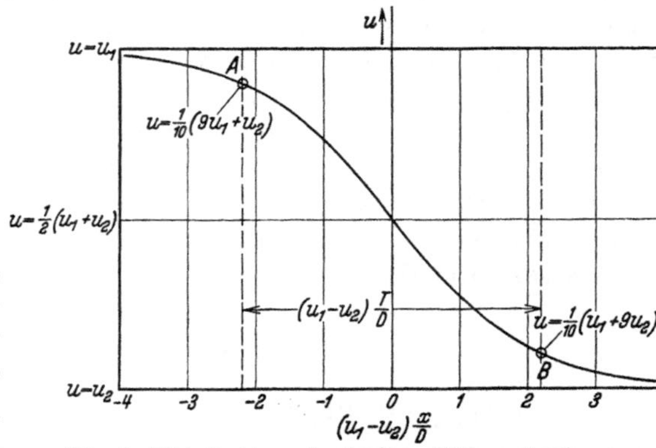

Fig. 4. Distribution of velocity within a shock wave.

$$T = \frac{2D}{u_1 - u_2} \log_e 9 = 4.4 \frac{D}{u_1 - u_2}$$

and substituting[1]

$$\frac{\varkappa J}{R} = \frac{1.6\,\mu}{(\gamma - 1)}, \qquad \gamma = 1.40, \qquad \varrho = 1.3 \times 10^{-3}, \qquad \mu = 1.9 \times 10^{-4},$$

which are the values of these constants for air, $D = 0.22\ \text{cm}^2.\ \text{sec.}^{-1}$

so that
$$T = \frac{1}{u_1 - u_2}\ \text{cm.} \tag{6.11}$$

In the case of a shock wave in which the change in velocity is only one tenth the speed of sound, *i. e.* 30 meters per second, the thickness of the transition layer is therefore only 1/3000 *i. e.* 1/300 of a millimeter. In all cases where a projectile exceeds the speed of sound by an appreciable amount, the shock wave may be expected to be less than 1/300 of a millimeter in thickness, so that the sharpness of the shadows of these waves in photographs of bullets in flight is very understandable.

It has now been shown that viscosity and conductivity are capable of explaining the maintenance of a shock wave of permanent type. It is worth noticing that though the whole of the conversion of mechanical energy into heat which occurs at the wave front is due to viscosity and conductivity, the actual magnitudes of μ and $\varkappa$ do not affect the total change in entropy as the gas passes through the wave. This may be compared with the case of dissipation of mechanical energy in the air

[1] The first relationship is derived from the kinetic theory of gases.

behind a flat plate placed perpendicular to a stream of fluid. The resistance of such a plate is practically independent of the viscosity of the fluid, but the whole of the energy is ultimately dissipated through the medium of viscosity. In the case of a shock wave, and probably also in the case of the flat plate, a decrease in viscosity merely decreases the linear dimensions of the region in which the changes in velocity take place.

It is worth noticing also that, as might be expected from the thermodynamic considerations advanced in connection with Rankine's solution of the shock wave, all waves of permanent type are waves of condensation. Since D is positive and $u_1 > u_2$, the value of x given by (6.9) necessarily increases as u decreases from u_1 to u_2. This is shown in Fig. 4. Hence the speed, relative to the wave, of the air approaching it is greater than that of the air which has passed through. In virtue of the continuity equation $\varrho_1 u_1 = \varrho_2 u_2$; this implies that the air increases in density as it passes through the wave.

CHAPTER II

STEADY FLOW THROUGH CHANNELS

1. Introduction. In Chapter I it has been shown that no state of steady flow or disturbance of permanent type can be set up in a gas when the motion is limited to one dimension except a shock wave. It is shown further that in the case of air (and the same is true for all gases) the shock wave is extremely thin so that calculations based on one-dimensional flow are still applicable for determining the changes in velocity and density on passing through it, even when the rest of the flow system is not limited to one dimension, provided that only the velocity component normal to the wave is considered. When the flow is not limited to one dimension, steady flow may be possible without the occurrence of shock waves, but no criterion has yet been put forward to enable us to distinguish cases in which they occur from those where they do not, except the general one that a shock wave cannot exist when the velocity is everywhere below that of sound. In many cases of steady flow through channels and jets, shock waves do not occur even when the speed of sound is exceeded. These will now be discussed and relevant experimental results described.

2. Bernoulli's Equation. When a gas is in steady motion the changes in density may, in general, be assumed to take place adiabatically, so that, if (p, ϱ), (p_0, ϱ_0) be the pressure and density at two points on the same stream-line,

$$p\,\varrho^{-\gamma} = p_0\,\varrho_0^{-\gamma} \tag{2.1}$$

where γ, the ratio of the specific heats, has the value of 1.405 for air. If q is the velocity at any point, the acceleration of a particle in the direction of motion is $q\,(dq/ds)$ where dq/ds represents the space rate of change in q in the direction of motion. Hence, neglecting gravity, the equation of motion in this direction is

$$q\frac{dq}{ds} = -\frac{1}{\varrho}\frac{dp}{ds} \tag{2.2}$$

The integral of this is

$$\frac{1}{2}q^2 = -\int\frac{dp}{\varrho} + C \tag{2.3}$$

where C is a constant for the stream-line.

Making use of the adiabatic relation (2.1), this becomes

$$C - \frac{1}{2}q^2 = \frac{\gamma}{\gamma-1}p_0\,\varrho_0^{-\gamma}\,\varrho^{\gamma-1} = \frac{\gamma}{\gamma-1}\frac{p}{\varrho} \tag{2.4}$$

If the gas is flowing from a state of rest out of a reservoir where its pressure is p_0 and its density ϱ_0, (2.4) takes the form

$$q^2 = \frac{2\gamma}{\gamma-1}\frac{p_0}{\varrho_0}\left[1-\left(\frac{p}{p_0}\right)^{\frac{\gamma-1}{\gamma}}\right] \tag{2.5}$$

or alternatively

$$q^2 = \frac{2\gamma}{\gamma-1}\frac{p_0}{\varrho_0}\left[1-\left(\frac{\varrho}{\varrho_0}\right)^{\gamma-1}\right] \tag{2.6}$$

If a, a_0 represent the velocity of sound in air defined by (p, ϱ) and (p_0, ϱ_0) respectively, then $a^2 = \gamma\,p/\varrho$ while $a_0^2 = \gamma\,p_0/\varrho_0$ so that (2.5) may be written

$$q^2 = \frac{2}{\gamma-1}(a_0^2 - a^2). \tag{2.7}$$

a will be called the "local velocity of sound". It is not the speed with which sound waves move relative to the channel but their speed relative to the air at any point. The maximum possible velocity attainable by a gas flowing under pressure from a reservoir is, therefore, $[2/(\gamma-1)]^{1/2}a_0$, or $2.22\,a_0$ if $\gamma = 1.405$. This velocity is attained when $p = 0$, *i. e.* $a = 0$ so that though the speed of this stream tends to an absolute maximum, the ratio of the velocity of the stream to the local velocity of sound increases indefinitely[1].

The absolute maximum speed is attained when the gas is expanded till its temperature and pressure are reduced to zero. The whole of the intrinsic energy of the gas is then converted into kinetic energy. The intrinsic energy E of unit mass of a gas can be found by calculating the work done against external pressure when it is expanded adiabatically to zero density. Thus

$$E = \int_{1/\varrho_0}^{\infty} p\,d\left(\frac{1}{\varrho}\right) = \int_{1/\varrho_0}^{\infty} p_0\,\varrho_0^{-\gamma}\left(\frac{1}{\varrho}\right)^{-\gamma} d\left(\frac{1}{\varrho}\right) = \frac{1}{\gamma-1}\frac{p_0}{\varrho_0} \tag{2.8}$$

In order to write down an energy equation it is convenient to imagine the reservoir itself to be retained at constant pressure by decreasing its volume. The intrinsic energy per unit mass of the gas which stays in the reservoir then remains unchanged as the gas flows out through the channel or orifice. The work done by the contracting walls of the

---

[1] See I **3**.

reservoir while unit mass of gas flows out is p_0/ϱ_0 so that the energy equation for flow into a vacuum is

$$\frac{1}{2} q^2 = \frac{p_0}{\varrho_0} + \frac{1}{\gamma - 1} \frac{p_0}{\varrho_0} = \frac{\gamma}{\gamma - 1} \frac{p_0}{\varrho_0}$$

which is identical with (2.5) when $p = 0$.

3. Flow Through a Channel of Varying Cross-Section. When a gas flows through a straight channel the cross-section of which varies gradually along its length, the dynamical equation of motion may be considered as one-dimensional, but the one-dimensional equation of continuity used in Chapter I no longer applies. It must be replaced in the case of steady motion by

$$S \varrho q = M \tag{3.1}$$

where S is the area of cross-section at any point, q the velocity along the channel and M the rate of discharge of mass. If we suppose that the stream originates from a reservoir containing air at pressure p_0 and density ϱ_0 we may substitute for q from Bernoulli's equation (2.5) and for ϱ from (2.1), thus obtaining the following equation between p and S:

$$M = S \left(\frac{2}{\gamma - 1} \right)^{1/2} a_0 \varrho_0 \left[\left(\frac{p}{p_0} \right)^{2/\gamma} - \left(\frac{p}{p_0} \right)^{\frac{\gamma + 1}{\gamma}} \right]^{1/2}$$

The curve giving $\frac{\gamma - 1}{2} \frac{M^2}{S^2 a_0^2 \varrho_0^2} = \left(\frac{p}{p_0} \right)^{2/\gamma} - \left(\frac{p}{p_0} \right)^{\frac{\gamma + 1}{\gamma}}$ as a function of p/p_0 is shown in Fig. 5.

It will be seen that for any given value of S the discharge has a maximum corresponding with the point A where $p/p_0 = 0.527$. If the pressure at the point concerned could be reduced beyond 0.527 the discharge would be reduced below its maximum value till at zero pressure the discharge would cease altogether. This apparently paradoxical result is due to fixing attention only on two states of the gas, namely, p_0, ϱ_0 and p, ϱ without considering in detail how the one is changed into the other. Suppose that the channel gradually contracts from a very large size down to a minimum area S_1 and then gradually expands again to an area S_2 (see Fig. 5).

If for any given rate of discharge, P_1 and P_2 are the points on the curve of Fig. 5 which correspond with the minimum section S_1 and the section S_2, then their ordinates are in the ratio $P_1 N_1/P_2 N_2 = (S_2/S_1)^2$. As the pressure at S_2 is gradually reduced, the point P_2 travels along the curve while the point P_1 moves ahead of it keeping the ratio $P_1 N_1/P_2 N_2 = (S_2/S_1)^2 = \text{constant}$. When the point P_2 reaches the position P_2' whose ordinate is $(S_1/S_2)^2 (A\,N)$, where $A\,N$ is the maximum ordinate of the curve, P_1 reaches its highest point A. No further continuous reduction in pressure at S_2 is then possible. At this stage the relationship between pressure and cross-section may be followed down the tube from the high pressure reservoir by starting at the end B of the curve, proceeding along to A where the pressure at the minimum section has its minimum value and then turning back along the curve to the point P_2'. If an attempt were made to reduce the pressure at S_2 still further it would be unsuccessful; any increase in the power of evacuating machinery designed to reduce pressure at S_2 would merely

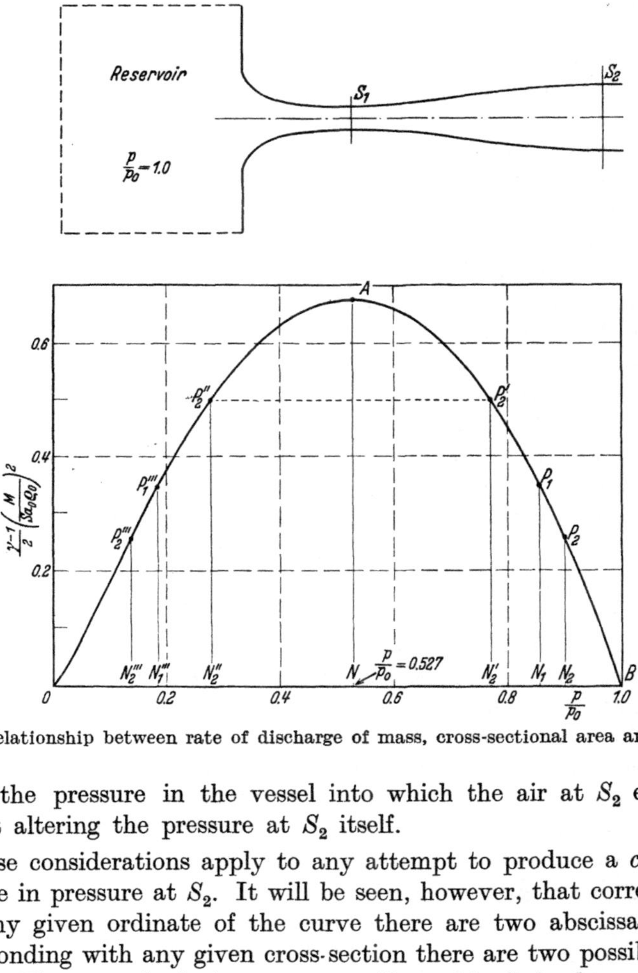

Fig. 5. Relationship between rate of discharge of mass, cross-sectional area and pressure.

reduce the pressure in the vessel into which the air at S_2 exhausted without altering the pressure at S_2 itself.

These considerations apply to any attempt to produce a *continuous* decrease in pressure at S_2. It will be seen, however, that corresponding with any given ordinate of the curve there are two abscissae so that corresponding with any given cross-section there are two possible values of p/p_0. The second abscissa corresponding with S_2 is shown in Fig. 5 at P_2'' for the case when the conditions at S_1 are represented by the point A; and to find whether it represents a physically possible condition, we must again follow the relationship between S and p/p_0 starting on the curve at B. It will be seen that to every point of the curve between B and P_2'' there corresponds a possible cross-section of the channel, the point P_2' corresponding with the point in the channel between the reservoir and the minimum section where the section is the same as that of the outlet, namely S_2.

On the other hand the pressure at the outlet cannot correspond with any other point on the low pressure side of the curve. Suppose for instance that it were possible for a still lower pressure represented by P_2''' to exist at the outlet; then the pressure at the minimum section would be found as before by erecting the ordinate $P_1''' N_1''' = (P_2''' N_2''')$ $(S_2/S_1)^2$ and the whole of the curve above the level of P_1''' corresponds with cross-sections smaller than any which exist in the channel, so that there could be no continuous decrease in pressure from that of the reservoir to that represented by P_2''' at the outlet. Similarly the pressure at the outlet cannot correspond with any point between P_2'' and A, because in that case the highest point A on the curve corresponds with a section which is not the minimum, so that there would be a length of the channel for which no corresponding points could be found on the curve of Fig. 5.

It appears, therefore, that as the pressure in the vessel into which the channel exhausts decreases, the pressure at the outlet decreases to the value given by the abscissa ON_2' (Fig. 5). No further continuous reductions are possible; but if the pressure outside the outlet is still further reduced till it is below the pressure represented by ON_2'' in Fig. 5, a second regime can be produced in which the pressure at S_2 is represented by ON_2''. No further reduction in pressure at S_2 can possibly be produced.

4. Velocity at Minimum Section. It was pointed out by Osborne Reynolds that in all cases where the second regime described above is established, the velocity of the stream at the minimum section is equal to the local velocity of sound. That this should be the case is to be expected because in the second regime, very small changes in cross-section in the neighborhood of the minimum section correspond with changes in pressure of a larger order of magnitude. The flow there is nearly the same as that in a channel with parallel walls where variations in pressure can only be maintained stationary when the velocity is equal to that of sound. To prove the proposition mathematically, we may express the conditions at the minimum section by the condition of $dS/ds = 0$ where ds is an element of length along the channel. Differentiating (3.1) logarithmically

$$\frac{dS}{S \cdot ds} + \frac{dq}{q \cdot ds} + \frac{d\varrho}{\varrho \cdot ds} = 0$$

so that at the minimum section

$$\frac{dq}{ds} = -\frac{q}{\varrho}\frac{d\varrho}{ds} \tag{4.1}$$

Substituting this value of dq/ds in the equation of motion (2.2) gives $-(1/\varrho)\,(dp/ds) = -(q^2/\varrho)\,(d\varrho/ds)$ so that $q^2 = (dp/d\varrho)$ or $q = (dp/d\varrho)^{1/2}$ which is the local velocity of sound. On the low pressure side of the minimum section, the speed is higher than that of sound while on the high pressure side it is lower than this speed.

5. Design of High Speed Wind Tunnels. The essential principle available for the design of high speed wind tunnels is contained in the foregoing discussion. A reservoir is maintained at high pressure by means of pumps and allowed to discharge into the atmosphere[1] through a channel or nozzle which first converges and then afterwards diverges. The diverging part is essential, as we have seen, if speeds higher than that of sound are to be obtained. The extent to which the conclusions of theory are verified in practice is illustrated in Fig. 6[2] where the results of a set of pressure measurements in a channel, whose section is shown inset, are given in the form of curves. The high pressure end was at atmospheric pressure p_0 while the low pressure end was connected with exhaust pumps. The pressure in the exhausted receiver is p_r and the abscissae of the curves represent p_r/p_0. The measurements were made at three points A, B and C, A being the point of minimum area, B farther down the channel and C very close to the outlet. The ordinates represent the pressures in cm. of mercury at the points A, B and C, while on a separate scale the rate of discharge is also given in grams per second.

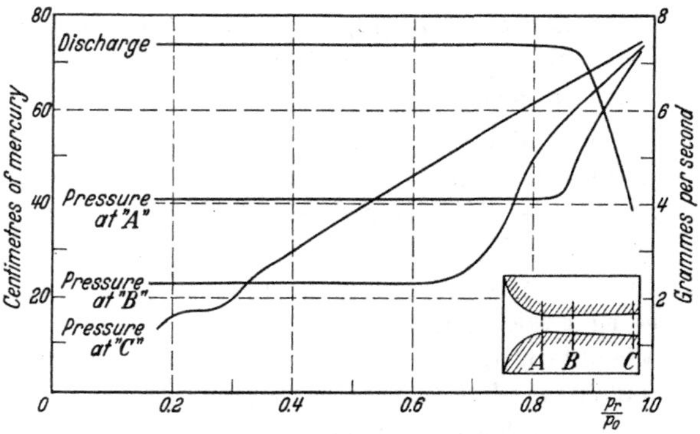

Fig. 6. Pressure measurements in a channel of varying cross-section.

It will be seen that the rate of discharge increases rapidly as p_r/p_0 decreases from 1.0 till at $p_r/p_0 = 0.86$ it becomes nearly constant. The pressure at the throat decreases rapidly till at $p_r/p_0 = 0.84$ it becomes nearly constant at 41 cm. of mercury. The pressure at C is practically the same as p_r, as can be seen by the fact that the corresponding line in Fig. 6 is nearly straight passing through the origin and tending to the pressure of 76 cm. when $p_r/p_0 = 1.0$, so that $p_0 = p_r = 76$ cm. of mercury. The regime in the range $p_r/p_0 = 0.86$ to $p_r/p_0 = 1.0$ is therefore qualitatively in accordance with theory.

[1] Or alternatively air is allowed to flow through such a nozzle into a chamber which is kept at a pressure lower than that of the atmosphere.

[2] Taken from a paper by T. E. Stanton, On the Flow of Gases at High Speeds, Proc. Roy. Soc., A, Vol. 111, p. 306, 1926.

At $p_r/p_0 = 0.83$, the pressure at the throat becomes constant and equal to 40.7 cm. of mercury which is $0.535\ p_0$, and no further reduction in pressure occurs there. This is in good agreement with the theory which gives the minimum pressure at the throat as $0.527\ p_0$. The theory, however, does not allow for the gradual reduction in pressure to 23 cm. of mercury which is observed at B. On the other hand, theory indicates that the pressure at B should, when the second regime is established, remain constant at some definite pressure which is lower than that at A. This is observed, the pressure at B becoming constant when $p_r/p_0 = 0.63$. The pressure at C appears to be trying to become constant at about $p_r/p_0 = 0.23$ but the pressure hole was so near the outlet that the flow could hardly have been even approximately uniform over the section at that point.

It is of interest to calculate the ratios of the effective areas at the points A, B and C in order that the observed curves may fulfil the conditions of adiabatic flow from which the curve of Fig. 5 was constructed. Taking for instance the measured pressures at $p_r/p_0 = 0.9$, the curves of Fig. 6 give $p_A = 56.8$ cm., $p_B = 64$ cm., $p_C = 69$ cm.; so that taking $p_0 = 76$ cm., $p_A/p_0 = 0.748$, $p_B/p_0 = 0.842$, $p_C/p_0 = 0.908$. We obtain from the curve of Fig. 5 the following values for $(\gamma - 1)/2\ S^2 a_0^2 \varrho_0^2$: 0.0540 at A, 0.0380 at B and 0.0240 at C. The ratios of the squares of the effective areas are, therefore, $(S_B/S_A)^2 = 0.054/0.038 = 1.42$, $(S_C/S_A)^2 = 0.054/0.024 = 2.25$. On the other hand, if the second regime is established at $p_r/p_0 = 0.23$, the values of p_B/p_0 and p_C/p_0 are $23/76 = 0.303$ and $16/76 = 0.21$ and the corresponding values of $(\gamma - 1)/2\ S^2 a_0^2 \varrho_0^2$ are 0.0536 and 0.0396 respectively, so that $(S_B/S_A)^2 = 0.0686/0.0536 = 1.28$, and $(S_C/S_A)^2 = 0.0686/0.0396 = 1.74$. Comparing these ratios with those deduced from the pressure measurements in the first regime, it will be seen that the effective areas of cross-section on the down stream side of the constriction, when the velocity there is greater than that of sound, are smaller than when the velocity is less than that of sound. This suggests the possibility that the boundary layer thickens more quickly at speeds above that of sound than at lower speeds. When the pressure p_r in the exhausted receiver is less than that necessary to lower the pressure in the throat to $0.527\ p_0$ but greater than that necessary to produce the second regime of stream-line flow, various types of flow are possible depending on the shape of the channel and the pressure in the receiver. If, after opening out to its maximum diameter the channel again contracts before reaching the receiver, a shock wave is likely to form. At such a surface there is an abrupt change from a speed greater to a speed less than that of sound. This is shown in Fig. 7 (on Plate VI at the beginning of this Division) which is a photograph, taken by Töpler's Schlieren method, of the flow in a contracting and expanding channel, the exit of which is slightly constricted by a diaphragm. The walls of this channel were artificially roughened, so that the wavelets produced in parts of the channel where the speed is greater than that of sound appear as lines or striations. It will be seen that these lines appear only in the portion of the channel from the throat to a point about half way down the expanding part. At this point they end in the shock wave and the remainder of the channel in which the speed is less than that of sound is clear. After passing through the wave the gas, which is then moving at a speed less than that of sound, can flow according to the first regime so that the pressure can increase as the channel expands. At the exit, however, the pressure will be less than what it would have been if the first regime had extended the whole way through the channel. As the pressure at the exit falls from the pressure which is just necessary to establish the speed of sound at the throat to the pressure necessary in order that the second regime may extend to the outlet, the shock wave moves along the tube from the throat to the exit and the minimum pressure occurs at the wave surface. The observed pressures in an expanding channel for various pressures at the exit are shown in Fig. 8. It will be seen that when the minimum pressure is less than $0.527\, p_0$, a sudden jump in pressure occurs at a point which moves down the tube with decreasing pressure at the exit.

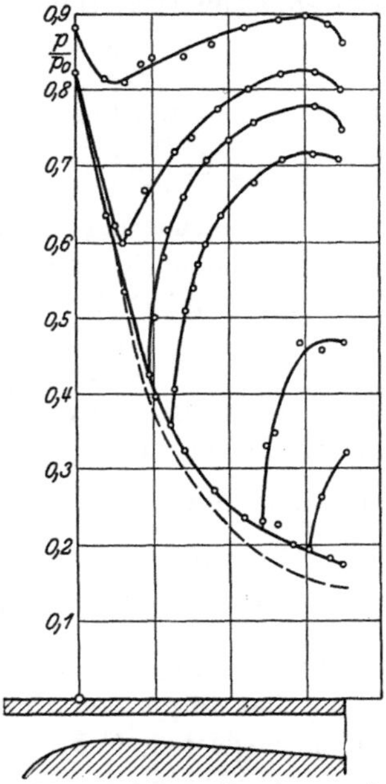

Fig. 8. Observed pressures in an expanding channel. (From Ackeret's article in the Handbuch der Physik. Vol. VII.)

The greatest wind velocity so far attained (1934) in a high speed wind tunnel seems to be about 3.5 times the speed of sound.

CHAPTER III

TWO-DIMENSIONAL FLOW AT SPEEDS LESS THAN THAT OF SOUND

1. Differential Equation for Irrotational Motion. When the components of velocity at the point (x, y) are u, v, the equation of continuity then is

$$\frac{\partial}{\partial x}(\varrho u) + \frac{\partial}{\partial y}(\varrho v) = 0 \tag{1.1}$$

If the fluid motion has a velocity potential φ so that $u = \partial\varphi/\partial x$, $v = \partial\varphi/\partial y$, (1.1) becomes

$$\frac{\partial^2\varphi}{\partial x^2} + \frac{\partial^2\varphi}{\partial y^2} = -\frac{1}{\varrho}\left[\frac{\partial\varrho}{\partial x}\frac{\partial\varphi}{\partial x} + \frac{\partial\varrho}{\partial y}\frac{\partial\varphi}{\partial y}\right] \tag{1.2}$$

The pressure at any point is given by Bernoulli's equation (see Chapter II)

$$\int \frac{dp}{\varrho} + \frac{1}{2} q^2 = \text{constant},$$

where $q^2 = u^2 + v^2 = (\partial\varphi/\partial x)^2 + (\partial\varphi/\partial y)^2$,

so that $$\frac{1}{2}\frac{\partial q^2}{\partial x} = -\frac{1}{\varrho}\frac{dp}{d\varrho}\frac{\partial\varrho}{\partial x}, \quad \frac{1}{2}\frac{\partial q^2}{\partial y} = -\frac{1}{\varrho}\frac{dp}{d\varrho}\frac{\partial\varrho}{\partial y}.$$

Now $dp/d\varrho = a^2$, where a is the velocity of sound at the point in question, and equation (1.2) thus becomes

$$\frac{\partial^2\varphi}{\partial x^2} + \frac{\partial^2\varphi}{\partial y^2} = \frac{1}{2a^2}\left[\frac{\partial\varphi}{\partial x}\frac{\partial q^2}{\partial x} + \frac{\partial\varphi}{\partial y}\frac{\partial q^2}{\partial y}\right] \tag{1.3}$$

which can be written also in the form

$$\frac{\partial^2\varphi}{\partial x^2} + \frac{\partial^2\varphi}{\partial y^2} = \frac{1}{a^2}\left[\left(\frac{\partial\varphi}{\partial x}\right)^2\frac{\partial^2\varphi}{\partial x^2} + 2\frac{\partial\varphi}{\partial x}\frac{\partial\varphi}{\partial y}\frac{\partial^2\varphi}{\partial x\,\partial y} + \left(\frac{\partial\varphi}{\partial y}\right)^2\frac{\partial^2\varphi}{\partial y^2}\right] \tag{1.4}$$

When considering the field of flow due to a body moving at speed U, it is convenient to express a^2 in terms of U^2 and q^2. Thus, for a gas obeying the adiabatic law, II (2.7) gives

$$q^2 = \frac{2}{\gamma-1}(a_0^2 - a^2) \quad \text{and} \quad U^2 = \frac{2}{\gamma-1}(a_0^2 - a_1^2)$$

where a_0, a_1 are the speeds of sound associated with the gas when at rest and when moving with speed U. Elimination of a_0^2 from these two equations gives

$$a^2 = a_1^2 - \frac{\gamma-1}{2}(q^2 - U^2)$$

and (1.3) can, therefore, be written in the form:

$$\left[\frac{\partial^2\varphi}{\partial x^2} + \frac{\partial^2\varphi}{\partial y^2}\right]\left[2a_1^2 - (\gamma-1)(q^2 - U^2)\right] = \frac{\partial\varphi}{\partial x}\frac{\partial q^2}{\partial x} + \frac{\partial\varphi}{\partial y}\frac{\partial q^2}{\partial y} \tag{1.5}$$

It is of interest to note than when u and v are small compared with a_1, so that terms containing squares and products of these quantities can be neglected, (1.5) reduces to

$$\frac{\partial^2\varphi}{\partial x^2} + \frac{\partial^2\varphi}{\partial y^2} = 0 \tag{1.6}$$

which is the equation for the potential flow of an incompressible fluid. We see, therefore, that at speeds small compared with that of sound, the flow of a gas is similar to that of an incompressible fluid.

Although no general solution of (1.5) has yet been proposed, it has been solved completely for a number of special problems at supersonic speeds. These solutions will be considered in Chapter IV. Approximate solutions have been obtained also for several problems when the speed of the fluid never exceeds the local speed of sound and two of these will now be discussed[1].

2. The Flow Past a Circular Cylinder. It has been indicated in the last section that, when a body moves through air at a speed which is small compared with that of sound, the effect of compressibility is merely to distort slightly the stream-lines from the positions they would have occupied if the air had been incompressible. Investigations of the

[1] See p. 250 for note on the work of A. Tchapliguine.

motion of a gas at low speeds can be based, therefore, upon the solution obtained for an incompressible, perfect fluid. Such calculations have been carried out analytically for two problems, namely, for the motions past a circular cylinder[1] and a sphere[2].

When the general equation (1.5) for steady motion in two dimensions is expressed in polar coordinates, it becomes

$$\nabla^2 \varphi \left[2 a_1^2 - (\gamma - 1)(q^2 - U^2)\right] = \frac{\partial \varphi}{\partial r}\frac{\partial q^2}{\partial r} + \frac{1}{r^2}\frac{\partial \varphi}{\partial \theta}\frac{\partial q^2}{\partial \theta} \tag{2.1}$$

In the first instance both q and U are assumed to be small compared with a_1 and hence a first approximation to (2.1) is $\nabla^2 \varphi = 0$. Using particular solutions of $\nabla^2 \varphi = 0$ of the type $\varphi = r^n \cos n \varphi$, the right hand side of (2.1) may be calculated as a function of r and θ, say $F(r, \theta)$. A second approximation to (2.1) may thus be obtained by solving the equation

$$\nabla^2 \varphi = \frac{1}{2 a_1^2} F(r, \theta) \tag{2.2}$$

The same method may be repeated to obtain higher approximations.

Janzen and Rayleigh applied this method to the flow past a circular cylinder and showed that however far the approximations are continued, the final solution gives a pressure distribution which is symmetrical about a radius at right angles to the direction of flow. There is, therefore, no resultant force experienced by the cylinder. The same conclusion is found to hold for a sphere.

Due to the excessive labour involved in passing from one approximation to the next, it is not possible to carry the above method very far in an analytical form. A mechanical method of solving (2.1) has been developed, however, by applying an analogy between the flow of a compressible fluid and the flow of electricity in a sheet of variable thickness. Successive solutions of (2.1) are obtained from electrical measurements of the current flowing through an electrolyte which is contained in a tank. The depth of the electrolyte is varied throughout the whole field in proceeding from one approximation to the next.

When this method was applied to the flow past a circular cylinder it was found that the approximations were convergent for $U/a_1 < 0.45$ but divergent for $U/a_1 > 0.45$. It appears that in this case failure of the solutions to converge occurs approximately when the maximum speed in the field reaches the local speed of sound. Similar conclusions were reached by purely numerical calculation in some other cases[3].

3. Prandtl and Glauert's Approximation to the Lift of an Airfoil. In Janzen's and Rayleigh's approximations the analysis ceases to be accurate as soon as U becomes comparable with a_1. This limits their

[1] Janzen, O., Phys. Zeitschrift, Vol. 14, p. 639, 1913.

[2] Rayleigh, (Lord), Phil. Mag., Vol. 32, p. 1, 1916; Scientific Papers, Vol. 6, p. 402.

[3] Taylor, G. I., Journal London Math. Soc., Vol. 5, p. 224, 1930; Zeitschrift angew. Math. Mech., Vol. 10, p. 334, 1930.

usefulness for aeronautical purposes. In the approximation of Prandtl and Glauert[1], U is not small compared with a_1 but the variations in q throughout the field are supposed small compared with U. The flow may be considered, therefore, as a uniform stream disturbed by a small perturbation. Such a disturbance can be caused by a thin airfoil, the slope of whose surfaces to the direction of the main stream is everywhere small. The approximation explores, therefore, the effect of compressibility on the flow around thin airfoils. The theory of thin airfoils is concerned with finding the distribution of vorticity along a line (the chord of the airfoil) which satisfies the condition that the normal component velocity at the surface of the airfoil shall be zero; accordingly the first step is to find the effect of compressibility on the field due to a single vortex filament. For this purpose, assume

$$\varphi = U\, r \cos\theta + A F(\theta) \tag{3.1}$$

where A is so small that all terms containing A^2 may be neglected. In this case

$$q^2 = U^2 \cos^2\theta + \left[-U \sin\theta + \frac{A}{r} F'(\theta)\right]^2 = U^2 - \frac{2\,A\,U F'(\theta)}{r} \sin\theta$$

where $F'(\theta) = dF(\theta)/d\theta$.

Remembering that by the initial assumption, $q^2 - U^2$ may be neglected compared with q, U or a_1 and by neglecting terms containing A^2 as a factor, (2.1) reduces to

$$2 a_1^2\, A\, F''(\theta) = 2\, A\, U^2 \sin\theta \frac{d}{d\theta}\left[F'(\theta) \sin\theta\right] + 2\, A\, U^2 \cos\theta \sin\theta\, F'(\theta)$$

or $$\frac{F''(\theta)}{F'(\theta)} = \frac{2\,U^2 \cos\theta \sin\theta}{a_1^2 - U^2 \sin^2\theta} \quad \text{so that} \quad F'(\theta) = \frac{\text{constant}}{a_1^2 - U^2 \sin^2\theta}$$

The radial and circumferential components of velocity u and v are:

$$\left.\begin{aligned} u &= U \cos\theta \\ v &= -U \sin\theta + \frac{B}{r\,[1-(U^2/a_1^2)\sin^2\theta]} \end{aligned}\right\} \tag{3.2}$$

where B is a constant. The circulation is

$$K = -\int \frac{B\,d\theta}{[1-(U^2/a_1^2)\sin^2\theta]} = -\frac{2\,\pi\,B}{\sqrt{1-U^2/a_1^2}}$$

so that

$$B = -\frac{K}{2\pi}\sqrt{1-U^2/a_1^2}$$

To find the lift on an airfoil which produces this circulation, first notice that to the degree of approximation required the pressure p is

$$\begin{aligned} p &= \text{constant} - (1/2)\,\varrho q^2 \\ &= \text{constant} - \frac{\varrho\, U K \sin\theta \sqrt{1-U^2/a_1^2}}{2\pi r\,[1-(U^2/a_1^2)\sin^2\theta]} \end{aligned} \tag{3.3}$$

The lift $L = -\int_0^{2\pi} [p \sin\theta + \varrho u\,(u \sin\theta + v \cos\theta)]\, R\, d\theta$, the integral

[1] Glauert, H., The Effect of Compressibility on the Lift of an Aerofoil, Proc. Roy. Soc., A, Vol. 118, p. 113, 1928.

being taken around a circle of radius R. Substituting (3.2) and (3.3) it is found that

$$L = \varrho\, U K$$

so that the relationship between circulation and lift is unaltered by compressibility.

To calculate the lift exerted by an incompressible fluid on a thin airfoil, it is necessary to find the distribution of vortices along the chord which will produce a component of velocity perpendicular to the direction of motion equal to $U \sin \psi$ where ψ is the small angle which

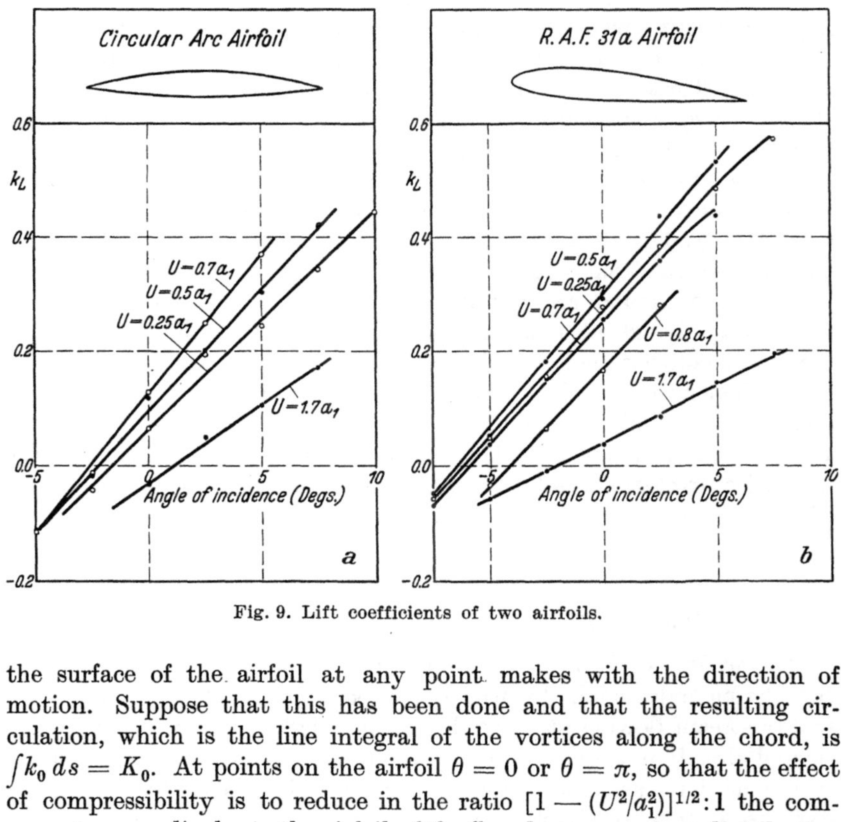

Fig. 9. Lift coefficients of two airfoils.

the surface of the airfoil at any point makes with the direction of motion. Suppose that this has been done and that the resulting circulation, which is the line integral of the vortices along the chord, is $\int k_0\, ds = K_0$. At points on the airfoil $\theta = 0$ or $\theta = \pi$, so that the effect of compressibility is to reduce in the ratio $[1 - (U^2/a_1^2)]^{1/2}:1$ the component perpendicular to the airfoil, of the flow due to any given distribution of vortices along the chord. Hence if the strength of every vortex element k_0 found for the incompressible case is now increased in the ratio $[1 - (U^2/a_1^2)]^{-1/2}:1$, the new vortex distribution will give the required distribution of normal velocity at the surface of the airfoil when the fluid is compressible. The circulation K is therefore equal to $K_0\,[1 - (U^2/a_1^2)]^{-1/2}$ and since $L = \varrho\, U K$, the lift is increased by compressibility in the ratio $[1 - (U^2/a_1^2)]^{-1/2}:1$.

Within certain limits this theoretical result is confirmed by experiments made in a high speed wind tunnel. Fig. 9a[1] shows the measured lift coefficient[2] k_L of a circular arc airfoil at angles of incidence ranging from -5^0 to $+7.5^0$ for speed $U = 0.25\,a_1$, $0.5\,a_1$, $0.7\,a_1$ and $1.7\,a_1$. At all angles of incidence within this range k_L increases as U increases from $0.25\,a_1$ to $0.7\,a_1$, but at higher speeds it decreases till at $U = 1.7\,a_1$, it is less than it was at $U = 0.25\,a_1$.

At low speeds when the effect of compressibility is negligible, it is known that the actual value of the lift coefficient is not so definitely related to the theoretical value given by the Joukowski theory as the slope of the lift curve. To compare Prandtl and Glauert's formula with observation, the slopes of the lift curve in Fig. 9a may, therefore, be measured. These are 0.0380 per degree at $U = 0.25\,a_1$, 0.0424 per degree at $U = 0.5\,a_1$, and 0.0484 per degree at $U = 0.7\,a_1$. The ratios of these slopes are 1:1.116:1.273. According to Prandtl and Glauert's theory they should be $(1-1/16)^{-1/2}:(1-1/4)^{-1/2}:(1-0.49)^{-1/2}$ or 1:1.120:1.357.

The observed increase in the slope of the lift curve from $U/a_1 = 0.25$ to $U/a_1 = 0.50$ is therefore almost exactly in accordance with theory, but the increase from $0.5\,a$ to $0.7\,a$ is less than that required by theory.

With the circular arc section shown in Fig. 9*a*, k_L increases up to $U = 0.7\,a_1$, though at a rate less than is required by theory; but with the more conventional airfoil section R.A.F. 31a (see Fig. 9b)[3] the lift coefficient rises only slightly from $U = 0.25\,a_1$ to $U = 0.5\,a_1$ and then begins to fall, so that at $U = 0.7\,a_1$ it is less than at $0.5\,a_1$ while at $U = 0.8\,a_1$ it is less still. This effect is probably due to failure of the theoretical irrotational solution of the equations of motion[4] which occurs with conventional sections at about $U = 0.6\,a_1$.

CHAPTER IV

TWO-DIMENSIONAL FLOW AT SUPERSONIC SPEEDS

1. The Motion Past Thin Airfoils. The simplification, introduced by considering only small disturbances imposed by a slender body on a steady stream flowing at a speed less than that of sound, has been considered already in connection with Prandtl and Glauert's approximations to the motion past thin airfoils. At supersonic speeds this conception leads to even more simple results.

---

[1] Stanton, T. E., British Aeronautical Research Committee, Reports and Memoranda No. 1130, 1928.

[2] It will be remembered that the lift coefficient k_L is taken relative to ϱV^2 while C_L, as used usually in other divisions of this work, is taken relative to $(1/2)\varrho V^2$.

[3] *loc. cit.*

[4] Taylor, G. I. and Sharman, C. F., British Aeronautical Research Committee, Reports and Memoranda, No. 1196, 1928, and Proc. Roy. Soc. A. 121, p. 194, 1928.

Referring to III (1.4), it will be seen that $\partial\varphi/\partial x$ is always large compared with $\partial\varphi/\partial y$ so that III (1.4) reduces to

$$\frac{\partial^2\varphi}{\partial x^2}+\frac{\partial^2\varphi}{\partial y^2}=\frac{U^2}{a_1^2}\frac{\partial^2\varphi}{\partial x^2}$$

or

$$\frac{\partial^2\varphi}{\partial y^2}-n^2\frac{\partial^2\varphi}{\partial x^2}=0 \qquad (1.1)$$

where $n^2=(U/a_1)^2-1$.

The general solution of (1.1) is of the form

$$\varphi=f_1(x-ny)+f_2(x+ny)$$

where f_1 and f_2 are arbitrary functions. This represents stationary sound waves making angles $\pm\, cot^{-1}\, n$ with the direction of the undisturbed stream. The angle $cot^{-1}\, n$, or $sin^{-1}\,(a_1/U)$, is known as the Mach angle.

Waves of this kind might be set up by the motion of a thin airfoil with sharp leading and trailing edges. We can conceive that the upper surface produces a sound wave propagating obliquely upward and that the lower one produces a similar wave moving obliquely downward. In Fig. 10 the upper wave is contained between the lines AB and DE while the lower wave is bounded by AC and DF. These lines all make the Mach angle with the direction of motion of the undisturbed flow. The two waves may be treated as plane sound waves in which the pressure p, density ϱ and velocity u within the wave due to the disturbance, are constant along lines like KL, MN which are parallel to the wave fronts. The velocity u must necessarily be taken at right angles to the wave front. As has been seen in I 2, the excess pressure within the wave will be $\triangle p=a_1\varrho u$ where a_1 is the velocity of sound and ϱ is the density.

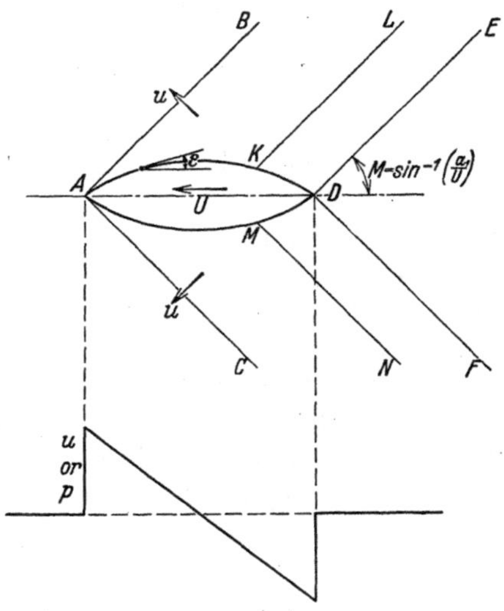

Fig. 10.
Waves produced by a thin airfoil.

If ε is the small angle which an element of the surface makes with the direction of motion then $U\varepsilon$ is the component of velocity of the surface normal to itself. This must equal the component of velocity of the air normal to the surface element which is seen to be $u\; cos\,(M-\varepsilon)$. Hence $u=U\varepsilon\; sec\; M$, since ε is small compared with M, and the excess pressure on the surface element is

$$\triangle p=a_1\varrho\,\varepsilon\, U\; sec\; M=\varrho\, U^2\,\varepsilon\,(U^2/a_1^2-1)^{-1/2}$$

If δ represents the inclination of the chord to the direction of motion and if ε_1, ε_2 are the slopes of the upper and lower surfaces to the chord, then the excess pressures $\triangle p_1$, $\triangle p_2$ at points on the upper and lower surfaces will be given by

$$\triangle p_1 = \varrho U^2 \left(\frac{U^2}{a_1^2} - 1\right)^{-1/2} (-\delta + \varepsilon_1)$$

$$\triangle p_2 = \varrho U^2 \left(\frac{U^2}{a_1^2} - 1\right)^{-1/2} (+\delta + \varepsilon_2)$$

When these pressures are integrated over the chord l, the following expressions are obtained for k_L, k_D, k_M, the coefficients[1] of lift, drag and moment about the leading edge[2].

$$k_L = 2\delta \left[\frac{U^2}{a_1^2} - 1\right]^{-1/2}$$

$$k_D = (2\delta^2 + \overline{\varepsilon_1^2} + \overline{\varepsilon_2^2}) \left[\frac{U^2}{a_1^2} - 1\right]^{-1/2}$$

$$k_M = \left[\delta + \frac{1}{l}\int_0^1 (-\varepsilon_1 + \varepsilon_2)\, x\, d\, x\right] \left[\frac{U^2}{a_1^2} - 1\right]^{-1/2}$$

where the bar indicates that the mean value is taken along the chord. In the case of a biconvex airfoil, these formulae are in good agreement with measured values obtained by Stanton in his high speed wind channel[3].

Busemann[4] has obtained recently a further approximation to the forces on an airfoil by calculating more accurately the relationship between the excess pressure $\triangle p$ and the local angle of incidence ε. His relationship is of the form $\triangle p = (C_1\varepsilon + C_2\varepsilon^2)\,\varrho U^2$, where

$$C_1 = \left[\frac{U^2}{a_1^2} - 1\right]^{-1/2}$$

as has been obtained above, and

$$C_2 = \left(\frac{U^2}{a_1^2} - 1\right)^{-1} \left[\frac{(\gamma + 1)(U^4/a_1^4)}{4(U^2/a_1^2 - 1)} - 1\right]$$

2. Oblique Shock Waves. In the above approximate calculations the velocity due to the disturbance is small compared with a_1, the speed of sound in the undisturbed stream. It has been seen in Chapter I that when this condition is not satisfied, shock waves may be formed. If the angle ε of the preceding section is not everywhere small, u is comparable with a_1 so that shock waves of finite amplitude may be expected.

[1] It will be remembered that the coefficients k_L, k_D, k_M are taken relative to ϱV^2 while C_L, C_D, C_M, as used usually in other divisions of this work, are taken relative to $(1/2)\,\varrho V^2$.

[2] Ackeret, J., Z.F.M., Vol. 16, p. 72, 1925.

[3] Taylor, G. I., British Aeronautical Research Committee, Reports and Memoranda, No. 1467, 1932.

[4] Forschung auf dem Gebiete des Ingenieurwesens, V.D.I., Vol. 4, p. 87, 1933.

Such waves may be seen in photographs of bullets in flight extending backward from the nose. Immediately in front of a round or flat nosed projectile the shock wave travels in the same direction and at the same speed as the body which produces it. The pressure behind the wave can, therefore, be deduced from I (5.1), (5.2) and (5.4). At all other points the shock wave due to a projectile moves obliquely to the direction of flight.

In discussing the flow, it is convenient to consider the shock wave as fixed, the air moving obliquely through it. Such a field of flow can be derived from the one-dimensional flow through a fixed shock wave, which was considered in Chapter I, by superimposing on the whole system a velocity parallel to the plane of the wave.

Let NN', Fig. 11, be part of a shock wave. In the one-dimensional problem the velocity components u_1, u_2 are represented by the vectors AN, NB which are perpendicular to NN'. Superimposing the velocity v, represented by CA or BD parallel to the wave, the pressures are unaltered but the velocities are now represented by the oblique vectors CN, ND. The speed of the approaching stream is now $U = [u_1^2 + v^2]^{1/2}$. The shock wave is oblique to the initial direction of motion of the gas and, if α and β are the angles ANC and BND, a stream-line is deflected through the angle $(\beta - \alpha)$ when it crosses the wave. The conditions in front of and behind the wave may be represented by U, p_1, ϱ_1 and V, p_2, ϱ_2 respectively.

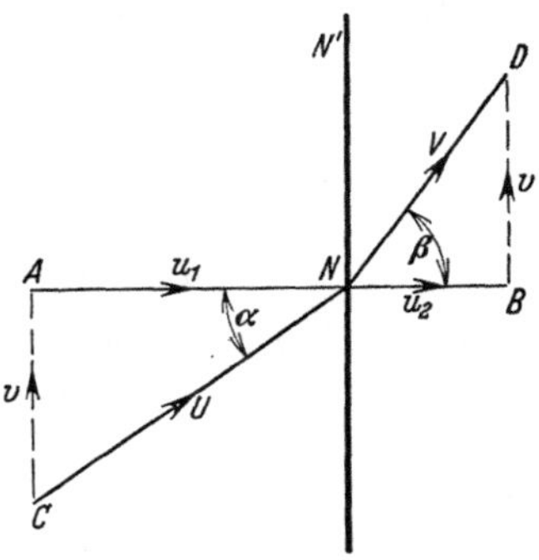

Fig. 11. Flow through an oblique shock wave.

In any steady flow which does not pass through a shock wave the velocity is connected with the pressure and density by Bernoulli's equation, so that the pressure and density corresponding with any given velocity can be found provided the pressure and density corresponding with zero velocity are known. If (p_0, ϱ_0) are the pressure and density at zero velocity for the stream defined by (U, p_1, ϱ_1), the ratio U/a_1, where a_1 is the velocity of sound in the air approaching the wave, depends only upon p_1/p_0. Thus II (2.5) may be written

$$U^2 = \frac{2}{\gamma - 1} \frac{\gamma p_1}{\varrho_1} \left[\left(\frac{p_0}{p_1} \right)^{\frac{\gamma-1}{\gamma}} - 1 \right]$$

and
$$a_1^2 = \frac{\gamma p_1}{\varrho_1}$$

so that
$$\left(\frac{U}{a_1} \right)^2 = \frac{2}{\gamma - 1} \left[\left(\frac{1}{y} \right)^{\frac{\gamma-1}{\gamma}} - 1 \right] \tag{2.1}$$

where $y = p_1/p_0$.

Although Bernoulli's equation does not hold in passing through a shock wave, it may be applied to the region behind the wave by making the necessary changes in the constant. The quantities (p_3, ϱ_3) are therefore defined as the pressure and density corresponding with zero velocity behind the wave. In subsequent work it will be found convenient to use the symbols $x = p_2/p_0$, $z = p_3/p_0$; it will be noticed that z is always less than 1.0.

In obtaining the equations which govern the flow through an oblique shock wave, we start from the equations obtained in Chapter I where we discussed the propagation of shock waves in one dimension.

In this case, since the shock wave is at rest, a slight modification is necessary in I (5.1), (5.2) and (5.4). These now take the form

$$\varrho_1 u_1 = \varrho_2 u_2 = m \tag{2.2}$$

$$p_2 - p_1 = m\,(u_1 - u_2) \tag{2.3}$$

$$\frac{1}{2}(p_1 + p_2)\left(\frac{1}{\varrho_1} - \frac{1}{\varrho_2}\right) = \frac{1}{\gamma - 1}\left(\frac{p_2}{\varrho_2} - \frac{p_1}{\varrho_1}\right) \tag{2.4}$$

Solving (2.4) for ϱ_2/ϱ_1, we obtain the Rankine-Hugoniot relationship in the useful form

$$\frac{\varrho_2}{\varrho_1} = \frac{(\gamma - 1)\,p_1 + (\gamma + 1)\,p_2}{(\gamma - 1)\,p_2 + (\gamma + 1)\,p_1} \tag{2.5}$$

Equations (2.2) and (2.3) give

$$p_2 - p_1 = \varrho_1 u_1^2 - \varrho_2 u_2^2 = \varrho_1 u_1^2\,(1 - \varrho_1/\varrho_2)$$

from which we obtain $u_1^2 = \dfrac{p_2 - p_1}{\varrho_1} \cdot \dfrac{\varrho_2/\varrho_1}{\varrho_2/\varrho_1 - 1}$

When we substitute the value of ϱ_2/ϱ_1 given by (2.5), this equation becomes

$$u_1^2 = \frac{p_1}{2\,\varrho_1}\left[(\gamma - 1) + (\gamma + 1)\frac{p_2}{p_1}\right]$$

Now $u_1^2 = U^2 \cos^2 \alpha$ and $p_1/\varrho_1 = a_1^2/\gamma$, where a_1 is the speed of sound in the approaching stream. Substituting these values in the above equation and applying the relationship given in (2.1) we obtain finally

$$\cos^2 \alpha = \frac{[(\gamma - 1) + (\gamma + 1)\,(x/y)]\,(\gamma - 1)}{4\gamma\left[(1/y)^{\frac{\gamma - 1}{\gamma}} - 1\right]} \tag{2.6}$$

as the equation connecting α with the pressure ratios x and y.

Since we have from (2.2)

$$\frac{\tan \beta}{\tan \alpha} = \frac{v/u_2}{v/u_1} = \frac{u_1}{u_2} = \frac{\varrho_2}{\varrho_1}$$

equation (2.5) gives

$$\frac{\tan \beta}{\tan \alpha} = \frac{(\gamma - 1) + (\gamma + 1)\,(x/y)}{(\gamma + 1) + (\gamma - 1)\,(x/y)} \tag{2.7}$$

which connects β with x and y when α is known. Equations (2.6) and (2.7) were first obtained by Meyer[1]. They govern the conditions on the two sides of the oblique shock wave, the directions of the streamlines ahead of and behind the wave being given by them when the pressure ratios x and y are stated.

[1] Forschungsarbeiten auf dem Gebiete des Ingenieurwesens, V.D.I., Heft 62, 1908.

These equations are represented conveniently by Figs. 12 and 13 which show contours of the pressure ratios x and y for varying values

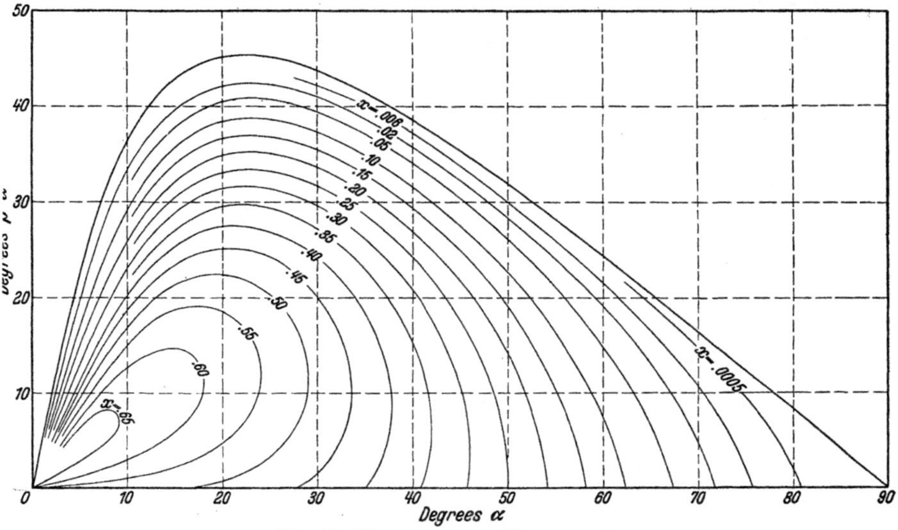

Fig. 12. The pressure ratio $x = p_2/p_0$.

of α and $(\beta - \alpha)$. It should be noted that α is the complement of the angle between the wave and the direction of the oncoming flow and

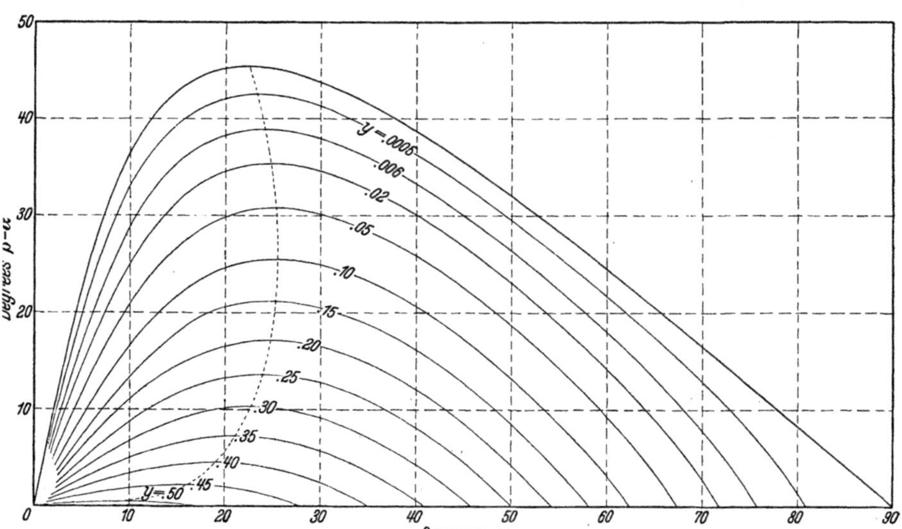

Fig. 13. The pressure ratio $y = p_1/p_0$.

$(\beta - \alpha)$, as has been already stated, is the angle through which a streamline is deflected when it passes through the shock wave.

The equation giving the pressure ratio z can be obtained by constructing the equation for β which is analogous to (2.6). Thus, substituting β for α, y/x for x/y and z/x for $1/y$ in (2.6) we find,

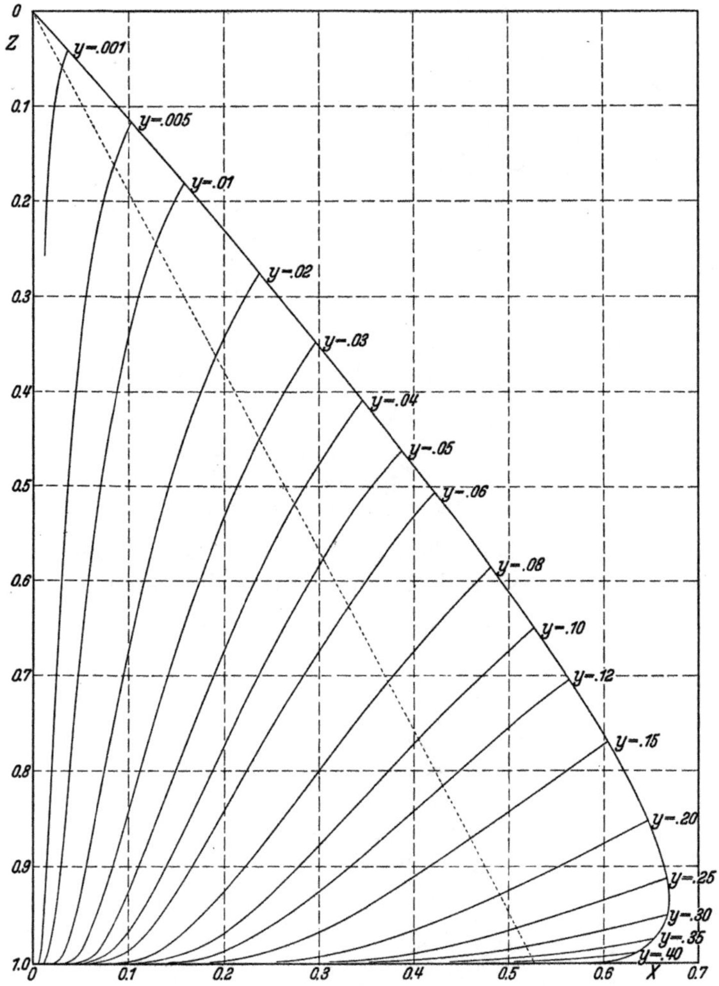

Fig. 14. The pressure ratio $z = p_3/p_0$ for varying values of x and y.

$$\cos^2\beta = \frac{[(\gamma-1)+(\gamma+1)(y/x)]\,(\gamma-1)}{4\gamma\left[(z/x)^{\frac{\gamma-1}{\gamma}}-1\right]}$$

This equation can be rearranged to give

$$\left(\frac{z}{x}\right)^{\frac{\gamma-1}{\gamma}} = 1 + \frac{\gamma-1}{4\gamma\cos^2\beta}\left[(\gamma-1)+(\gamma+1)\frac{y}{x}\right] \tag{2.8}$$

from which z may be calculated when x, y and β are known.

Fig. 14 shows the variation of z with x for constant values of y. When $p_2/p_3 = x/z > 0.527$, the speed of the gas behind the shock wave is less than the local speed of sound. The broken line shown in the diagram thus divides the figure into two portions; to the left the conditions are such that the speed of the gas behind the wave is greater than the local speed of sound, while to the right of the line the speed of the gas is less than this speed.

It is interesting to note that if we are given the shape of a shock wave moving into still air (as, for example, is obtained by the photography of bullets in flight) then y, which depends only on U/a_1, has a constant value over the whole wave. All the conditions at the wave are defined, therefore, by one of the contour lines in Fig. 13 and, α being known, the values of $\beta - \alpha$ may be read directly off the diagram. The pressure ratios x and z can then be found by using Figs. 12 and 14.

3. Pressure in a Pitot-Tube or at the Nose of a Projectile. The limiting curve on the right of Fig. 14 corresponds with the condition $\alpha = \beta = 0$. This condition is obtained in practice when a Pitot-tube is placed in the stream. In using such a tube for measuring the speed of the air flow, the pressure p_3 in the Pitot-tube and the static pressure p_1 in the undisturbed stream are observed. From the ratio p_3/p_1 or z/y, the value of U/a_1 can be calculated as follows.

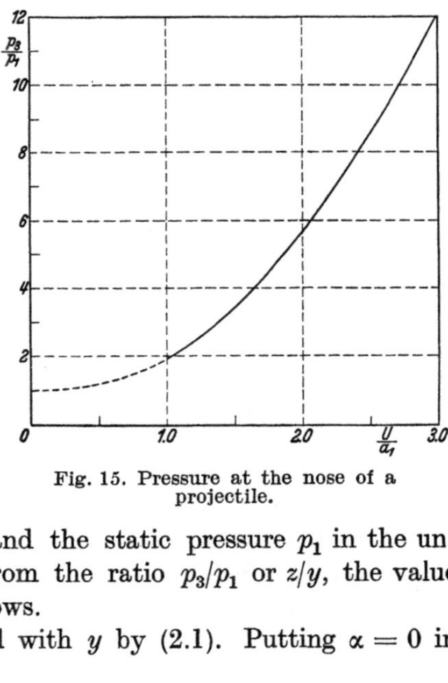

Fig. 15. Pressure at the nose of a projectile.

The quantity U/a_1 is connected with y by (2.1). Putting $\alpha = 0$ in (2.6) we obtain the expression

$$\frac{x}{y} = \frac{1}{\gamma^2 - 1}\left[4\gamma y^{\frac{1-\gamma}{\gamma}} - (\gamma+1)^2\right] = \frac{2\gamma}{\gamma+1}\frac{U^2}{a_1^2} - \frac{\gamma-1}{\gamma+1} \qquad (3.1)$$

Putting $\beta = 0$ in (2.8) and multiplying the resulting equation by $(x/y)^{\frac{\gamma-1}{\gamma}}$ we obtain also

$$\left(\frac{z}{y}\right)^{\frac{\gamma-1}{\gamma}} = \left(\frac{x}{y}\right)^{\frac{\gamma-1}{\gamma}}\left[\frac{(\gamma+1)^2}{4\gamma} + \frac{\gamma^2-1}{4\gamma}\frac{y}{x}\right] \qquad (3.2)$$

The connection between p_3/p_1 and U/a_1 is shown in Fig. 15 by a continuous line[1]. This curve gives directly the pressure in atmospheres

[1] These formulae were obtained by Lord Rayleigh, Proc. Roy. Soc., A, Vol. 84, 1910; Scientific Papers, Vol. 5, p. 609.

at the nose of a projectile which moves at a supersonic speed. When the speed of the projectile is less than that of sound no shock wave is formed in front of the moving body and the Pitot pressure p_3 is then equal to p_0. The pressure on the nose of a projectile in atmospheres is measured by p_0/p_1 or $1/y$, and this pressure is shown by the broken curve in Fig. 15.

4. The Flow Past a Wedge and a Cone at Supersonic Speeds. The results for oblique shock waves, which were obtained in 2, may be applied directly to give the flow past a wedge at supersonic speeds. For, if OS (Fig. 16) is the side of a wedge which makes an angle θ_s with the direction of the approaching stream, all that is necessary is to find a shock wave OW which will deflect the undisturbed flow through the angle θ_s. After passing through the wave the stream will flow parallel to the surface of the wedge and the conditions will be uniform over the region between the wedge and the wave. The pressure ratios x and y, which exist with such waves, are found from the contours of Figs. 12 and 13 by taking lines which make $\beta - \alpha = \theta_s$.

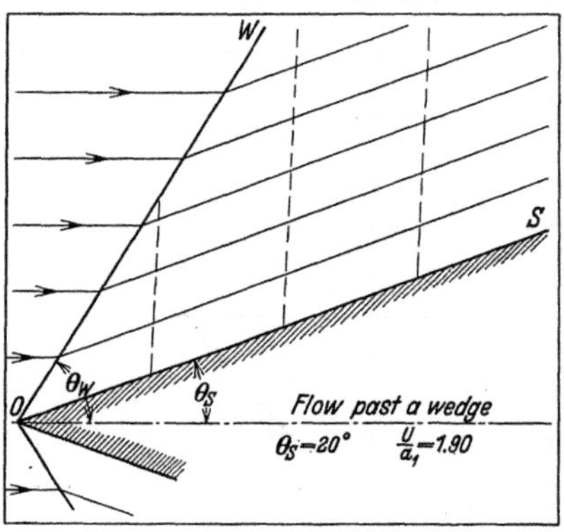

Fig. 16.

Fig. 13 shows that, since a y contour is intersected by a horizontal line at two points, there are two possible waves for one value of U/a_1; it is found, however, that only the wave with the smaller value of θ_w can occur in practice. Fig. 13 shows also that for each wedge angle there is a minimum value of U/a_1 below which the present type of solution does not exist. Photographs reveal that, when this value of U/a_1 is not exceeded, the wave front is neither straight nor attached to the tip of the wedge. The distance between the wave and the wedge increases as U/a_1 decreases till for $U/a_1 = 1.0$, the wave recedes to a great distance ahead of the disturbing body.

The problem of the flow past a cone at supersonic speeds is not so simple as that of the wedge because, in order to satisfy the condition of continuity, the stream-lines must follow curved paths after passing through the shock wave. The velocity and pressure cannot be uniform in the region behind the wave as they are in the case of a wedge. It is necessary, therefore, to solve the hydrodynamic equations in this region and to find a shock wave which is capable of changing the stream of air from its uniform state to a condition which satisfies the solution obtained.

Calculations have been carried out along these lines[1] for three cones of semi-vertical angle 10°, 20°, and 30° when the flow behind the wave is irrotational and the pressure, velocity and density are functions of θ only. Excellent agreement was found when the calculated surface pressures were compared with observations of pressure made in a high speed wind channel and when the calculated position of the shock wave was compared with the position obtained from photographs of projectiles in flight. Fig. 17 shows the flow past a cone of 20° semi-vertical angle at $U/a_1 = 1.65$. The broken lines shown in Figs. 16 and 17 represent stationary wavelets caused by small disturbances in the flow. The angle between the wavelet and the direction of flow is the local Mach angle.

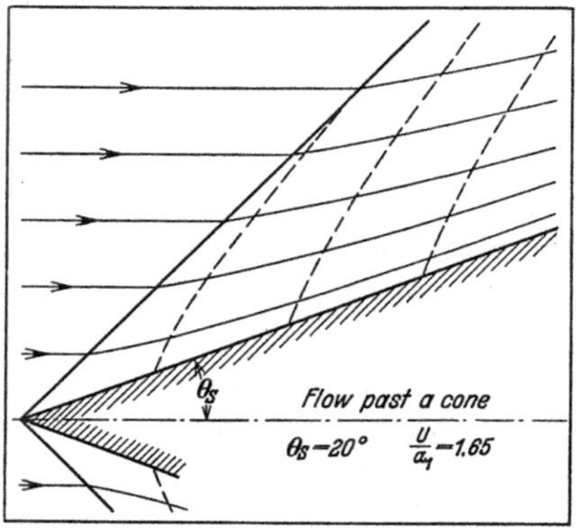

Fig. 17.

Fig. 18 (on Plate VI at the beginning of this Division) shows a photograph[2] of a 1.57 inch diameter projectile in flight at a speed of 1.576 a_1. The head of the projectile is a cone of semi-vertical angle 20°. The wavelets, which are clearly seen along the head and parallel portion, are due to slight irregularities on the surface. Those on the conical part may be compared with the theoretical wavelets of Fig. 17.

5. The Two-Dimensional Flow Around a Corner. The expansion of a uniform two-dimensional stream of gas flowing around a corner at a supersonic speed has been investigated by Prandtl[3] and Meyer[4]. The system considered may be described briefly as follows. A uniform stream of gas flows parallel to a rigid boundary AO at a supersonic speed (Fig. 19). Expansion commences when the stream reaches the radial line OM, which makes, with the direction of flow, the Mach angle M_1 and continues up to a radial line ON. After passing ON the stream flows parallel

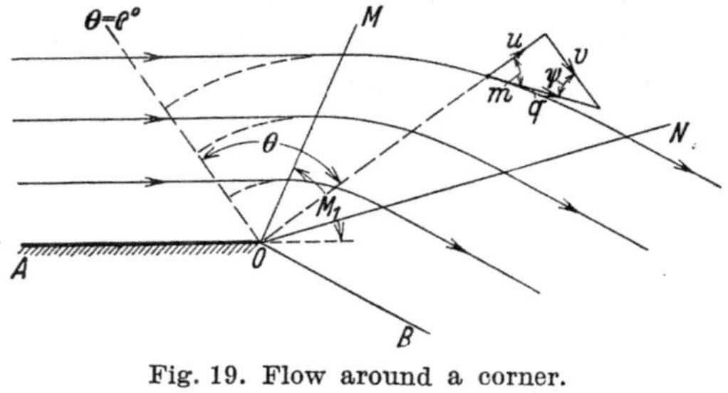

Fig. 19. Flow around a corner.

[1] TAYLOR, G. I. and MACCOLL, J. W., Proc. Roy. Soc., A, Vol. 139, pp. 278—311, 1933.

[2] This photograph was obtained by Mr. T. Harris of the Research Department, Woolwich, and is reproduced here by kind permission of the Ordnance Committee.

[3] Phys. Zeitschrift, Bd. 8, p. 23, 1907.

[4] Forschungsarbeiten auf dem Gebiete des Ingenieurwesens, V.D.I., Heft 62, 1908.

to OB. OB may be either a rigid boundary or the free surface of the gas jet. It is proposed to investigate the expansion of the stream in the region between OM and ON. A similar type of flow in three dimensions can be seen at the shoulder of the projectile in Fig. 18.

The dynamic equations for any two-dimensional fluid motion, expressed in polar coordinates, are:

$$u\frac{\partial u}{\partial r} + \frac{v}{r}\frac{\partial u}{\partial \theta} - \frac{v^2}{r} = -\frac{1}{\varrho}\frac{\partial p}{\partial r}$$

$$u\frac{\partial v}{\partial r} + \frac{v}{r}\frac{\partial v}{\partial \theta} + \frac{u\,v}{r} = -\frac{1}{\varrho}\frac{\partial p}{r\,\partial \theta}$$

where u, v are the components of velocity along and perpendicular to the radius vector through the point considered.

The equation of continuity is

$$\frac{\partial}{\partial r}(\varrho\, u\, r) + \frac{\partial}{\partial \theta}(\varrho\, v) = 0$$

In the present problem we assume that the velocity, density and pressure are constant along a radius. Thus the above equations reduce to

$$v = \frac{d\,u}{d\,\theta} \tag{5.1}$$

$$\frac{v}{r}\left(\frac{d\,v}{d\,\theta} + u\right) = -\frac{1}{\varrho}\frac{d\,p}{r\,d\,\theta} \tag{5.2}$$

$$\varrho u + \frac{d}{d\,\theta}(\varrho\, v) = 0 \tag{5.3}$$

Equation (5.1) is the condition for irrotational flow and shows that this type of motion is implied when we assume that conditions are constant along a radial line.

Since $d\varrho/d\theta = (1/a^2)\, dp/d\theta$, where a is the local speed of sound, (5.2) may be written

$$\frac{v}{r}\left(\frac{d\,v}{d\,\theta} + u\right) = -\frac{a^2}{\varrho}\frac{d\,\varrho}{r\,d\,\theta} \tag{5.4}$$

Substituting in (5.3) the value of $d\varrho/d\theta$ given by (5.4) we have,

$$\left(\frac{d\,v}{d\,\theta} + u\right)\left(1 - \frac{v^2}{a^2}\right) = 0 \tag{5.5}$$

as the general equation to be satisfied by the flow in the region of expansion.

Equation (5.2) shows that $(dv/d\theta + u)$ cannot be zero for all values of θ as the pressure would then be constant throughout the entire field. For (5.5) to be satisfied, we must have, therefore,

$$v = a \tag{5.6}$$

Thus, in two-dimensional flow around a corner, the component of velocity perpendicular to the radius vector must always equal the speed of sound at the local conditions of pressure and density.

By the adiabatic law we have

$$v^2 = a^2 = \frac{\gamma\, p}{\varrho}$$

and Bernoulli's equation

$$\frac{1}{2}(u^2 + v^2) + \frac{\gamma}{\gamma - 1}\frac{p}{\varrho} = \frac{\gamma}{\gamma - 1}\frac{p_0}{\varrho_0} = \frac{1}{2}c^2 \tag{5.7}$$

(where c is the velocity which the gas would attain if allowed to flow in steady motion into a vacuum) then becomes,

$$u^2 + \frac{\gamma + 1}{\gamma - 1}v^2 = c^2$$

Substituting for v from (5.1) we have,

$$\left(\frac{d\,u}{d\,\theta}\right)^2 = \lambda^2 (c^2 - u^2)$$

where $\lambda^2 = (\gamma - 1)/(\gamma + 1)$.

Integration of this equation gives

$$\lambda\,\theta = \sin^{-1}(u/c)$$

when the constant of integration is made to vanish by measuring θ from the radius at which $u = 0$.

The components of velocity are thus given by the equations

$$\left.\begin{aligned} u &= c \sin \lambda\,\theta \\ v &= du/d\theta = c\,\lambda \cos \lambda\,\theta \end{aligned}\right\} \tag{5.8}$$

It will be noted that the flow is pure radial when $\theta = \pi/2\lambda$. For air $\gamma = 1.405$, thus making $\lambda = 0.410$, so that θ is then 219.3^0. The maximum angle through which a two-dimensional stream of air can be deflected by expansion is, therefore, $219.3^0 - 90^0 = 129.3^0$.

To obtain the relationship between the pressure ratio p/p_0 and θ, we have, by Bernoulli's equation,

$$q^2 = u^2 + v^2 = \frac{2\gamma}{\gamma - 1}\frac{p_0}{\varrho_0}\left[1 - \left(\frac{p}{p_0}\right)^{\frac{\gamma-1}{\gamma}}\right] \tag{5.9}$$

while equations (5.8) give

$$\left.\begin{aligned} q^2 &= c^2\,[\sin^2 \lambda\,\theta + \lambda^2 \cos^2 \lambda\,\theta] \\ &= c^2\left[1 - \frac{1}{2}(1 - \lambda^2)(1 + \cos 2\lambda\,\theta)\right] \end{aligned}\right\} \tag{5.10}$$

Comparing these two expressions and remembering that $c^2 = [2\gamma/(\gamma - 1)]\,p_0/\varrho_0$ and $(1/2)(1 - \lambda^2) = 1/(\gamma + 1)$ we obtain finally,

$$\left(\frac{p}{p_0}\right)^{\frac{\gamma-1}{\gamma}} = \frac{1}{\gamma + 1}(1 + \cos 2\lambda\,\theta) \tag{5.11}$$

as the required relationship between p/p_0 and θ.

The local Mach angle is given by $m = \sin^{-1}(v/q)$, and is thus the angle between the radius vector through the point and the direction of motion. Hence ψ, the complement of m, is given by the equation

$$\tan\psi = \frac{u}{v} = \sqrt{\frac{\gamma + 1}{\gamma - 1}}\tan\left[\sqrt{\frac{\gamma - 1}{\gamma + 1}}\,\theta\right] \tag{5.12}$$

Equations (5.11) and (5.12) can be used to calculate the values of θ and ψ which correspond to different pressure ratios p/p_0. The angle ν turned through by the stream (relative to conditions at $\theta = 0$) is $\theta + m - \pi/2$ or $\theta - \psi$ and thus, if suffixes 1 and 2 refer to the conditions at the beginning and end of any expansion, the total deflection of the stream during the expansion will be given by $\nu_2 - \nu_1 = (\theta_2 - \psi_2) - (\theta_1 - \psi_1)$.

The following table gives θ, ψ, ν and q/c calculated for several values of p/p_0:

p/p_0	θ (deg.)	ψ (deg.)	ν (deg.)	q/c
0.527	0.00	0.00	0.00	0.410
0.50	17.30	16.90	0.40	0.426
0.40	39.15	35.05	4.10	0.482
0.30	55.55	45.70	9.85	0.542
0.20	72.10	54.15	17.95	0.609
0.10	92.85	62.40	30.45	0.696
0.05	108.65	67.40	41.25	0.760
0.01	135.55	74.30	61.25	0.857
0.00	219.30	90.00	129.30	1.000

6. Two-Dimensional Flow Past a Curved Surface. Provided no shock waves are formed to cause divergence from the adiabatic law in the region of flow considered, the solution obtained in the last section may be generalized to explain the expansion or contraction of a two-dimensional stream flowing at a supersonic speed along a curved surface. This extension is made possible by the fact that disturbances cannot be propagated forward into the fluid and consequently the conditions at any stage of an expansion or contraction depend only on the angle turned through by the stream. The latter quantity being known from the shape of the surface, it follows that the conditions of flow can be obtained at all points on the surface when the conditions are known at one point.

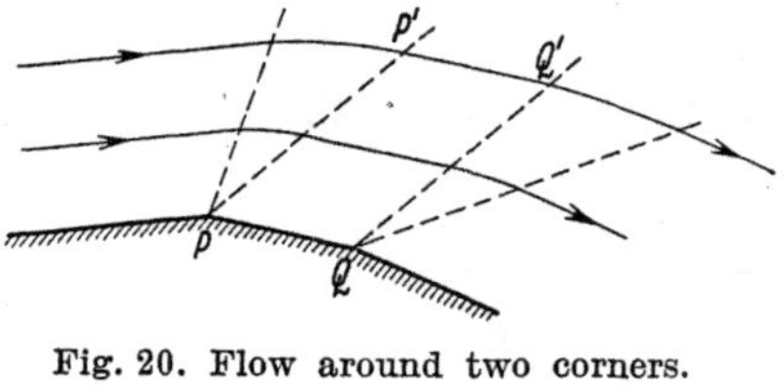

Fig. 20. Flow around two corners.

These considerations will be more clearly appreciated by examination of the flow around two corners P and Q (Fig. 20). After expanding around the corner P, the gas flows in a uniform stream between PP' and QQ'. Expansion around Q then takes place in exactly the same way as if P and Q were coincident. It will be evident therefore that by considering the motion past a set of suitably chosen corners, we can approximate to the flow along a surface of given shape. From this it will be clear also that the conditions at any point on the surface must depend only upon the angle turned through by the stream.

Using this extension of the problem treated above, Prandtl[1], Ackeret[2] and Busemann[3] have solved graphically several two-dimensional types

[1] Stodola Festschrift, Zürich, p. 499, 1929.
[2] Schweiz. Bauzeitung, Bd. 94, p. 179, 1929.
[3] Handbuch der Experimentalphysik, Bd. 4 (i), pp. 428—433, 1931.

of supersonic flow. Their solutions are obtained by means of certain characteristic curves which are, in fact, hodographs of the motion along a stream-line when expansion takes place around a corner.

Fig. 21a shows two characteristic curves R and L. Referred to the origin O, curve R gives the velocity in magnitude (q) and direction (ν) when expansion accompanies an increase of ν in the clockwise sense. When expansion accompanies an increase of ν in the anti-clockwise sense, the characteristic curve is of the form L. In a particular problem, the given conditions of flow are satisfied by a suitable choice of the initial radial line $\nu = 0$.

As an example of the use of these characteristic curves let us investigate the conditions when an extremely thin

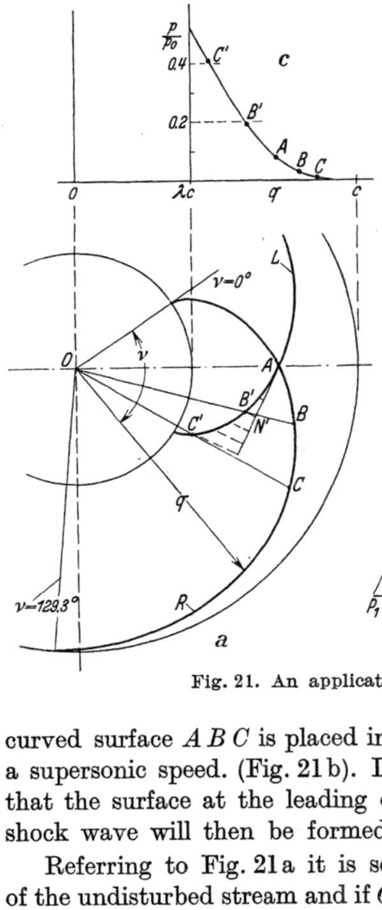

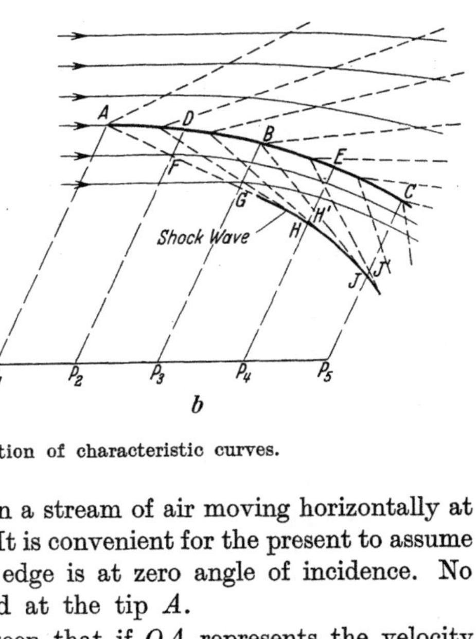

Fig. 21. An application of characteristic curves.

curved surface ABC is placed in a stream of air moving horizontally at a supersonic speed. (Fig. 21b). It is convenient for the present to assume that the surface at the leading edge is at zero angle of incidence. No shock wave will then be formed at the tip A.

Referring to Fig. 21a it is seen that if OA represents the velocity of the undisturbed stream and if $OB'B$, $OC'C$ are parallel to the tangents at B, C, then the velocities at these points on the *upper* surface are represented by OB, OC while the velocities on the *lower* surface are given by OB', OC'. The pressures on both surfaces may be calculated from Bernoulli's equation by inserting the derived values of q, or by reading the values of p/p_0 from the curve shown in Fig. 21c.

It can be shown that a wavelet, which makes the local Mach angle with the direction of flow through the point considered, is at right angles to the characteristic curve through the corresponding point on that curve. Wavelets can thus be drawn in (see Fig. 21 b). Stream-lines can also be traced throughout the whole field of flow since the direction of motion is the same at all points along a wavelet and is governed by the surface angle near the origin of the wavelet.

In the problem discussed above we have assumed that no shock wave is formed at the leading edge of the surface. This restriction is, however, not necessary provided that the wave present at the nose of the body is not very intense. When the angle between the slope of the surface at A and the direction of the approaching stream is small, the conditions in passing through the shock wave follow the adiabatic law closely, as may be verified from the Rankine-Hugoniot relationship by assuming the pressure change to be small. The decrease in speed due to a sudden, but small, change in direction of motion near the leading edge will, therefore, be given by the characteristic curve. The above methods employing characteristic curves can thus be applied to investigate the flow past slender bodies, such as thin airfoils, which have sharp leading and trailing edges.

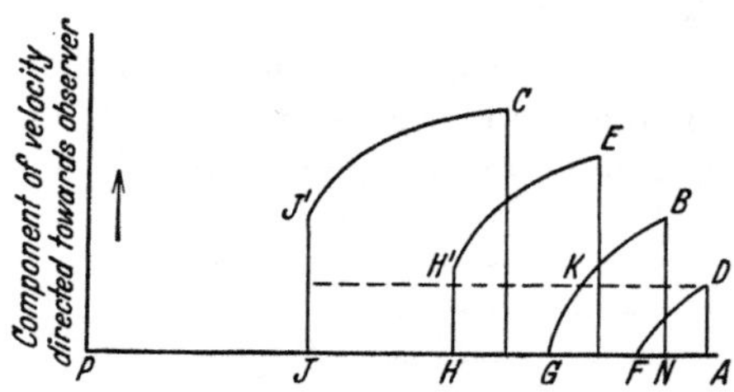

Fig. 22. Formation of a shock wave.

It will be noted that, due to compression of the flow along the lower surface of ABC, the disturbance lines converge in this region and form a shock wave at some distance from the surface. If this distance is sufficiently great, the shock wave will not affect the surface pressure. It is of particular interest, however, to trace the connection between the present conception of the mode of formation of the shock wave with that previously put forward and shown in Fig. 2. Let $P_1, P_2, \ldots P_5$ be successive positions of an observer relative to the moving surface ABC and suppose that the observer is viewing the disturbance produced by it in the direction P_1A which is perpendicular to the approaching wave front, *i. e.* the angle AP_1P_2 is the complement of the Mach angle. In Fig. 21 b the successive positions of the observer's line of sight are shown as the parallel lines P_1A, P_2D, P_3B, P_4E, P_5C.

Fig. 22 shows a number of positions of the disturbance as it approaches the observer. The ordinates denote the components of velocity in the direction of the observer's line of sight. The curves in this figure can be obtained as follows.

The velocity at any point in the disturbed area is given by the corresponding vector in Fig. 21 a. The velocity of the surface itself is represented in Fig. 21 a by AO, so that the velocity of the disturbed air relative to the observer is represented by the vector from A to the appropriate point on the characteristic curve. Thus the velocity relative to the observer, of the air close to B (Fig. 21b), is represented by the vector AB' (Fig. 21a) which has a component AN' directed toward the observer at P_3. It will be seen, therefore, that the ordinates of Fig. 22 can be found from the characteristic curve by taking the components of velocity parallel to the tangent at A. The formation of the wave as it approaches the observer can thus be followed in Fig. 22.

When the observer is between P_1 and P_3, say at P_2, the approaching disturbance increases continuously along the line of sight from the first infinitesimal effect associated with the wave produced at the forward edge A of the moving sheet till the maximum value is attained at the surface of the sheet. This is shown in Fig. 22 as FD. By the time the observer is looking along the line P_3B, the wave front in the line of sight has become much steeper as will be seen in the curve GB of Fig. 22. When the observer is looking along the line P_4E the disturbance has taken the form $HH'E$ (Fig. 22) and there is now a finite shock wave present at the forward edge of the disturbance.

The sequence represented by the curves FD, GKB, $HH'E$, $JJ'C$ of Fig. 22 could also have been viewed from the standpoint of the conception of shock wave formation illustrated in Fig. 2. In this way for instance, it will be seen that the disturbance represented by FD in Fig. 22 would become steeper as it proceeded toward the observer, as shown at GK, and finally a discontinuity would set in before the condition represented by $HH'E$ was realized.

Bibliography.

General.

The following general works deal with the mechanics of compressible fluids and contain numerous references to original papers. The bibliography in Prof. Bateman's paper is very complete.

Bateman, H.: National Research Council, Bulletin No. 84, Report of the Committee on Hydrodynamics, Part IV, (Washington, 1932).

Prandtl, L.: Abriß der Strömungslehre, Section IV, (Braunschweig, 1931).

Ackeret, J.: Handbuch der Physik, Volume 7, Chapter 5, (Berlin, 1927).

Busemann, A.: Handbuch der Experimentalphysik, Volume 4, Part I — "Gasdynamik", (Leipzig, 1931).

Pistolesi, E.: Aerodinamica, (Turin, 1932).

Supplementary List of Original Papers.

The following short list of original papers supplements the references made in the text of the present article.

Stokes, G. G., On a Difficulty in the Theory of Sound, Phil. Mag., (3), **33**, p. 349, 1848; Math. and Phys. Papers, Vol. 2, pp. 51—55.

REYNOLDS, O., On the Flow of Gases, Phil. Mag., (5), 21, pp. 185—199, 1886; Scientific Papers, Vol. 2, pp. 311—320.

RAYLEIGH, Lord, Aerial Plane Waves of Finite Amplitude, Proc. Roy. Soc., A. 84, pp. 247—284, 1910; Scientific Papers, Vol. 5, pp. 573—610.

TAYLOR, G. I., The Conditions necessary for discontinuous Motion in Gases, Proc. Roy. Soc., A. 84, pp. 371—377, 1910.

STEICHEN, A., On the Motion of a Gas in Two Dimensions, Journal Indian Math. Soc., 3, pp. 7—15, 53—65, 1911; Dissert. Göttingen, 1909.

RAYLEIGH, Lord, On the Discharge of Gases under High Pressures, Phil. Mag., (6), 32, pp. 177—187, 1916; Scientific Papers, Vol. 6, pp. 407—415.

v. KÁRMÁN, Th. and MOORE, N. B., Resistance of Slender Bodies Moving with Supersonic Velocities, with Special Reference to Projectiles, Trans. Amer. Soc. Mech. Eng., 54, pp. 303—310, 1932.

Note on the Work of A. Tchapliguine.

Prof. D. RIABOUCHINSKY (Comptes rendus de l'Académie des Sciences, Paris, **194**, p. 1215, 1932) has drawn attention to the work on two-dimensional flow at speeds below the speed of sound carried out thirty years ago by A. TCHAPLIGUINE. In this work, which deals with the motion in regions bounded by straight walls and free streamlines, it is assumed that the gas can separate from the rigid boundary in a manner similar to that considered in the Helmholtz-Kirchoff theory of the flow of a perfect incompressible fluid. Unfortunately an account of TCHAPLIGUINE's theory could not be included in the present article but full details will be found in several papers by B. DEMTCHENKO (Comptes rendus, Paris, **194**, p. 1218 and p. 1720, 1932; Publications Mathématiques de l'Université de Belgrade, Vol. 2, p. 85, 1933).

DIVISION I

EXPERIMENTAL METHODS—WIND TUNNELS

EDITOR'S PREFACE

In common with the advances in all branches of science, the developments of theoretical aerodynamics must constantly be checked and the direction of new avenues of research guided by appeal to experimental investigation. From the beginning of the special developments which have marked the advances in aeronautics during the present century, theoretical advance and experimental test have gone hand in hand.

However, the conditions of airflow about a model in a constrained or limited stream of air are not, in all respects, the same as those for an airplane flying free. There results the need for various corrections depending on the shape and dimensions of the flow relative to the model and on the conditions of flow at the point of observation, jet free or constrained by rigid walls.

In Part 1 of the present Division, some historical account is given of the early beginnings of aerodynamic research, together with brief descriptions of the various types of wind tunnels which have come into use during the past quarter century, and including a discussion of the economic and energy relations inherent in these various types.

The major theoretical discussion of the Division is then presented in Chapter III, wherein are developed formulae and practical methods relating to the corrections referred to above and which must be applied to wind tunnel observations in order to render them comparable with results in free flight.

Again the dependence of the results of aerodynamic performance on the so-called Reynolds number and the impossibility in experimental work with models in atmospheric air of approaching values of this number representing conditions full scale, has in recent years, brought into special prominence the significance of experimental work carried on in air under high pressure and under conditions which permit of a nearer approach to full scale Reynolds numbers.

In the same way the recent rapid increase of flying speeds and the outlook toward still further advance in this direction, together with the high speeds now reached by the tips of propeller blades, all conjoin to emphasize the importance of more extended information regarding the phenomena of flow and the force reactions involved under these high velocities.

Part 2 of this Division presents some theoretical discussion bearing on the design and use of wind tunnels intended for use under these extreme conditions, together with descriptive matter relating to recent installations at the Langley Memorial Laboratory, U.S.A.

Supplementary to the descriptive material presented in the Division, numerous references are given in the Bibliography to experimental aerodynamic installations in various countries, wherein may be found descriptive matter in detail regarding the form and construction of such installations and of their instrumental equipment.

W. F. Durand.

EXPERIMENTAL METHODS—WIND TUNNELS

PART 1

By

A. Toussaint,

Saint-Cyr-l'École

CHAPTER I

CLASSIFICATION OF METHODS—USE OF NATURAL WINDS

1. Classification of Experimental Methods Used in Aerodynamic Research. The most natural method of classification of the methods of aerodynamic research is based on the nature of the movement impressed, either on the body under investigation or on the fluid in which the experiment is carried on. We therefore distinguish the two *principal categories.*

(1) Methods in which the object is stationary and the fluid is in motion.

(2) Methods in which the object moves with reference to a stationary fluid.

The *first* category comprises two secondary divisions:

(*a*) Methods utilizing the natural movement of the air in aerodynamic problems or the natural flow of a stream for hydrodynamic research.

(*b*) Methods utilizing an artificial current of air produced by some type of fan. Such installations are commonly known as Wind Tunnels or Wind Channels.

The *second* principal category comprises three secondary divisions:

(*a*) Methods involving a straight line movement of the body as in the case of the study of the motion produced by the fall or the rise of a body in a resisting medium; or again, methods involving the use of an aerodynamic carriage carried on a cable or on a railway line, and, by analogy, methods commonly employed in hydraulic testing channels and naval tanks.

(*b*) Methods comprising the circular movement of the body in installations either for air or water research. Of the same general character are methods involving the use of a periodic movement as of a pendulum.

(*c*) Methods involving any movement of the body whatever, such as those which utilize an airplane or an airship in motion as a flying laboratory, and by means of special instruments set up on board, provide for a measure of the motion relative to the air or, otherwise, the character and amount of the stresses to which the various parts may be subject through the action of the fluid pressures.

2. Methods in which the Object is Stationary and the Fluid in Motion. (Methods Utilizing Natural Movements of the Air.) These methods are no longer in common use. They involve numerous difficulties arising chiefly from the irregularities of the wind, both in magnitude and direction and which are difficult of measure.

For example the ordinary anemometer which might give, with sufficient accuracy the magnitude of the velocity, is in general seriously influenced by the variations in direction, especially if it can turn only in a horizontal plane. In such case, the variations in the direction in a vertical plane will affect more or less seriously, the indications of the instrument. Furthermore the vanes intended to indicate the direction of the wind are in general imperfect and lack the needed sensitiveness. The influence of their proper inertia renders impossible suitable response to very rapid changes in the direction of the wind.

However the more modern wind measuring or indicating devices using a hot wire do not have these disadvantages and would replace advantageously, for certain researches, the older instruments used in early experiments with natural wind.

In spite of such difficulties, however, these methods are worthy of attention because, if judiciously employed with all the modern improvements, they might well contribute, in particular, to a study of the soaring flight of birds.

Following this thought, note will be made at a later point of the experiments of M. Idrac. Measuring the velocity of the wind V and its ascending component W in regions where soaring birds are found, and knowing furthermore the weight P and the wing surface S of these birds, he was able to determine the ratio $C_x/C_z = W/V$ and the coefficients $C_z = P/(1/2)\,\varrho\, S\, V^2$ and C_x. The average values found were

$$\frac{C_x}{C_z} = \frac{1}{16}$$

with $C_z = 1.40$, whence $C_x = 0.0875$.

These values are remarkable, as well for the size of C_x/C_z as for that of C_z.

Similar experiments could be carried out with the aid of suitably designed aerodynamic balances carrying lifting surfaces and exposed to the natural winds in countries or regions where conditions favor soaring flight. The purpose of such experiments would be to determine the over all effect of irregular wind upon a lifting wing rather than to measure the aerodynamic characteristics of such a wing. It is known, indeed, that the action of an ascending wind or even of an oscillating wind (Katzmayr effect) upon a fixed wing may give sustentation without drag or even with a drag negative. These properties, demonstrated in artificial currents of air might be advantageously sought in the natural winds of various regions.

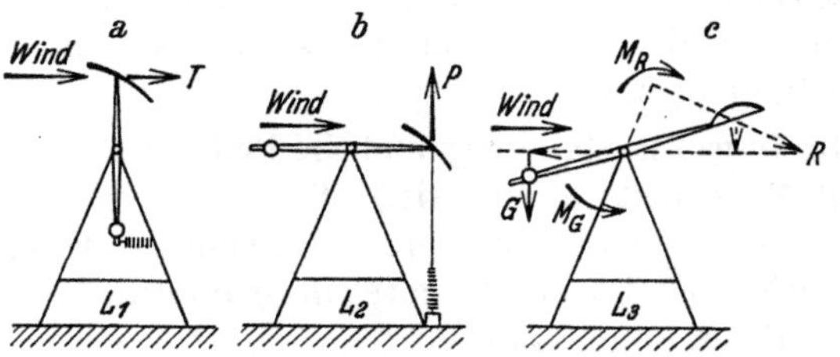

Fig. 1.

The utilization of natural winds might be further investigated through the examination of the aerodynamic forces acting on elevated constructions or on air motors.

In any case the experimental means to be employed vary with the particular conditions and the study of such means can best be made through a description of some of the principal applications of this general method of procedure. To this end the following experiments may be briefly noted.

1) Lilienthal 1889
2) Wellner 1893
3) Stanton 1907—1908
4) Pannell 1915—1916
5) Idrac 1919—1922
6) Huguenard, Magnan and Planiol 1924.

Lilienthal, in 1889 made use of natural winds for the measurement of the aerodynamic properties of lifting wings of various forms. These, fixed at a suitable angle, were carried on a horizontal axis under spring control, the whole supported on a rigid structure.

Under these conditions, the arrangement of Fig. 1a would serve to measure the horizontal component of the wind force and that of Fig. 1b, the vertical component. Again with the arrangement of Fig. 1c, the surface takes a position of equilibrium such that the moment due to the wind forces with reference to the point of support is equal to that of gravity. These measurements will then serve to give the center of pressure on the surface. The surfaces tested in this manner were 1.8 meters span and 0.40 meters chord. Such simple apparatus could not give results of great accuracy; nevertheless, the experiments of Lilienthal served to indicate the special properties of curved or arched surfaces.

In 1893, Wellner made use of natural winds for the measurement of the components of the air forces on surfaces slightly inclined to the direction of the wind. To this end, five different pieces of apparatus were employed, of which two were for the measure of the velocity and direction of the wind and the three others for the measure of the forces on the surfaces.

For anemometer, Wellner made use of a plate form as in Fig. 2a, assuming $R = 0.125\,S V^2$ at 15^0 C and 760 m/m.

For a wind vane he made use of the arrangement of Fig. 2b composed of two vanes, one for indicating inclinations vertically and the other for inclinations horizontally. These two vanes were mounted with a universal joint in such manner that they were in neutral equilibrium in calm air. The apparatus for the measure of the components of the wind forces was of the general character of a disc anemometer, the surface being fixed at a suitable angle on an articulated or guided system which turned face to the wind through the action of a wind vane. The inclination of the articulated system or the pointer on the spring control indicated the magnitude of the components of the resultant wind force. See Fig. 3.

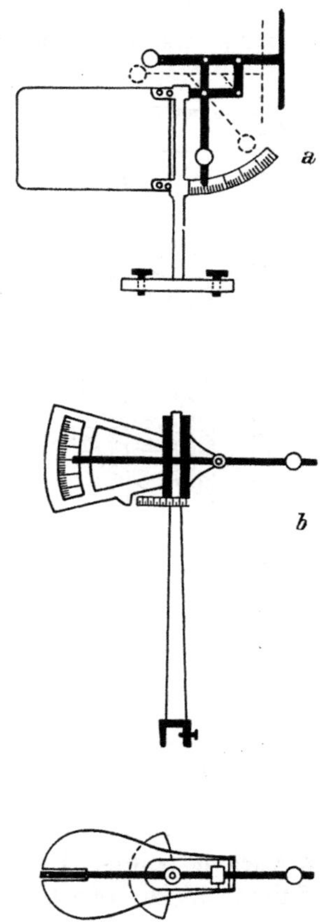

Fig. 2.

These aerodynamic balances comprise very ingenious applications of systems of articulated parallelograms. As mechanical devices they are well-known and are of advantageous use in that they permit the determinations of the applied forces independently of the couples which complete the knowledge of any system of forces acting on a system. However, such devices being exposed as a whole to the wind, their indications, as regards the resultant forces on the wing, require correction 1) for the aerodynamic action on the balance itself and 2) for the interaction between the balance and the wing under test. Any modern research seeking to make use of similar devices would need to make sure of the elimination of these two sources of error.

It may be further noted that with devices of this general character, the direct observations will give the components, horizontal and vertical, rather than parallel and normal to the direction of the wind. For the lift force the difference is not great, but for the resistance or drag, there must be added or subtracted, the component of the sustentation in the direction of the wind.

In 1908 Stanton made measurements of the wind pressure on plane surfaces going to dimensions of about 10 feet square. These surfaces were fixed at the top of a tower of about 50 feet height. The wind velocity was determined by means of a Dines anemometer (combination of Pitot tube and Dines tube). The surface was supported on a movable frame and the aerodynamic pressure, transmitted by a steel rod to a steel diaphragm, was balanced against a calibrated spring. The diaphragm formed furthermore the bottom of an air cylinder which communicated with a water manometer. This latter was connected with a second air cylinder, similar to the first, in such manner as to compensate for changes in barometric pressure. The manometer indicated the displacement of the diaphragm and a direct calibration served to give the corresponding forces in play. Simultaneous readings were made on the anemometer and on the manometer and from a number of such sets of readings, coefficients were derived for large plates normal to the wind as follows:

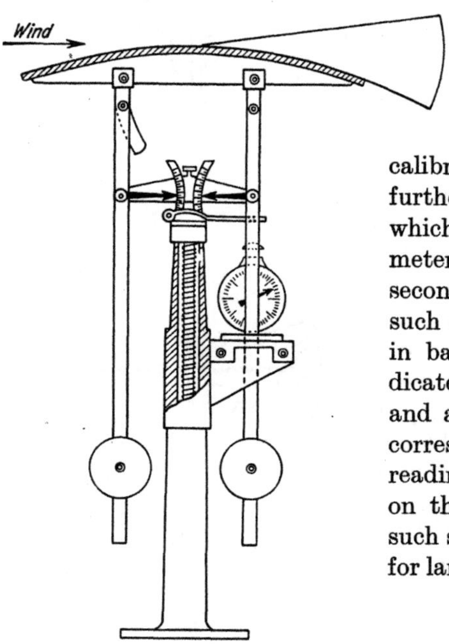

Fig. 3.

Size	*K*
5 ft. × 5 ft.	0.08
10 ft. × 5 ft.	0.079
10 ft. × 10 ft.	0.081

These overall measurements were verified by local measurements at various points. It was thus shown that the pressure at the center of the front face of a large plate was the same as that measured on a plate of small dimensions; but that the depression on the after face of the large plate was somewhat greater (11% in the case of two plates 1.0 foot and 2 in. on the side). For a large plate 10 ft. × 10 ft. the ratio of the depressions as compared with those for a small plate reached 1.18. A similar result was found for the overall forces on a girder structure (0.10/0.084 = 1.18).

In 1909 at the N. P. L. Teddington, England, two whirling arm aerodynamic towers were constructed of the dimensions:

Height	62 ft.
Base	14 × 14 ft.
Top	8 × 8 ft.

They were surmounted by a turning platform 20 ft. × 4 ft. The two towers were separated by a distance of 350 ft. and the observing station was placed between the two, the pressures being transmitted thence by means of lead pipes[1].

One cause of uncertainty in these measurements arose from the variations in the velocity of the wind at any one instant of time at various points in the system. Dines found variations in the ratio from 0.75 to 1.25 for points distant no more than 10 feet. Stanton found variations even at distances of 2 feet.

Sir Benjamin Baker also made experiments in natural winds at the Forth bridge. Use was made of a panel of wood 16 ft. × 16 ft. and of a plane surface about 13 in. × 13 in. The mean of the daily maxima of pressures was greater for the large panel.

In 1915—16, Pannell at the N. P. L. was led to carry out measurements on spheres in natural winds. It is known that the coefficients of resistance for a sphere are subject to considerable changes in value for slight changes in velocity or in the nature of the surface. Likewise the size of the tunnel, the regularity of the current of air, its degree of turbulence, etc. also have an important influence on the values of this coefficient.

These experiments were made on one of the 62 foot towers provided with a turning platform. A special balance was constructed. The arm was horizontal and it was brought to the zero position (with the wind acting on the sphere) by means of a micrometer screw, as with torsion balances. The zero was determined in calm days by placing the arm of the balance parallel to the wind. The wind velocity was read by means of a standard Pitot tube connected to a U form manometer, the tube being partially orientable in the wind. The platform was brought to a suitable position and numerous readings were made, simultaneously of wind velocity and force. The sphere used was 6 in. diameter.

The results were arranged as a function of the velocity and a mean of 10 observations was utilized for each point marked down. The velocities varied from 20 to 50 feet per second. The resultant points were considerably scattered but in general lie below the curve found for spheres in the wind channels of the N. P. L. It may be concluded that the turbulence of natural winds is similar or perhaps less than that of the air in the wind channels of the N. P. L.

During the years 1919—21 M. Idrac undertook the study, "on the spot", that is in regions specially suited to soaring flight, of different forms of the internal energy of the wind susceptible of being utilized by soaring birds.

Two types of internal energy in the wind may be distinguished—*static* and *dynamic*. The first depends on the condition of the air as

[1] Stanton, The Experimental Wind Towers. Report British A.R.C., p. 17, 1909—10.

regards pressure and temperature; the second depends on the irregularities in the velocity of the wind, both in magnitude and direction, and especially on the magnitude of the vertical component of this velocity.

M. Idrac was able to carry out the experimental measures needed to characterize these sources of energy, in different regions and especially in Africa with direct reference to the great soaring birds—Kites and Vultures (1919 at Constantine; 1919 autumn in Senegal, in French Guinea and the French Soudan, 1921 spring in Senegal, 1921 autumn, Ivory Coast and Senegal).

To carry out these measures, M. Idrac made use of special apparatus designed in view of the difficulties of experimentation in the African jungle and by a single observer. The chief characteristics of this experimental equipment may be summarized as follows:

a) Measurement of temperatures. Use was made of a thermometric antenna based on the change in electrical resistance of platinum wire with temperature. The antenna was composed of a spiral of platinum wire of resistance 300 ohms contained in a blackened tube open at the two ends to permit the passage of the air. This antenna was carried by a kite and connected by wire to one arm of a Wheatstone bridge with "invar" resistances. The deviations of the galvanometer give then an indication of the temperature at the height of the antenna. The characteristics of the assemblage were chosen in such manner that the warming of the wire itself by the passage of the current was negligible and assurance was found that the passage of a current of air from 0 to 20 m. per sec., dry and at constant temperature, had no effect on the indications of the galvanometer. This feature differentiates this form of antenna from that employed in the so-called *hot-wire* type of anemometer.

Such an arrangement comes to temperature equilibrium with the air in less than 3 seconds and its sensitiveness makes possible the recognition of changes of 1/40 degree Cent.

b) Measurement of pressure. The barometric antenna, for such use, must be of great precision. It was composed of a thermos receptacle, of which the mouth directed downward communicated with a tube with three branches in which a conducting liquid preserved a level independent of the slope of the antenna and of the aerodynamic effects on the lateral branches. Differences of pressure, then, between the interior of the receptacle and the external air cause a change of level of the liquid which is evaluated electrically (change in the resistance of a circuit comprising a galvanometer, the resistance of the liquid and an adjustable resistance). The sensitiveness at a fixed point was less than 5 cm. height. Practically, on the kite, less than 1 meter could not be counted upon.

c) Changes and Irregularities in the Velocity of the Wind. The anemometric antenna was composed of a kite [1.80 $\times$ 1.0 (meter)] with two triangular wings and two triangular prismatic cells. The tension on the cable was transmitted to a recording dynamometer and the preliminary calibration had shown that the pull was sensibly proportional to the square of the velocity.

d) Changes or Irregularities in the Direction of the Wind. The antenna for this purpose was also formed of a kite the movements of which were followed by means of a sighting device connected with a registering pencil. The inertia of the kite, however, permitted the registration only of changes in direction non-instantaneous in character.

e) Study of the Vertical Component of the Wind Velocity. For velocities below 4 m. per sec. use was made of sounding balloons (use of kite not practicable). This method consists simply in determining by observation the vertical ascent during a specified period of time.

In the absence of an ascending current, the vertical velocity v_0 of the balloon is known. If the vertical air velocity is v, there results the equation,

$$h = (v + v_0)\,(t_2 - t_1)$$

whence v, knowing h, v_0 and $(t_2 - t_1)$.

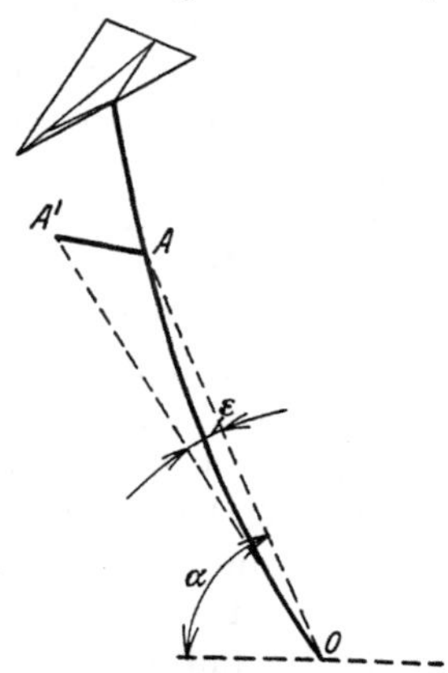

Fig. 4.

The value of h was found by measuring the angle subtended by a base of 1 meter carried by the balloon (thus giving the distance) together with the angle of elevation above the horizon measured with a suitable instrument (Lunette de Rochon).

This method gives an accuracy of measurement of the vertical velocity of about 0.10 m. per sec. when the balloon is at a distance of about 1000 meters.

For vertical velocities greater than 4 m. per sec., use was preferably made of a kite with streamers (see Fig. 4). To the kite line, a few meters below the kite, a streamer of silk muslin was attached which places itself sensibly at an angle ψ with the direction of the wind, which angle ψ depends on the velocity of the wind and on the weight of the streamer. ($\psi = 4^0 - 50'$ for $v = 4$ m. sec; $\psi = 1^0 - 50'$ for $v = 10$ m. sec.). The inclination of the streamer φ_1 with the horizon is measured by the use of the Rochon lunette and there results the inclination of the wind,

$$\varphi = \varphi_1 + \psi.$$

M. Idrac also made use to the same end of the kite itself. To this end he established theoretically the conditions for which the angle α or height of the kite above the horizon could become a simple function of the angle φ of the wind with the horizon. The vertical component may then be deduced if there is given the wind velocity V.

M. Idrac was able to determine, through these means that the African soaring birds hold themselves preferably in zones of ascending air which are formed above the ground and which are subject to movements of displacement without apparent law, but which the birds seem able to recover. The origin of such ascending currents may be due, at least in part, to solar heat (convection vortices).

In more recent years, M. M. Huguenard, Magnan and Planiol have measured the aerodynamic characteristics of a glider with flexible wings under the action of natural winds. The glider was suspended from its center of gravity at a height of 4 or 5 meters above the ground and the horizontal and vertical components of the aerodynamic resultant were registered on suitable dynamometric apparatus. At the same time an anemometer measured the velocity of the wind.

It may be added that these same authors have made notable studies on the internal structure of natural winds, studies in which use was made of anemometric devices of their own invention[1].

Finally as regards the action of natural winds on stationary structures, mention may be made of the researches of Dryden and Hill.

A cylinder 10 ft. diameter and 30 ft. high (representing a factory chimney) was exposed to the wind. The distribution of aerodynamic pressures on a number of sections suitably chosen, was measured. In addition the overall tipping couple due to the action of the wind was measured. In accordance with these measurements, the safety of such structures should be secured through the use of a coefficient $F/(1/2)\,\varrho\, S\, V^2 = 0.78$.

References.

DRYDEN, H., and HILL, G., Wind Pressure on Circular Cylinders and Chimneys, Journal of Research U.S., Bureau of Standards, Vol. 5, p. 653, September 1930.

EIFFEL, G., Résistance de l'air, 1910, p. 194.

IDRAC, Etudes expérimentales sur le vol à voile, 1921—22, L'Aérophile Janvier, Février, Mars 1922.

HUGUENARD, MAGNAN et PLANIOL, Sur l'étude aérodynamique des ailes d'oiseaux et des voilures souples, 1924, C. R. Acad. des Sciences 7-I-1924 et Bulletin du Service Technique de l'Aéronautique No. 21, 1924; No. 74, 1931.

LILIENTHAL, O., Der Vogelflug als Grundlage der Fliegekunst 1889, Revue de l'Aéronautique 8ème. Année p. 22, 1895.

PANNEL, Experiments on the Resistance of Spheres, British A.R.C. R. and M. 190, p. 19, 1916.

STANTON, The Experimental Wind Towers, Report of the British A.R.C. p. 17, 1909—10.

— Experiments on Wind Pressure, 1907—1908, Proceedings of the Institution of Civil Engineers, Vol. C. L. XXI.

WELLNER, Expériences sur la résistance de l'air dans le cas de surfaces courbes en mouvement rectiligne, 1893, Revue de l'Aéronautique 7ème. Année, p. 57, 1894.

[1] Bulletins du S. T. Ae No. 32, 1926, nos 73 and 74, 1931, etc.

CHAPTER II

METHODS USING AN ARTIFICIAL AIR FLOW—CLASSIFICATION OF WIND TUNNELS, DESCRIPTION, ENERGY LOSSES

1. Methods Using Artificial Fluid Currents. The experimental equipment employed in methods of this type are generally known under the name of "Wind Tunnels". Aerodynamic laboratories make use of installations of this type of the most varied forms and dimensions.

It may be said that such installations now replace almost entirely all other experimental methods. They permit of observation, continuous, methodical and rapid and are therefore susceptible of a high order of experimental efficiency. Almost the total vast number of aerodynamic coefficients relating to the action of air in motion on a body of given form have been measured in laboratories of this character. It follows that the present Division of this work should be devoted, in major part, to a study of equipment of this general type.

2. Classification of Wind Tunnels.

(*a*) *From the point of view of the mode of action on the fluid.*

1) Wind tunnels with the point of observation on the downstream side of the fan.

2) Wind tunnels with the point of observation on the suction or upstream side of the fan.

3) Wind tunnels with the circulation of air guided and continuous (closed circuit).

(*b*) *From the point of view of the conditions of the fluid vein at the point of observation.*

1) Wind tunnels with the flow constrained by solid walls.

2) Wind tunnels with the flow free; that is, limited only by the air surrounding the moving stream. This surrounding air is at rest at a small distance from the moving stream.

Each of the types under classification (*a*) may have indifferently its air stream *constrained* or *free*. In the present study the general classification (*a*) will be followed. However, it must be noted that the needs of modern research lead to the consideration of another criterion of classification—that furnished by the influence of the laws of *kinematic similitude.*

From this point of view there may be distinguished:

(1) Wind tunnels which may operate under very high Reynolds numbers for researches relating to the effect of similitude, due, chiefly, to the influence of the viscosity of the air.

(2) Wind tunnels operating at very high velocities for researches relating to the effect of similitude, due, chiefly, to the influence of compressibility of the air.

Discussion and descriptive matter relating to these special types will be found in Part 2 of the present Division.

In advance of a study of these various forms of wind tunnel and of the equipment appropriate thereto, certain general characteristics must be noted and defined. Such are: the "coefficient of utilization", efficiency, losses of energy from various causes, and the degree of turbulence.

Certain other questions, common to these various types of installation, will, however, be reserved for consideration at a later point. Such are

Honeycomb constructions

Direction vanes

Fans

Influence of the limitation of the air stream; corresponding corrections.

Influence of the gradient of static pressure in the air stream; corresponding corrections.

3. Coefficient of Utilization—Efficiency—Losses. Under steady conditions of operation, the air which circulates in a wind tunnel traverses an irreversible cycle characterized by:

(*a*) Transformations of energy, kinetic and potential, which, theoretically, are realized without corresponding losses.

(*b*) Losses of energy in the air, suffered in various parts of the circuit.

(*c*) A receipt of energy on the part of the air during its passage through the fan.

The energy communicated to the fluid by the fan is, under the conditions of steady flow, equal to the sum of the losses suffered.

If we denote the efficiency of the fan by η_f and the power which it receives from the motor by P_m, then evidently

$$\eta_f P_m = \text{(losses)} \tag{3.1}$$

The coefficient of utilization of the tunnel, or *tunnel coefficient* as it may be conveniently termed, is then the ratio of the kinetic energy of the column of fluid at the experimental section, to the energy given to the air by the fan[1].

Let the velocity of the air, assumed constant over the experimental section, be denoted by V_0 and the area of the section by S_0. Further, let the density of the air be denoted by ϱ_0, again assumed constant at all ordinary velocities, for which the effects of compression may be neglected. Then since $\varrho_0 S_0 V_0$ measures the mass m_0 passing per unit time, the kinetic energy at the experimental section will be

$$P_u = \frac{m_0 V_0^2}{2} = \frac{1}{2} \varrho_0 S_0 V_0^3 \tag{3.2}$$

Then in accordance with the preceding definition, the tunnel coefficient will be

$$C_t = \frac{P_u}{\eta_f P_m} = \frac{P_u}{\Sigma\,\text{(losses)}} \tag{3.3}$$

[1] This ratio is sometimes referred to as an *efficiency*. Since its value is normally greater than 1, however, it seems preferable to characterize it as a *coefficient*.

By analogy the overall coefficient of the assemblage of tunnel and fan will be

$$C_{t\,m} = \frac{P_u}{P_m} = \eta_f\, C_t \tag{3.4}$$

The corresponding efficiencies η_1 and η_2 may likewise be defined as follows:

Thus the efficiency of the tunnel will be the ratio

$$\eta_1 = \frac{P_u - \Sigma\,(\text{losses})}{P_u} = 1 - \frac{1}{C_t} \tag{3.5}$$

And similarly the efficiency of the assemblage of fan and tunnel will be the ratio

$$\eta_2 = \frac{P_u - P_m}{P_u} = 1 - \frac{1}{C_{tm}} \tag{3.6}$$

The difference $P_u - \Sigma$ (losses) represents in effect the energy apparently disposable in the fluid. The expression "apparently disposable" signifies that in reality it could not all be recovered (by an air motor for example) due to the additional losses which would be introduced by the presence of any such device. In the same way the difference $P_u - P_m$ represents the energy apparently disposable in the fluid, taking into account the fact of losses at the fan.

Preferably, use is made of the coefficients C_t and $C_{t\,m}$ for the characterization of wind tunnels from the point of view of the power required to obtain a given velocity of flow at the experimental section.

It will be seen that these coefficients depend chiefly on the losses suffered by the air in the circuit. These losses are due in major part, to the following causes.

a) Surface friction of the air along the walls of the tunnel.

b) Variations of section area, abrupt or gradual.

c) Changes in direction of flow, guided or non-guided.

Certain of these losses can be calculated with a sufficient degree of approximation, as will be seen in certain special cases. However, such determinations are more commonly based on experiment with a reduced size model of the tunnel.

These remarks apply equally to the efficiency of the fan, for a discussion of which reference may be made to Division L XI.

4. Degree of Turbulence of Wind Tunnels. In all wind tunnels, under normal regime, the Reynolds number at the experimental section greatly exceeds the critical value $R \simeq 2000$, below which the flow may be laminar in type. Thus, even in the absence of any other perturbing cause, and especially aside from the presence of a peripheral layer of air directly disturbed as a result of surface friction, it is to be expected that there will exist a certain general turbulence over the cross-section of flow at the point of observation.

The influence of this turbulence on aerodynamic resistance is appreciable principally for cylinders, spheres, ellipsoids, fusiform bodies, either cylindrical or of revolution, and lifting wings. The differing values, frequently found among the results obtained in different laboratories, may, for such forms, be imputed to the differing degrees of turbulence of the air stream in these various laboratories. It is then desirable to be able to define quantitatively the *degree of turbulence* of an air stream in a wind tunnel.

In the present state of our knowledge regarding the laws which govern the turbulent flow of fluids, and for the special case of wind tunnels, the following definition, proposed by Messrs. H. L. Dryden and A. M. Kuethe[1] may be adopted.

The turbulence at a given point is taken to be the ratio of the square root of the mean square of the deviations of the speed from its mean value to the mean value.

The variations of the local velocity with reference to the mean velocity are measured during intervals of time which depend on the inertia and other characteristics of the anemometer used for this purpose. For measurements of this character, the authors noted above recommend the use of a suitably compensated hot wire anemometer[2].

The preceding definition which takes account only of the variations in the local velocity, is applicable over the zone of constant mean velocity, to the exclusion of the peripheral zones or the zones directly influenced by the wake of objects upstream. Practically, this definition is then suited to the case of wind tunnels comprising an experimental section of constant mean velocity and sufficiently removed from the influence of honeycombs upstream. Under these conditions, it is noted that, in general, the turbulence varies but little over the zone of constant mean velocity.

The direct measures of the degree of turbulence, through the use of a hot wire anemometer, were carried out by the authors noted above, for three types of wind tunnel. In these same tunnels they carried out, furthermore, measurements on the resistance of spheres of varying size. It is known that the values of the coefficient of resistance of spheres present notable variations over a certain zone of variation of the Reynolds number R, defined as follows

$$R = \frac{VD}{\nu}$$

V = velocity of flow,
D = diameter,
ν = kinematic viscosity of air.

---

[1] DRYDEN, H. L. and KUETHE, A. M., Effect of Turbulence in Wind Tunnel Measurements, U.S. N.A.C.A. Technical Report No. 342, 1930.

[2] DRYDEN, H. L., and KUETHE, A. M., The Measurement of Fluctuations of Air Speed by the Hot-Wire Anemometer, U.S. N.A.C.A. Technical Report No. 320, Washington, 1929.

In these experiments verification was found of results developed previously by Wieselsberger[1], showing that representative curves of values of the coefficient $C = \frac{F}{(1/2)\,\varrho\, S\, V^2}$ as a function of the Reynolds number R, are displaced stepwise, as it were, relative to each other, according to the degree of turbulence in the tunnel in which the test is made.

Messrs. Dryden and Kuethe found that the mean displacement of these curves $C = f(R)$ for the mean value $C = 0.3$, corresponded to the regularly decreasing values of the Reynolds numbers with increase in the degree of turbulence as measured directly with the hot wire anemometer.

Other comparisons, more complex in character were also made by these authors between the degree of turbulence thus measured and the coefficients for fusiform bodies of revolution. From this study the following general conclusions may be drawn:

The results obtained in different wind tunnels regarding the aerodynamic resistance of bodies of certain forms, even with all things the same otherwise, can only be compared under conditions of equal turbulence. This latter may be characterized by the Reynolds number for which the coefficient of resistance of a sphere has the value 0.30. To this end use may be made of the following table deduced from the experiments of Dryden and Kuethe.

Values of R for $C = 0.30$	270,000	240,000	214,000	190,000	172,000	160,000	145,000	132,000
Degree of Turbulence in per cent	0.50	0.75	1.00	1.25	1.50	1.75	2.00	2.25

However, in the case of forms for which the aerodynamic resistance is due chiefly to surface friction over a well polished surface, the results obtained in a wind tunnel with a moderate degree of turbulence are better applicable (by extrapolation) to the true size, than those obtained in a tunnel with small turbulence. This remark applies, for the same Reynolds number, to the resistance of fusiform bodies, wing profiles at small angles of incidence, etc.

Thus for applications in practice, the turbulence in wind tunnels should tend toward the production, on the model under experiment, of tangential resistance corresponding to a turbulent surface friction similar to that which governs corresponding resistances full scale.

It must be further noted that the turbulence proper to any given conditions as regards velocity, size and viscosity of an air stream, will

[1] Wieselsberger, C., Zeitschrift für Flugtechnik und Motorluftschiffahrt, p. 140, 1914.

be increased by the influence of the turbulent wake of any obstacle placed upstream in the flow. Thus the presence even of one or of several cylindrical wires upstream from a body of the forms noted, will prematurely determine the effect of turbulence on the resistance of such body. Thus for example a multicelled honeycomb structure, such as is often interposed just upstream of the experimental section, will cause an additional turbulence at this point.

5. Wind Tunnels with the Experimental Section Downstream from the Fan. A wind tunnel of this type comprises essentially (Fig. 5) a motor A driving a fan B which drives the air along a passage $CDEF$ in such manner as to give at F, the experimental section, a velocity V_0 as great as possible and distributed uniformly (as nearly as may be) in magnitude

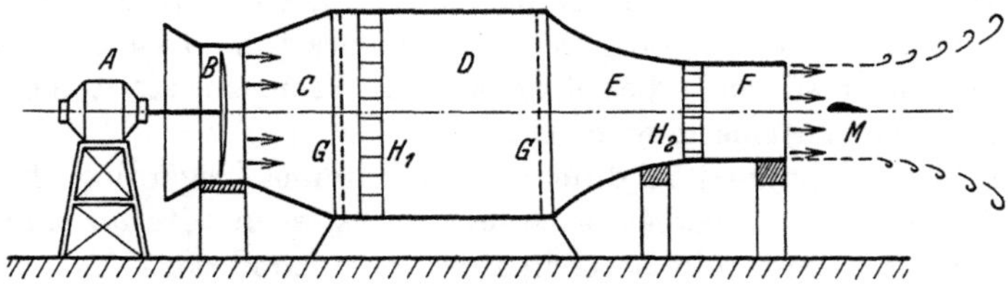

Fig. 5.

and direction over this section. To this end the air leaving the fan is first expanded in a chamber D of large size. Within this chamber are placed any special "regulating" devices such as grilles or perforated screens as at G, or honeycomb structures such as at H_1. These devices are intended to control or suppress movements of rotation in the air as a whole as it leaves the fan and to redistribute the velocity over the cross-section through their retarding action on zones of excess velocity. Following the chamber D, the air stream enters the convergent passage E which further aids in giving a uniform distribution of velocity at the experimental section F. A second honeycomb H_2 may contribute further to this regulation of the distribution of velocity. The model M may be subjected to experiment either in an open or closed stream. It is readily seen that the degree of turbulence in an air stream of this general character may be expected to be, in general, very high.

It is further seen that the power P_m required at the fan in order to obtain a given kinetic energy P_u at F, must be equal to this power P_u plus whatever losses the air stream may suffer in its flow along $CDEF$. With the preceding notation we shall then have:

$$\eta_f P_m = P_u + \Sigma\,(\text{losses})$$

and hence the coefficient C_t will have the value

$$C_t = \frac{P_u}{\eta_f P_m} = 1 - \frac{\Sigma\,(\text{losses})}{\eta_f P_m} < 1$$

And in consequence what we have called the efficiency will be negative. In a tunnel of this form the entire structure operates as a retarder of the air flow delivered by the fan. This arrangement is therefore relatively inefficient and expensive, and cannot be considered as suited either to large dimensions or high velocities.

Wind tunnels of this type were employed in 1899 by H. Maxim and Col. Renard and in 1909—10 by M. Rateau.

6. Wind Tunnels with the Experimental Section Upstream from the Fan. A wind tunnel of this type may be formed as in Fig. 6 of a motor *A*

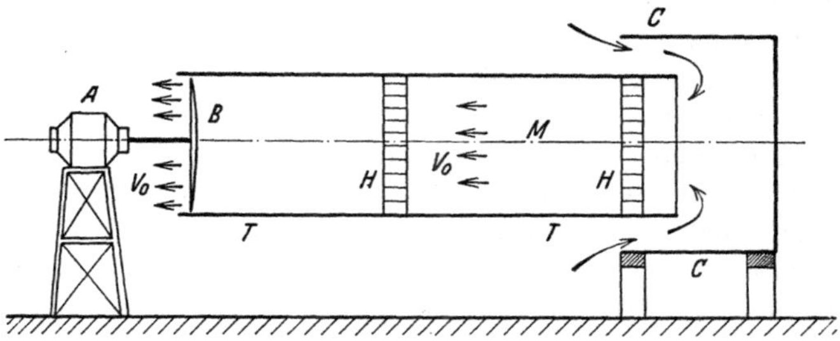

Fig. 6.

driving a fan *B* which draws the air along a tube *T* and drives it out at the fan end. The admission of the air into the tube *T* may be controlled by the addition of some form of entrance *C* and the experimental

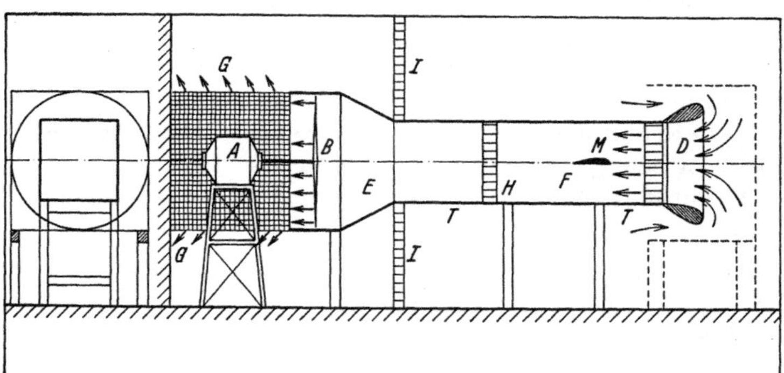

Fig. 7.

section may be preceded and followed by honeycomb structures such as *H*. Such a primitive form of tunnel will again have a coefficient $C_t < 1$ and cannot be considered as suited either to high velocities or large size.

The first wind tunnel in the aerodynamic laboratory of Koutchino (Moscow, 1910) was of this type.

The operation and the efficiency of such a form of tunnel may be improved by the changes indicated in Fig. 7. The entrance end of the tube T is provided with a rounded guide form as at D in order to secure an inflow of the air, progressive and without the parasitic vortices formed in rounding a sharp corner. Beyond the experimental section the channel T is replaced by a divergent conical section E so that the fan B operates in a section of much greater size than the tube T. Beyond the fan the passage may be prolonged by a section slatted or grilled, thus permitting a peripheral diffusion of the air delivered from the fan. This diffusion may be carried still further by allowing the air to return to the entrance end through a perforated screen I occupying the remaining section of the room containing the tunnel. These screens, sometimes called *distributors* have usually a thickness equal to the square root of the area of one perforation and the total surface of the perforations should be about one third the surface of the screen. These arrangements are effective in reducing variations of wind velocity over the experimental section. They aid by reducing local velocities above the mean and in suppressing local accelerations or rotations of the stream as a whole.

Examples of wind tunnels of this general type are found in the laboratories of the English N. P. L., in the Belgian laboratory at Rhode St. Genèse near Brussels and elsewhere in Europe and the United States.

The tunnel coefficient of this type of tunnel is improved by the increase of the section at and beyond the fan. In effect the loss of energy in the stream at the point of discharge being equal to the kinetic energy of the stream itself, any decrease in the velocity of the stream at the point of discharge will decrease this loss. However, in order to make this advantage effective it is necessary to give to the diffuser E proportions such that the loss due to the effects of boundary layer separation may be kept within proper limits. In addition the various obstacles in the air circuit, screens, honeycombs, etc. all tend to increase the losses of energy. It results in general that the coefficient for this form of tunnel is often less than unity and in some cases not above 0.6 or 0.7.

7. Wind Tunnels of the Eiffel Type. A notable improvement in wind tunnels of the general type of the preceding section was introduced by M. G. Eiffel.

In a tunnel of this improved form, the wind channel comprises the following three major features (see Fig. 8):

1) A convergent collector C.

2) An experimental section of constant cross-section D.

3) A divergent diffuser E of limited cone angle.

The fan, as in this general type, is located beyond the experimental section and at or near the mouth of the diffuser.

Collector. The mouth of the convergent collector C should preferably be formed with stream lined guide surfaces as shown (see Fig. 8) thus giving a smoothly rounded entrance. The external air, without sensible initial velocity, is then gradually accelerated as a result of the depression within the entrance due to the action of the fan, and the air in entering rounds the boundary of the collector without forming parasitic vortices, as would be the case with a viscous fluid rounding a sharp corner. The essential property of the flow in a collector of this type, is the realization of a uniform distribution of velocity in the sections beyond the collector. To this end the form of the curved entrance might be governed by the purpose of impressing on the entering air a smooth and continuous acceleration in the velocity of flow. But practically, a curved generatrix smooth and continuous and joining a cone of about 60^0 apex angle gives excellent results.

Whatever may be the conditions, however, regarding the distance between the collector and the walls of the room containing the structure,

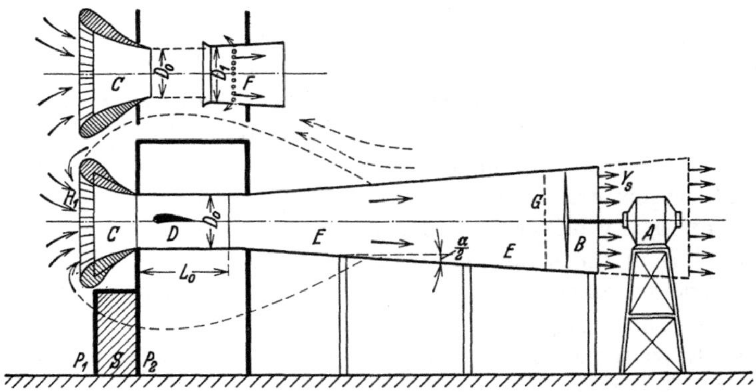

Fig. 8.

uniformity of flow of the air over the experimental section, in velocity and direction, can be realized only through the addition of some form of honeycomb structure H_1 at the upstream end of the collector. It has been noted in certain relatively recent installations that even a location of the entrance to the collector geometrically centered with reference to the walls of the room does not avoid certain irregularities in the distribution of the velocity such as local vortices, rotation as a whole, inequalities in the velocity, etc.

The turbulence due to a honeycomb structure thus located is much less than that due to a similar structure directly upstream from the experimental section. If the turbulence produced by such a structure is considered as bearing a direct relation to its aerodynamic resistance (due chiefly to friction) it is then easy to see that a honeycomb placed in a section of low velocity will have a resistance much less than when located at or near the experimental section. The turbulence caused by the honeycomb will then be diminished by this fact and furthermore the wake turbulence carried downstream from the honeycomb will be effectively reduced in amount in passing along this accelerated part of the flow. In any case the initial turbulence due to the air entering the collector is small or negligible, since the return velocity of the air is very small as a result of the large dimensions of the return section, at least with any reasonable dimensions of the room in which the tunnel is located.

Experimental Section. In the experimental section, the air stream may be either constrained or free. In either case, however, it is useful or necessary to construct a closed space surrounding this section. Under the conditions of steady flow, this closed space or experiment room will be maintained in pressure equilibrium with the stream of air flowing through it. For tunnels with the fan at the downstream end of the flow (working by aspiration), this mean pressure in the stream is necessarily less than that in the room containing the tunnel (Bernoulli's law) and under these conditions it is essential that the experiment room be provided with an air-lock with doors P_1, P_2 in order to permit passage to and fro during the operation of the tunnel.

For the operation of such a tunnel with a free air stream, this closed experiment chamber is evidently indispensable. For operation with constrained flow (continuous channel) it is to be recommended in order to avoid the many inconveniences resulting from the necessity otherwise of preventing any inflow of air from the exterior over the experimental section while at the same time realizing stuffing box fittings practically without friction at points of passage for connections between the model and the balances employed for the measurement of the various forces in play. In particular in the case of the use of the so-called wire balance, this latter difficulty is practically insurmountable.

Furthermore, the existence of an experiment room, air tight, suitably proportioned, well lighted, warmed in cold weather, etc., is a significant element contributing to comfort and experimental efficiency. The dimensions to be given to this room should take account of the possibility of the installation of apparatus for observation and measurement in various locations about the air stream. The external contours of this structure may be joined to the collector and to the diffuser with continuous forms in such way as to minimize disturbance to the return flow of air, especially in cases where the velocity of such return flow is appreciable as a result of small dimensions of the containing chamber.

A final controlling reason for the fitting of such an air-tight chamber in any case is found in the fact that in a modern wind tunnel it should be possible to test a given model at will under either set of conditions, free or constrained air stream. This possibility often leads to interesting comparisons and extends the domain of possible investigation. To this end the walls of the tunnel at the experimental section are readily made in such way as to be demountable or remountable as the case may require.

The length of the experimental section should be made in view of the length of models to be subjected to test. For the usual range of experiments in view of aeronautic applications, a length of 1 to 1.5 D_0 will give a reasonable proportion. A length of D_0 may rather be recommended for a free air stream in order to reduce the energy losses at this point. With a guided stream a length of 1.5 D_0 may be adopted without inconvenience.

As a result of the uniformity of distribution of velocity due to this form of collector, a length greater than as indicated above is without experimental significance, and is on the whole harmful by reason of the increase of perturbed peripheral zones and of resulting losses of energy.

The Divergent Diffuser. The rôle of the divergent diffuser is to gradually reduce the velocity of the air stream while at the same time increasing its pressure. However, this transformation of kinetic energy into potential (pressure) energy cannot be realized exactly in accordance with Bernoulli's law, for the flow of a fluid in such a diffuser always entails losses due to friction and to expansion which we shall evaluate at a later point.

In the case of a tunnel with constrained air stream, the upstream section of the diffuser is made equal to the downstream section of the experimental section. With a free air stream, however, the entrance section of the diffuser should be made of somewhat greater dimensions in view of the inflow of air through the walls of the experiment room around doors, windows etc., which practically cannot be expected to be perfectly air tight. A further reason for an increase in the diameter of the diffuser entrance is found in the slight spread to which the air stream is subject in passing through the experiment room and the resulting formation of parasitic currents due to impact with the edge of the diffuser. For these reasons a diameter of diffuser entrance 1.05 to 1.10 D_0 may be recommended. Furthermore it is found of advantage to provide this entrance with a curved flare of moderate proportions, somewhat as indicated in Fig. 11. Finally the presence of openings F in the wall of the diffuser at a moderate distance from the entrance (0.50 to 0.60 D_0) and opening into the experiment room are found to aid in the reduction of pulsations, more or less violent, which are often observed in wind tunnels of this type[1].

Such leaks of air where the difference of pressure is small affect only slightly the tunnel coefficient.

The generatrix of the diffuser should preferably be rectilinear. Various tests have shown that a diffuser of this form leads to a higher tunnel coefficient than one of parabolic form and the construction is simpler.

[1] Jacobs, Eastman, N., Investigation of Air Flow in Open-Throat Wind Tunnels U.S. N.A.C.A. Technical Report No. 322, 1929.

The angle of divergence of the diffuser is of great importance from the viewpoint of tunnel coefficient. In fact according to Fliegner the losses due to expansion of the air stream in the diffuser are proportional to the sine of the angle of divergence. However, if with a view to the diminution of these losses a very small angle is adopted, the length of the diffuser required to give the desired relation between the sections at entrance and discharge will be unduly increased as well as the size and cost of the structure.

Practically, the angle very commonly employed is 7^0. It results, for a desired ratio between the sections of discharge and entrance (S_d/S_e) that the length of the diffuser will be given by the equation

$$\frac{L}{D_e} = \frac{\sqrt{n}-1}{2\,\tan 3^0,\, 30'} \simeq 8.2\left(\sqrt{n}-1\right)$$

where D_e is the diameter of the entrance section if circular or the side if square.

If the fan shaft is to be provided with a shell of stream-line form of considerable section, the form of the diffuser may be suitably modified in order to preserve the desired law for the increase of section areas. Beyond the plane of the fan, the diffuser may be advantageously extended a moderate distance.

In order to reduce size and first cost it would be desirable to increase the angle of the diffuser cone. However, this leads soon to larger energy losses by reason of the separation of the boundary layer. To avoid such separation, it has been proposed to provide the interior of the diffuser with longitudinal partitions along the lines of flow, or again to provide a peripheral depression permitting the aspiration of the boundary layer thus decreasing the losses which would otherwise be incurred.

8. Evaluation of the Losses in a Tunnel with Experimental Section Upstream from the Fan. The energy losses in a tunnel of this type with air stream constrained may be evaluated by the use of the following method proposed by M. Margoulis[1].

The schedule of these losses for the various parts of the system is as follows:

a) In the collector, losses by friction on the walls and on the surface of the cells of the honeycomb.

b) In the experimental section, losses by friction on the walls (assuming the absence of any honeycomb in this section) and with a free airstream, losses from the suppression of the walls.

c) In the divergent diffuser, c_1 losses by friction on the walls and on the surfaces of partitions (if present), c_2 losses by expansion (Fliegner effect).

[1] Margoulis, M., Résumé des principaux travaux exécutés pendant la guerre au Laboratoire aérodynamique Eiffel, Souffleries Aérodynamiques, p. 177.

d) At the discharge end of the diffuser, losses corresponding to the kinetic energy of the stream leaving the fan.

The return velocity of the air toward the collector is supposed to be negligible by reason of the greatly increased area of return section and the corresponding losses are considered as negligible likewise.

The evaluation of the various losses by friction is based on the knowledge of the value of the coefficient of surface friction for the flow of turbulent air in tubular passages of various forms, of which the walls are smooth and well polished. This coefficient is of the form

$$C_f = \frac{F}{S_f\,(1/2)\,\varrho\,V^2} = c\,\varphi\,(R)$$

where F = force or drag due to friction,
S_f = surface exposed to the flow,
V = velocity,
$R = D\,V/\nu$ = Reynolds number,
c = non-dimensional coefficient.

The form $\varphi\,(R) = R^n\ (n < 0)$

is frequently used for $\varphi\,(R)$.

Thus experimental results for friction with turbulent flow in smooth pipes are often represented by the Blasius formula

$$C_f = 0.079\,(R)^{-0.25} \tag{8.1}$$

For application to wind tunnels the values of R are in general very high and extrapolation to these values with the preceding formula hardly seems permissible. It leads in fact to values of C_f which tend toward zero when R increases beyond limit. It seems preferable to employ for the function $\varphi\,(R)$, a form with two terms such as that of Dr. Lees or of Jakob and Erk, that is,

$$C_f = a + b\,R^n = 0.0018 + 0.1526\,R^{-0.35} \tag{8.2}$$

which leads to the limit value 0.0018 when R increases beyond limit.

It should be noted that the values of C_f calculated by the aid of the preceding formulas can be applied only to the case of *smooth continuous surfaces*. As a result of the dimensions and the modes of construction of modern wind tunnels, this condition is not always met with in all portions of the air circuit. It is therefore often necessary to employ values of C_f for non-polished or non-smooth or even for rough surfaces. The values available for such purposes are numerous but not always consistent. In default of more definite information, values as derived from the preceding formulas may be employed by multiplying them by an arbitrary coefficient of increase such as 1.50 or 2.00 according to the degree of roughness of the walls[1].

[1] Handbuch der Physik, Mechanik der flüssigen und gasförmigen Körper, Strömung in Röhren und Gerinnen, p. 151.

The computation of the loss by friction in a section cylindrical in form of diameter D_0 and length L_0 would then be given in the form:

$$\triangle E_0 = 4 C_{f0} (\varrho/2) V_0^2 \frac{1}{D_0} \int_0^{L_0} d L = 4 C_{f0} (\varrho/2) V_0^2 \frac{L_0}{D_0} \tag{8.3}$$

For a convergent collector or a divergent diffuser, this relation may readily be extended to conical forms giving

$$\triangle E_c = 4 (\varrho/2) V_0^2 D_0^4 \int_0^{L_c} C_f \frac{d L}{D^5}$$

Introducing the expression for C_f and the ratio n of the sections of entrance and discharge, and $\alpha/2$ for the half angle of the cone, this reduces to

$$\triangle E_c = \frac{0.5}{sin\,(\alpha/2)} (\varrho/2) V_0^2 \left[a \frac{n^2 - 1}{n^2} + \left(\frac{1.093\, b}{R_0^{0.35}} \right) \frac{n^{1.825} - 1}{n^{1.825}} \right]$$

for the usual values of the ratio n this can be written more simply

$$\triangle E_c = \frac{1}{2\, sin\,(\alpha/2)} (\varrho/2) V_0^2 \frac{n^2 - 1}{n^2} \left[a + 1.09\, b\, (R_0)^{-0.35} \right]$$

The expression within the brackets is, however, only slightly different from C_{f0} so that practically we may write

$$\triangle E_c = \frac{C_{f0}}{2\, sin\,(\alpha/2)} (\varrho/2) V_0^2 \frac{n^2 - 1}{n^2} \tag{8.4}$$

The loss by expansion in the divergent diffuser is evaluated by Fliegner's formula as follows:

$$\triangle E_e = (\varrho/2) V_0^2 \left(\frac{n - 1}{n} \right)^2 sin\, \alpha \tag{8.5}$$

And finally the kinetic energy at discharge will be

$$\triangle E_k = (\varrho/2) V_0^2 \frac{1}{n^2} \tag{8.6}$$

The losses due to friction on the surfaces of honeycombs may be computed in the same manner by using the coefficient of friction corresponding to the Reynolds number which characterizes the flow within the individual cells. The preceding relations permit the formulation of the following interesting conclusions regarding the proportions of the various elements constituting a wind tunnel of this general type.

1) The losses by friction in the convergent collector decrease as the angle α is increased, but these losses are in general very small and in consequence the form of the collector is without notable influence on the value of the tunnel coefficient.

2) The losses in the experimental chamber are proportional to the ratio L_0/D_0; it is therefore desirable to give to this ratio the minimum value consistent with experimental requirements.

3) The sum of the losses within and at the discharge of the diffuser has the expression

$$\Sigma \triangle E_D = (\varrho/2)\, V_0^2 \left[\frac{C_{f0}}{2\, sin\,(\alpha/2)} \cdot \frac{n^2-1}{n} + \left(\frac{n-1}{n}\right)^2 sin\,\alpha + \frac{1}{n^2}\right] \qquad (8.7)$$

The values of n and of α corresponding to the minimum values of the losses are given as a function of C_f in the form

$$\left.\begin{aligned} cos\,\alpha\, sin\,(\alpha/2)\, tan\,(\alpha/2) &= \left(\frac{n+1}{n-1}\right)\frac{C_{f0}}{4} \\ n &= 1 + \frac{1}{sin\,\alpha}\left[1 - \frac{C_{f0}}{2\, sin\,(\alpha/2)}\right] \end{aligned}\right\} \qquad (8.8)$$

The first of these gives, in particular, the optimum value of the angle α for given values of the ratio n and of C_{f0}.

Suppose for example a project for a wind tunnel comprising the following characteristics

$$D_0 = 2 \text{ meters},$$
$$V_0 = 40 \text{ meters/sec.},$$
$$n = 4.$$

The value of C_f corresponding to the Reynolds number $R_0 = V_0 D_0/\nu$ will be $C_{f0} = 0.0025$ for smooth walls and sensibly $2C_{f0}$ for walls only moderately smooth or with rough spots. The solutions corresponding to the first of (8.8) are respectively $\alpha = 3^0 \cdot 7$ and $\alpha = 5^0 \cdot 3$.

These values are definitely smaller than the value 7^0 referred to above, which value, while it does not correspond to the highest value of the tunnel coefficient, does have, however, the advantage of reducing the length and the cost of the structure.

Tunnel losses may also be evaluated by direct measures on a reduced size model of the wind tunnel. Such measurements have been carried out by Costanzi, Castellazzi, Eiffel and others and as regards wind tunnels of the type under present consideration, these experimental measures confirm the conclusions noted above.

Certain further results due to experimental measurement on reduced size models may also be here noted.

a) Influence of the Room Containing the Tunnel. The model of a tunnel of the type tested in open space, *i. e.* without surrounding walls gave a coefficient of 2.43. Tested again with surrounding walls the coefficient dropped to 1.18 or otherwise to about half value.

b) Influence of the Suppression of the Walls at the Experimental Section in Order to Give a Free Stream. A free stream of length $L_0 = 1.95\, D_0$, according to Castellazzi, leads to a loss in the coefficient of 40% of its original value of 2.43 with guided stream. A similar result is indicated by Norton and Warner[1], that is a reduction of 50% for a free stream of length $L_0 = 2D_0$.

[1] Warner, Edward P; Norton, F. H. and Herbert, C. M., The Design of Wind Tunnels and Wind Tunnel Propellers, U. S. N. A. C. A. Technical Report No. 73, 1919.

These results may furthermore be expressed by the value of the supplementary loss due to the suppression of the walls. This loss may be put in the form $k\,(L_0/D_0)\,(\varrho/2)\,V_0^2$ and for tunnels similar to those above noted, k may be taken equal to about 0.16. This loss is considerable in comparison with that due to friction on the walls and it would be preferable, therefore, to limit the length of the free stream to about one diameter.

Special Arrangements at the Discharge of the Diffuser. M. Lapresle has suggested an arrangement comprising a discharge from the diffuser into an "aspiration chamber" into which the air is drawn by several fans. This arrangement permits the subdivision of the fan motive power into several groups of moderate dimensions. It permits also some reduction in the length of the diffuser.

A large number of wind tunnels have been established in accordance with the general Eiffel type among which may be noted tunnels in the Eiffel laboratory, Auteuil, Paris (free stream), in the laboratory of the Aerotechnical Institute, St. Cyr (stream free or guided), in the laboratory of the Aeronautical Research Service, Paris (free stream), in the laboratory of the Belgian Aerotechnical Service, Brussels (free stream), and many others in Europe and in the United States.

9. Wind Tunnels with Continuous Closed Circuit. A wind tunnel of this type comprises essentially an air circuit composed of a passage, $CDEF$ Fig. 9, and with section constant or variable. The fan B induces

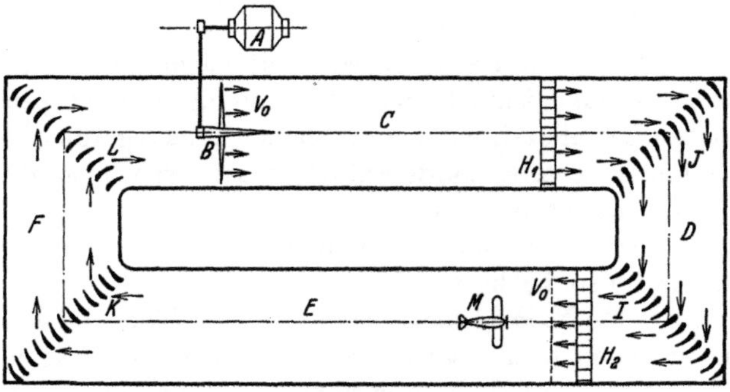

Fig. 9.

the circulation in the circuit and honeycombs and guide vanes may be disposed at suitable points in order to improve distribution of velocity over the experimental section. Changes in direction are furthermore aided by guide vanes at points such as $IJKL$.

Fig. 9a represents the usual form of profile for guide vanes of this character. Vanes disposed in this manner are to be preferred to curved elbow turns in which the air flow, as a result of centrifugal forces, will not distribute itself uniformly over the section. With such an elbow turn, therefore, it is necessary to guide the air by intermediate partitions or by series of vanes and the losses of energy in such case are very considerable. See Fig. 10.

The experimental section E may be either with airstream constrained or free. In the case of a free stream, the experiment room is formed by the room containing the tunnel and may, therefore, be made as large as desired.

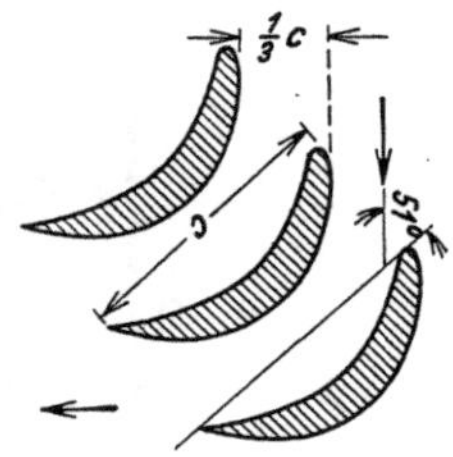

Fig. 9a.

For an air-circuit of uniform section, the tunnel coefficient is necessarily less than unity and an important improvement in this type of tunnel is, therefore, obtained by the use of a circuit with variable section (see Figs. 11, 12) comprising a convergent collector upstream of, and a divergent diffuser downstream of the experimental section. The

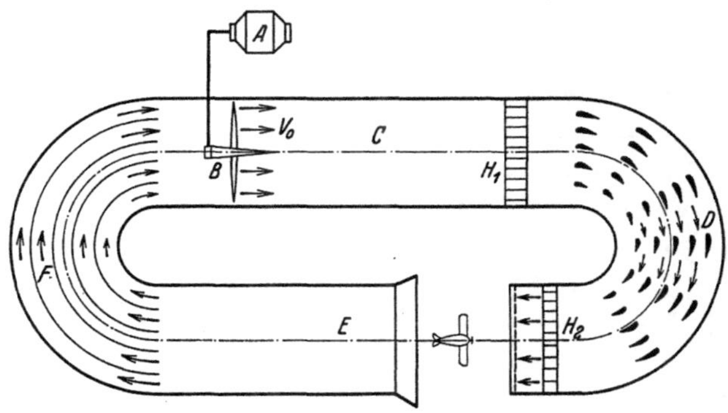

Fig. 10.

fan is then placed in a divergent section of the air return, which is extended to join the entrance to the collector. Under these conditions it

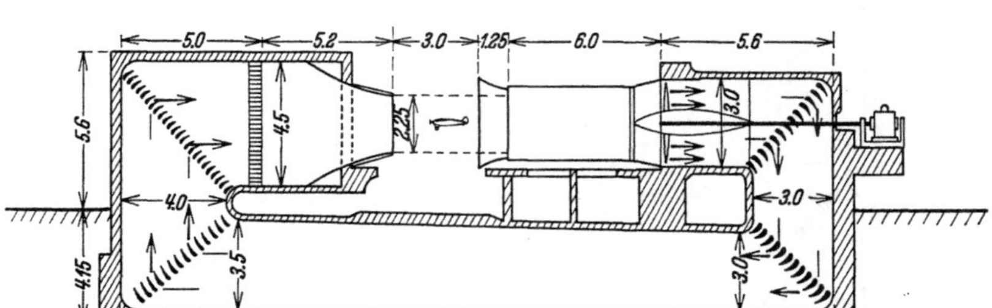

Fig. 11. Göttingen Wind Tunnel. Vertical Section, Dimensions in meters.

is possible to obtain a value of the tunnel coefficient much greater than unity. Furthermore, the turbulence caused by the guide vanes, by the fan and by the honeycomb of the collector is much reduced. As in the preceding case, the experimental section may be either with a constrained or free air stream, in the latter case the surrounding room at atmospheric pressure constituting the experiment chamber.

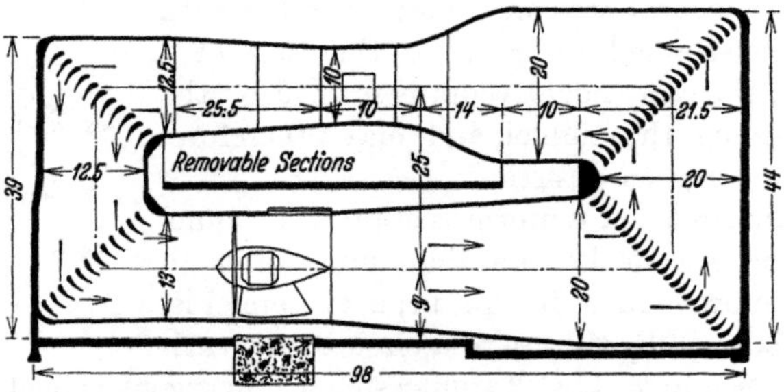

Fig. 12. Wind Tunnel at the Calif. Inst. of Technology. Vertical Section, Dimensions in feet.

Tunnels of this type have been installed at Göttingen, at the California Institute of Technology, Pasadena, and elsewhere, see Bibliography at end of Division.

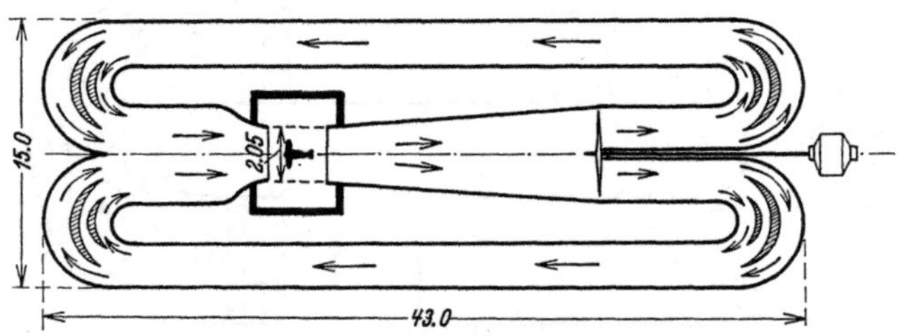

Fig. 13. Wind Tunnel at the Aerodynamic Institute of Warsaw. Plan View, Dimensions in meters.

Another type of wind tunnel with constrained stream comprises two returns for the air as indicated in Fig. 13. The central portion comprises

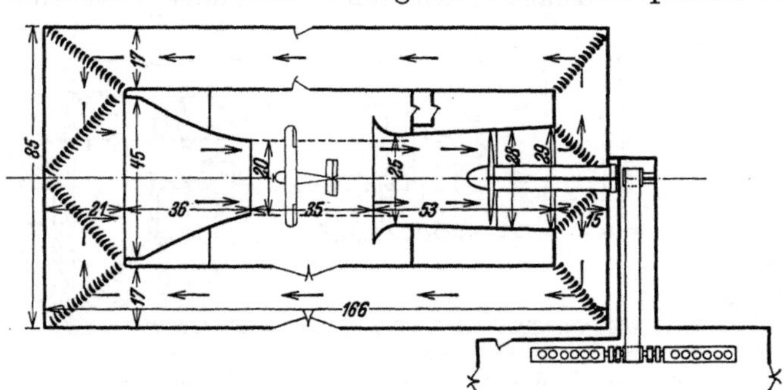

Fig. 14. Propeller Tunnel U.S. N.A.C.A. Plan View, Dimensions in feet.

the essential elements of a wind tunnel as in the preceding forms with collector and diffuser. Downstream of the fan, the section of the diffuser is divided into two equal parts corresponding to the two air returns which may be either cylindrical or divergent as in the previous cases.

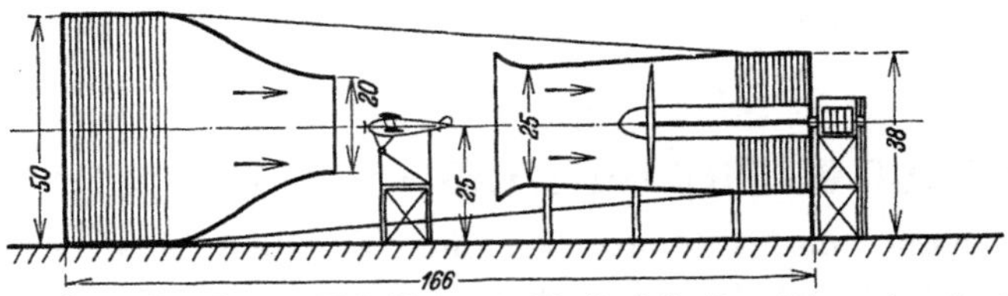

Fig. 15. Propeller Tunnel U.S. N.A.C.A. Vertical Section, Dimensions in feet.

The same general conditions as regards honeycombs, guide vanes and formation of turbulence apply equally to this type of tunnel.

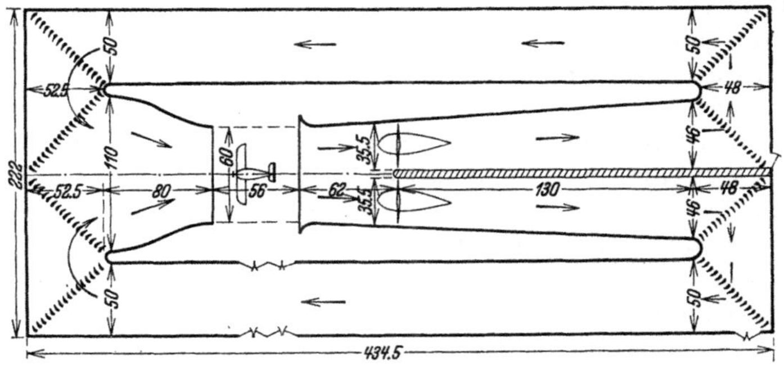

Fig. 16. Full Scale Tunnel, U.S. N.A.C.A. Plan View, Dimensions in feet.

Wind tunnels of this general type are found at the Central Aerodynamic Institute at Rome, at the Aerodynamic Institute of Warsaw, and at the Langley Field Laboratory of the N.A.C.A. See Figs. 14, 15, 16.

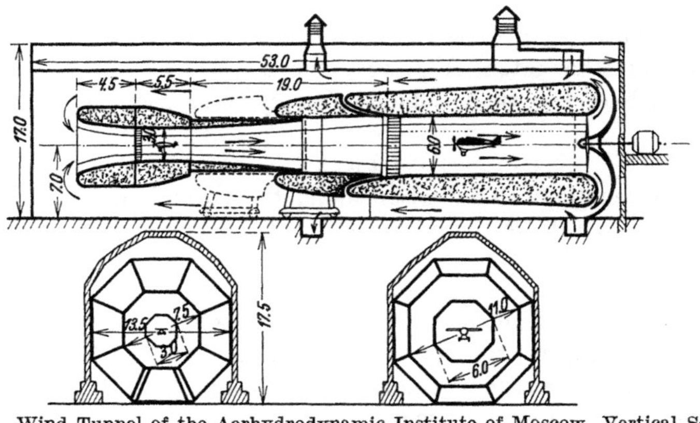

Fig. 17. Wind Tunnel of the Aerhydrodynamic Institute of Moscow. Vertical Section, Dimensions in meters.

And finally a variant of this type comprising an annular air return as in Fig. 17 is found in the Central Aerhydrodynamic Institute of

Moscow, in the Variable Density Tunnel at Langley Field[1], and at the N. P. L., Teddington, at the Polytechnic School of Bucharest and elsewhere, see Bibliography at end of Division.

CHAPTER III

INFLUENCE OF THE DIMENSIONS OF THE AIR STREAM[2]

1. General Considerations. The limited extent of air stream used in wind tunnels and the nature of the free surfaces or of the rigid walls limiting this stream, impose certain restrictions on the flow of the air about the object under investigation. In order to deduce, from measurements in a wind tunnel, coefficients applicable in unlimited free air, due regard must be had of these constraints, and suitable corrections made to the direct experimental values. These corrections are likewise essential in order to properly compare the results obtained in different wind tunnels.

The general examination of the restrictions due to the limitation of the air stream, may be carried out by the following method due to Prof. L. Prandtl.

It is assumed that the potential φ_l which characterizes the motion of the fluid under constraint is equal to the corresponding potential φ_0 of the indefinite free fluid, augmented by a complementary potential φ_c.

$$\varphi_l = \varphi_0 + \varphi_c \tag{1.1}$$

The complementary potential φ_c will then be determined by expressing the conditions on the limiting surface of the fluid stream. Two principal cases may first be considered.

1) The case of a stream constrained or limited by rigid walls.

2) The case of a free stream or fluid jet.

Case of a Stream Guided or Constrained. In this case the constraint requires, for the potential φ, that the component of velocity normal to the walls of the tunnel shall be everywhere zero. We must therefore have

$$\frac{\partial \varphi_l}{\partial n} = 0 \tag{1.2}$$

where ∂n denotes differentiation along the normal to the walls.

From (1.1) this will give, at the walls,

$$\frac{\partial \varphi_0}{\partial n} = -\frac{\partial \varphi_c}{\partial n} \tag{1.3}$$

Thus the complementary potential φ_c, having in view the influence of the walls limiting the stream, must satisfy the condition (1.3). Let us apply what precedes to the case of a lifting wing. For the wing the potential φ_0 characterizing the motion in free air may be itself decomposed into two parts; one corresponding to translation with the velocity U_0

[1] See Part 2 of the present Division.

[2] See also Division E IV, Part C.

and of which the expression is $U_0 x$ (taking the axis of x parallel to U_0) and the other corresponding to the bound and free vortices representative of the wing in the approximate wing theory. Denote this latter by φ'. Then we shall have
$$\varphi_0 = U_0 x + \varphi' \tag{1.4}$$

Then since at the walls the velocity U_0 will have no component normal to the walls, it follows from (1.3) and (1.4) that, at the walls,
$$-\frac{\partial \varphi_c}{\partial n} = \frac{\partial \varphi'}{\partial n} \tag{1.5}$$

The determination of φ_c reduces therefore to a knowledge of the potential φ' which corresponds to the velocity induced by the ensemble of the vortices representative of the wing. This potential φ' satisfies in itself at ∞ both upstream and downstream, certain special conditions which contribute to the simplification of the problem. At the same time the complete calculation is often very difficult, except in certain special cases.

Case of a Free Stream. In this case the conditions require that the pressure over the limiting surface of the stream must be equal to the constant pressure which prevails throughout the fluid at rest surrounding the stream. This pressure p_0 corresponds to the general velocity U_0 of the undisturbed stream. Thus at every point of the free surface, the velocity resulting from the composition of U_0 with the added velocities u, v, w (due to the vortex systems) must satisfy the relation
$$p + \frac{\varrho}{2}\left[(U_0 + u)^2 + v^2 + w^2\right] = p_0 + \frac{\varrho}{2} U_0^2 \tag{1.6}$$
with $p = p_0$.

Whence
$$u^2 + v^2 + w^2 + 2 U_0 u = 0.$$

If the squares of the added velocities be neglected, this reduces to
$$u = 0 \tag{1.7}$$

Hence, the conditions of constraint in this case require that the added velocities shall have no component in the direction of U_0. Thus on any stream-line s at the limiting surface the resultant velocity $\partial \varphi_l / \partial s$ must satisfy the relation,
$$\frac{\partial \varphi_l}{\partial s} = U_0 = \text{constant} \tag{1.8}$$
which, by integration gives
$$\varphi_l - U_0 s = C \tag{1.9}$$

From (1.1) this gives
$$\varphi_0 - U_0 s = -\varphi_c \tag{1.10}$$

For the case of a lifting wing, the potential φ_0 being decomposed in accordance with (1.4), the condition of (1.10) becomes (assuming that the inclination of s to the direction of U_0 is negligible)
$$\varphi' + U_0 x - U_0 s \cong \varphi' = -\varphi_c \tag{1.11}$$

Thus the determination of the complementary potential φ_c again becomes reduced to that of φ' as in the preceding case. At ∞ upstream, this potential φ' is constant or zero since at this distance the influence of the wing vortex systems vanishes. At ∞ downstream, there is only the influence due to the free vortices and the value of φ' here corresponds to plane (two-dimensional) motion.

Along a generatrix of the cylinder limiting the stream, the component of the induced velocities parallel to the direction of the flow, can only come from the bound vortices representative of the wing. This component $u = \partial\varphi'/dx$ takes the same values for all symmetrical points $\pm x$. It follows that the value of φ' for $x = 0$ is one half of the value at ∞ downstream. It will be sufficient, therefore, to know the value of φ' at ∞ downstream to be able to deduce that corresponding at the wing and the problem is thus brought into the domain of plane motion. However, as already noted, it remains, in general, difficult of final resolution.

2. Wind Tunnels with Circular Section—Lifting Systems, Monoplane or Multicellular. *a) Air Stream Constrained.* Let us consider a wing of span $2b$, and surface S under test in a wind tunnel with constrained stream of section $S_0 = \pi R^2$. It is further assumed that the wing is centered with reference to the walls of the tunnel. In general, the distribution of the circulation Γ along the span b will not be uniform and will depend on the characteristics of the wing under test. However, to simplify the problem, the case of *uniform distribution of circulation* will first be considered. In this case the vortex system will consist of a vortex Γ_0 extended along the span $2b$ and of two free vortices of the same strength Γ_0 extending from the tips of the wing downstream to ∞. See Figs. 18, 19.

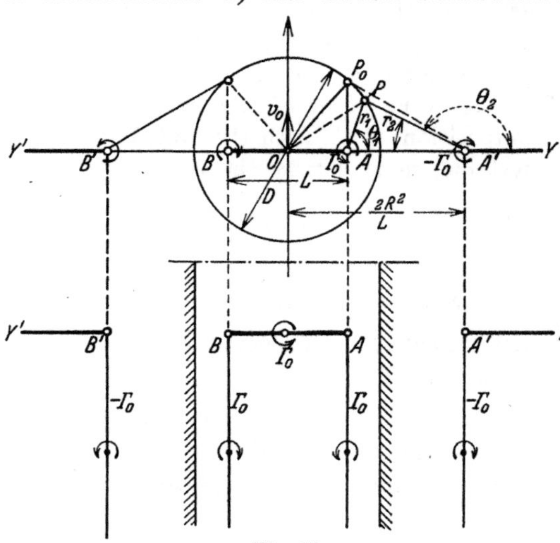

Fig. 18.

The constraint due to the rigid walls of the tunnel imposes the condition that the component of the velocity normal to the walls shall be zero and we have seen that this condition corresponds to the addition to the potential φ_0 (motion in unlimited air) of a complementary potential φ_c corresponding to a fictitious motion satisfying the condition of (1.5).

This we may write: $$\left(\frac{\partial \varphi_c}{\partial n} + \frac{\partial \varphi'}{\partial n}\right)_{\text{wall}} = 0 \tag{2.1}$$

This relation expresses the fact that the motion resulting from the addition of the potentials φ_c and φ' cannot have any component radial velocity on the cylindrical surface of section S_0. The potential φ' corresponds to the motion resulting from the vortex system representing the wing. For the transverse sections located at a sufficient distance downstream (theoretically at ∞) the influence of the bound vortex is negligible and the only induced velocities are those due to the free vortices A and B. For such sections then, the circumference of radius R must be a stream-line of the motion resulting from the free vortices A and B and of the fictitious motion corresponding to the potential φ_c. These conditions are fulfilled by the complementary potential φ_c corresponding to two vortex images A' and B', of circulation equal and of sign opposite to that of A and B and located at reciprocal distances[1]

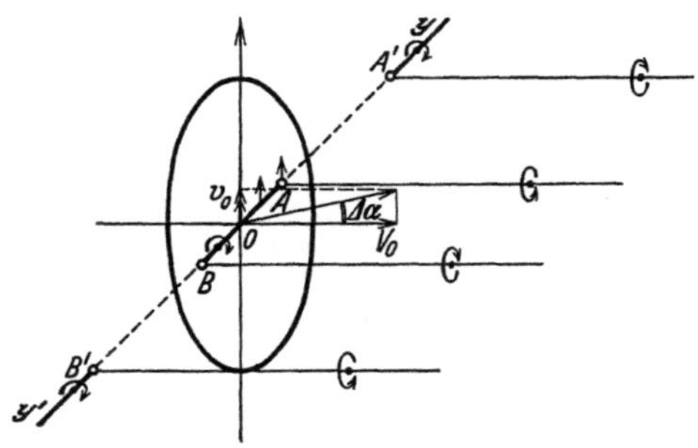

Fig. 19.

$$OA' = OB' = \frac{R^2}{OA} = \frac{R^2}{OB} = \frac{R^2}{b}$$

Thus in the case of a guided stream of circular section the constraint of the stream is equivalent to the perturbations due to two vortex images A' and B' as above defined. These vortices are to be considered as issuing from A' and B' and extending to infinity downstream.

The perturbations due to these vortices A' and B' will have the effect of causing, at the wing AB, an addition to the general motion with velocity U_0, of an induced complementary velocity v, variable along the span. Practically the value of v at the mid section of the wing can be adopted as the measure of the mean perturbation.

For the two vortices A' and B', this induced velocity is upward and is expressed by the equation

$$v_0 = 2 \times \frac{\Gamma_0}{4\pi OA'} = \frac{\Gamma_0 b}{2\pi R^2} = \frac{\Gamma_0 b}{2 S_0} \tag{2.2}$$

We have furthermore,

$$L = C_L S\,(\varrho/2)\,U_0^2 = 2\varrho\, U_0 \Gamma_0 b$$

whence $$\Gamma_0 b = \frac{C_L}{4} S U_0$$

and $$v_0 = U_0 C_L \frac{S}{8 S_0} \tag{2.3}$$

[1] See Note at end of Section.

Finally, the influence on the wing of the rigid walls guiding the stream is equivalent to that of a fictitious ascending current causing a change in the angle of incidence of amount

$$\triangle \alpha_0 \simeq \frac{v_0}{U_0} = C_L \frac{S}{8\, S_0} \qquad (2.4)$$

L. Rosenhead[1] has shown that the approximation made above in considering only the induced velocity v_0, at the mean section may be avoided in the following manner.

At a point y of the wing $A\,B$, the ascending velocity induced by the vortex images A' and B', has the value

$$v = \frac{\Gamma_0}{4\pi}\left[\frac{1}{R^2/b - y} + \frac{1}{R^2/b + y}\right] \qquad (2.5)$$

or putting $\xi = y b/R^2$

$$v = \frac{\Gamma_0 b}{4\pi R^2}\left[\frac{1}{1-\xi} + \frac{1}{1+\xi}\right] = C_L \frac{S}{8\, S_0}\, U_0 \left(\frac{1}{1-\xi^2}\right) \qquad (2.6)$$

or

$$v = C_L \frac{S}{8\, S_0}\, U_0 \,(1 + \xi^2 + \xi^4 + \ldots) \qquad (2.7)$$

The local induced angle at any section with abscissa y will then be

$$\frac{v}{U_0} = C_L \frac{S}{8\, S_0} \cdot \frac{1}{1 - y^2\,(b/R^2)^2} \qquad (2.8)$$

and the mean corrections for drag and for incidence will be:

$$\triangle C_D = \frac{1}{(\varrho/2)\, S\, U_0^2} \cdot \frac{1}{2b} \int_{-b}^{+b} \frac{v}{U_0}\, L\, d\,y$$

$$= C_L^2 \frac{S}{8\, S_0} \cdot \frac{1}{2b} \int_{-b}^{+b} \frac{d\,y}{1 - y^2\,(b/R^2)^2}$$

or $$\triangle C_D = C_L^2 \frac{S}{16\, S_0}\, \frac{R^2}{b^2}\, log \frac{1 + (b/R)^2}{1 - (b/R)^2} \qquad (2.9)\ [2]$$

$$= C_L^2 \frac{S}{8\, S_0}\left[1 + \frac{1}{3}\left(\frac{b}{R}\right)^4 + \frac{1}{5}\left(\frac{b}{R}\right)^8 + \frac{1}{7}\left(\frac{b}{R}\right)^{12} + \ldots\right] \qquad (2.10)$$

---

[1] Rosenhead, L., The Effect of Wind Tunnel Interference on the Characteristics of the Aerofoil. Proc. Royal Society, Series A, Vol. 129 No. a 809, p. 135.

[2] $$\int_{-b}^{+b} \frac{2\,d\,y}{1 - y^2\,(b/R^2)^2} = \underset{(A)}{\int_{-b}^{+b} \frac{d\,y}{1 - y\,(b/R^2)}} + \underset{(B)}{\int_{-b}^{+b} \frac{d\,y}{1 + y\,(b/R^2)}}$$

But $A = B$ and hence

$$\int_{-b}^{+b} \frac{d\,y}{1 - y^2\,(b/R^2)^2} = \int_{-b}^{+b} \frac{d\,y}{1 - y\,(b/R^2)}$$

whence we derive

$$\triangle\alpha = \frac{\triangle C_D}{C_L} = C_L \frac{S}{8 S_0}\left[1 + \frac{1}{3}\left(\frac{b}{R}\right)^4 + \frac{1}{5}\left(\frac{b}{R}\right)^8 + \frac{1}{7}\left(\frac{b}{R}\right)^{12} + \ldots\right] \quad (2.11)$$

For the usual values of the ratio b/R, the corrections differ but slightly from those computed by the approximate formula.

This disturbing upward current increases then the experimental angle of incidence by the angle $\triangle\alpha$ and the lift L is inclined forward by the same angle (see Fig. 20). It results that the horizontal component D is decreased by the corresponding component of L, that is by

$$\triangle D = L \sin \triangle\alpha \cong L \triangle\alpha \quad (2.12)$$

Since $\triangle\alpha$ is normally small, the vertical component is not sensibly changed.

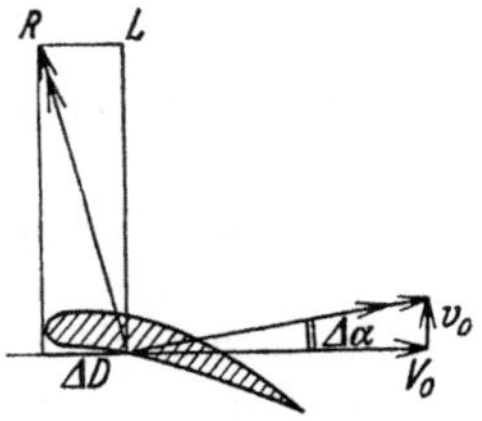

Fig. 20.

It is desirable to extend these results to the more general case of a *variable distribution of circulation* along the span. In this case, the vortex system comprises the bound vortex of variable circulation Γ together with a sheet of free vortices of strength $(d\Gamma/dy)\,dy$ for an element dy situated at the distance y from the middle section. See Fig. 21.

As in the previous discussion, we shall match to a free vortex $M\,\xi$ located at a distance y, a vortex image $M'\xi'$ located at the distance R^2/y, and of which the strength is $-(\partial\Gamma/\partial y)\,dy$. The complementary potential will then correspond to two sheets of vortex images, located on one side and the other of the section of the tunnel reckoned from the points A' and B', conjugate to the points A and B and extending laterally to ∞ and indefinitely downstream.

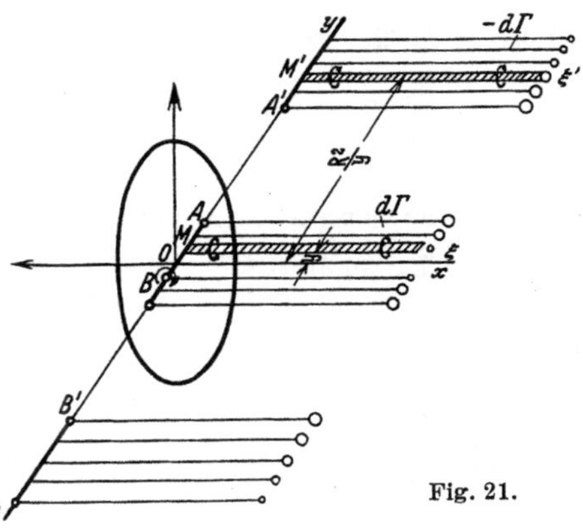

Fig. 21.

In other words, the constraint due to the rigid walls of the tunnel is equivalent to the action of two wing images extending indefinitely from A' and B' and of which the circulation at each point is equal and of opposite sign to that of the conjugate image—point on the wing.

The influence of these "image-wings" (as we may call them) and of the corresponding sheets of free vortices is manifest at the wing by a complementary induced velocity of which the value at a point y' of the wing is given by the relation[1]

$$v(y') = \frac{-1}{4\pi}\int_{-\infty}^{+\infty}\left[-\left(\frac{d\Gamma}{dy}\right)\frac{dy}{R^2/y - y'}\right] = \frac{1}{4\pi}\int_{-b}^{+b}\frac{d\Gamma}{dy}\cdot\frac{dy}{R^2/y - y'} \qquad (2.13)$$

The complementary induced angle at any point y' will then be

$$\triangle\alpha = \frac{v(y')}{U_0}$$

and the corresponding induced resistance will be given for any wing AB by the relation $\triangle D = \varrho\int_{-b}^{+b}\Gamma\, v(y')\, dy$

The calculation of $\triangle\alpha$ and of $\triangle D$ may then be made as soon as the law of distribution of the circulation Γ along the span is known. This calculation has been made by Prandtl and by Rosenhead for the particular case of an elliptical distribution of the circulation defined by the equation $\Gamma = \Gamma_0\sqrt{1-(y/b)^2}$

For such a distribution, the induced velocity at a section y of the wing will have the value

$$v(y) = C_L\frac{U_0 S}{8 S_0}\left[1+\frac{3}{4}\left(\frac{yb}{R^2}\right)^2+\frac{5}{8}\left(\frac{yb}{R^2}\right)^4+\frac{35}{64}\left(\frac{yb}{R^2}\right)^6+\ldots\right] \qquad (2.14)$$

where R is the radius of the air stream. We have then,

$$\triangle C_D = \frac{1}{(\varrho/2)SU_0^2}\int_{-b}^{+b}\frac{v_y}{U_0}\frac{dL}{dy}dy = \frac{1}{(\varrho/2)SU_0^2}\int_{-b}^{+b}\frac{v_y}{U_0}\varrho\Gamma U_0\, dy \qquad (2.15)$$

And the corresponding induced drag has the value

$$\triangle D = \frac{C_L^2 S^2(\varrho/2)U_0^2}{8 S_0}\left[1+\frac{3}{16}\left(\frac{b}{R}\right)^4+\frac{5}{64}\left(\frac{b}{R}\right)^8+\ldots\right] \qquad (2.16)$$

or again

$$\triangle C_D = C_L^2\frac{S}{8 S_0}\left[1+\frac{3}{16}\left(\frac{b}{R}\right)^4+\frac{5}{64}\left(\frac{b}{R}\right)^8+\ldots\right] = C_L^2\frac{S}{8 S_0}(1+\delta) \qquad (2.17)$$

The usual values of the ratio $(b/R) = (2b/\text{Diam.})$ = span of wing to the diameter of the tunnel, rarely exceeds the values 0.60 or 0.70, and hence the factor $(1+\delta) = 1+\frac{3}{16}\left(\frac{b}{R}\right)^4+\frac{5}{64}\left(\frac{b}{R}\right)^8+\ldots$

is rapidly convergent and the value given above for $\triangle C_D$ differs but little from that found above for the case of a uniform distribution. An approximate calculation for any system of wings whatever, but suffi-

[1] Throughout this Division, frequent use is made of a symbol in parenthesis to denote some special limitation, location or other condition limiting the application of the preceding symbol. Thus $v(y')$ = vertical velocity at the point y'. Also in general, z will be used in the sense of the complex $x + iy$, and w in the sense of the complex $u - iv$.

ciently small, placed at the center of a guided air stream leads to the same expression for the induced complementary resistance, so that the law of variation of Γ along the span does not exercise a large influence.

Practically, the influence of the walls of the tunnel on the wing is equivalent to that of a fictitious ascending current producing a change in the angle of incidence of the amount

$$\triangle\alpha \cong \frac{\triangle C_D}{C_L} = C_L \frac{S}{8 S_0}\left[1 + \frac{3}{16}\left(\frac{b}{R}\right)^4 + \dots\right] = C_L \frac{S}{8 S_0}(1+\delta) \qquad (2.18)$$

and the corrections to be applied to the experimental results in a stream under constraint in order to obtain the corresponding values for unlimited flow are the following:

1) For a given lift coefficient C_L the experimental values of the angle of incidence must be *increased* by the angle

$$\triangle\alpha = C_L \frac{S}{8 S_0}(1+\delta) \qquad (2.19)$$

2) For a given lift coefficient C_L, the experimental values of the drag coefficient C_D must be *increased* by the amount

$$\triangle C_D = C_L^2 \frac{S}{8 S_0}(1+\delta) \qquad (2.20)$$

The following table gives the values of the factor $(1+\delta)$ for the usual values of the ratio b/R

$b/R =$	0.30	0.40	0.50	0.60	0.70
$(1+\delta) =$	1.0015	1.0050	1.0130	1.0255	1.0500

Strictly speaking, the distribution of lift along a wing placed in an air stream as here assumed is modified by the velocity due to the system of "image-wings". However, for usual values of b/R, this modification is negligible[1].

b) Model in Free Air Stream. The method of treatment in this case is entirely similar to that for the case with constrained or fixed boundary air stream.

It has been seen that the constraints due to a free stream correspond to a fictitious movement of which the potential φ_c must satisfy, on the limiting surface of the stream, the condition

$$\varphi_c + \varphi' = 0 \qquad (2.21)$$

In other words, the motion resulting from the addition of the potentials φ_c and φ' must not show, on the limiting surface, any component of velocity along the direction of flow.

For the transverse sections far enough downstream, the influence of the bound vortex is negligible and the potential φ' corresponds to

[1] MILLIKAN, C. B., On the Lift Distribution for a Wing of Arbitrary Plan Form in a Circular Wind Tunnel, Trans. of Amer. Soc. of Mech. Engrs. Applied Mechanics Division, 1932—33.

the free vortices only. The condition is then satisfied by associating with each free vortex $M\xi$ with circulation $(d\Gamma/dy)\,dy$ the following (see Fig. 22a and 22b):

1) A vortex image $M'\xi'$ of the same circulation as $M\xi$ and situated at the distance OM' conjugate to OM, *i. e.* $OM' = R^2/OM$.

2) An axial vortex Ox' with inverse circulation $-(d\Gamma/dy)\,dy$.

For the wing as a whole, the axial vortices mutually cancel, two by two, and there remains simply the "image-wings" $A'y$ and $B'y'$ with their sheets of free vortices, as in the case of the guided air stream. However, in the present case, the velocities induced at the wing will be directed downward, since the circulation of the vortex-images is of the same sign as that of the corresponding free vortices on the wing.

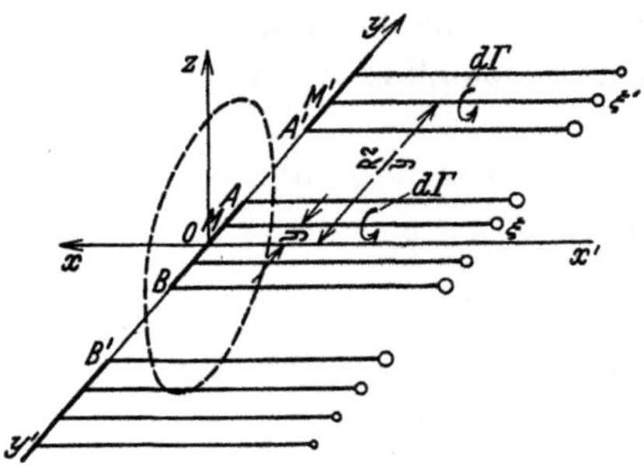

Fig. 22a.

It results that the effect in a free stream is equivalent to the action of a complementary descending flow, and quantitatively *equal and of opposite sign* to that due to the constraint in a guided stream.

The corrections to be made in the case of experimental results in a free stream in order to obtain the corresponding values for unlimited flow consist then in subtracting from the experimental values the values of $\Delta\,\alpha$ and of $\Delta\,C_D$, calculated as in (2.19) and (2.20).

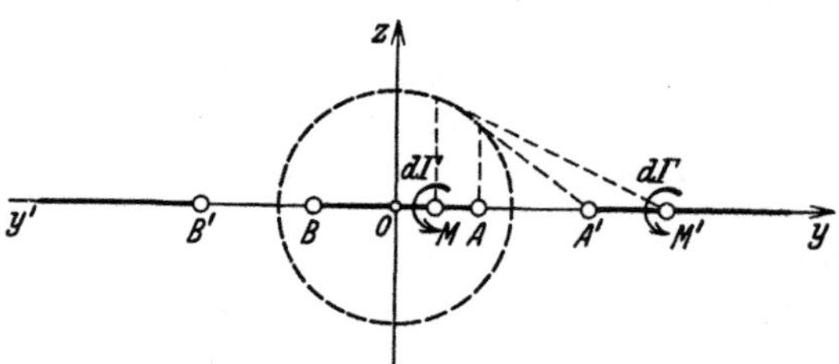

Fig. 22b.

It is to be noted that the preceding method assumes that the bound vortex representative of the wing does not itself directly influence the air stream flow. In reality it will be seen at a later point (corrections for plane flow) that this influence is not always negligible. In particular, in the case of a free air stream, the distance between the wing under test and the end section of the free stream should be as short as possible in order that the incurvation of the stream due to the vortex-images of the bound vortex may be negligible. This condition is generally realized in experimental work on reduced scale models. Furthermore, the ratio of the chord of the wing to the diameter of the stream is generally very small.

In general, no attempt is made to apply these supplementary corrections, especially as they are of relatively still smaller importance for experiments in a fixed boundary air stream.

In the same order of ideas, the corrections noted above might also be completed by taking account of the gradient of static pressure. However, for models of wings or of airplanes relatively small or of over all length moderate, this correction is negligible.

From the preceding discussion it follows that the experimental results relative to the same wing system tested in a wind tunnel with constrained stream and with free stream, both of the same diameter and all conditions equal otherwise, will differ between themselves by twice the corrections calculated as above. See Fig. 23. An especially interesting experimental verification, as regards the value of these corrections, could therefore be obtained by tests of various wings or wing systems carried out in a wind tunnel permitting of use at will, either with air stream constrained or free.

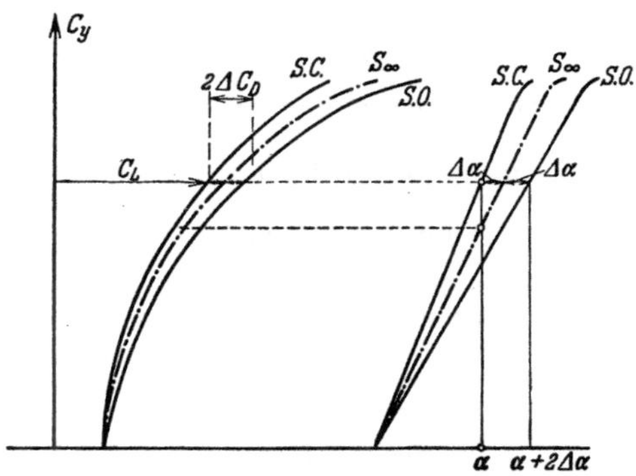

Fig. 23. Corrections for ΔC_D and $\Delta\alpha$. S.C. = stream constrained. S.O = stream open. S ∞ = Indefinite flow.

It is surprising that such tests are either non-existent or very rare because existing wind tunnels in general are arranged for operation in either one way or the other, but not both.

However, experimental verification of corrections of this character are very numerous and very conclusive as regards the application of the Prandtl corrections to wings of different proportion or of different dimensions tested in the same tunnel. Especial reference may be made to the experiments of Higgins[1] with regard to tests in a constrained air stream and to the tests in the Göttingen Laboratory in a free stream.

Note to Section 2. Consider, in a section far down stream, Fig. 18 the vortex A of circulation Γ_0 located at the distance $OA = b$. The potential function for this vortex has the form

$$f(A) = -\frac{i\Gamma_0}{2\pi} \log(z - b) \tag{1}$$

For the image vortex A' of circulation $-\Gamma_0$ located at the distance $OA' = R^2/b$, the complex potential will be similarly

$$f(A') = +\frac{i\Gamma_0}{2\pi} \log\left(z - \frac{R^2}{b}\right) \tag{2}$$

[1] Higgins, George J., The Effect of the Walls in Closed Type Wind Tunnels, U. S. N. A. C. A. Technical Report No. 275, 1927.

The potential of the motion resulting from the sum of $f(A)$ and $f(A')$ will be

$$F(A, A') = -\frac{i\Gamma_0}{2\pi} \log \frac{z-b}{z-R^2/b} \tag{3}$$

For a point P on the circumference of radius R, we shall have,

$$z-b = r_1 e^{i\theta_1}$$

and

$$z - R^2/b = r_2 e^{i\theta_1}$$

and the expression for $F(A, A')$ for the point P may be written

$$F(A, A') = \frac{-i\Gamma}{2\pi} \log \frac{r_1}{r_2} + \frac{\Gamma}{2\pi}(\theta_1 - \theta_2) \tag{4}$$

The equation of the stream-line passing through P will be

$$\psi_{(P)} = \frac{-\Gamma}{2\pi} \log \frac{r_1}{r_2} = \text{const.} \quad \text{or} \quad \frac{r_1}{r_2} = \text{const.} \tag{5}$$

It is easily seen that the circumference of radius R satisfies this condition for we have $\quad \frac{r_1}{r_2} = \frac{b}{R} = \sin O P_0 A$

3. Wind Tunnel with Square or Rectangular Section—Constrained Air Stream. Let us consider a wing of span $2b$ of area S, tested in the center of a constrained air stream, rectangular in section of width h and height k.

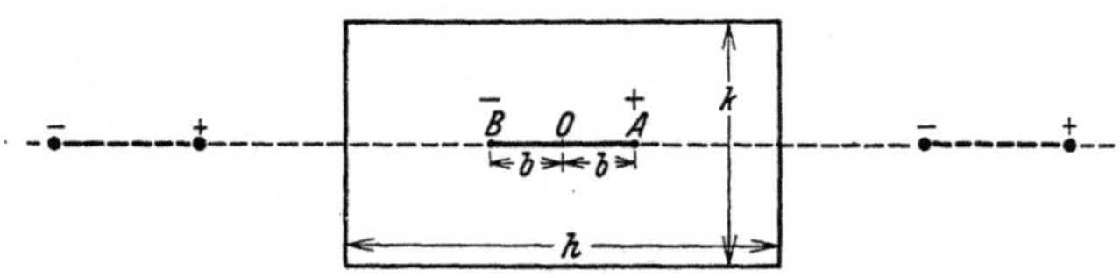

Fig. 24.

We assume that the circulation is uniform over the span and in consequence the equivalent vortex system will be composed of a bound vortex AB of circulation Γ_0 and of two free vortices of the same circulation Γ_0. See Fig. 24.

In applying the method of vortex images, it is easy to see that the constraint due to the walls of the tunnel may be replaced by a doubly infinite series of "wing-images" of which the free vortices constitute, downstream, horizontal ranges of alternating vortices.

Taking for example the horizontal range in the plane of the wing, it is seen that this range is composed 1) of a file of vortices such as A of the same circulation Γ and 2) of a file of vortices such as B of circulation $-\Gamma$. The interval between two files of the same sign is equal to the width h of the tunnel section. Taking the initial vortices as those located at $\quad A, (z_0 = +b)$ and $B, (z_0 = -b)$
the potential function at any point $z = (x + iy)$ is given by the equation[1]

$$F = \frac{\Gamma}{2i\pi} \log \frac{\sin(\pi/h)(z-b)}{\sin(\pi/h)(z+b)} \tag{3.1}$$

[1] VILLAT, H., Leçons sur la Theorie des Tourbillons, p. 56, Paris, 1930.

Let us now consider a horizontal range of order n. The origin-vortices corresponding to A_0 and to B_0 will have for coordinates

$$z_0 = b + ink \quad \text{and} \quad z_0' = -b + ink$$

The corresponding potential will then be

$$F(n) = \frac{(-1)^n \Gamma}{2 i \pi} \log \frac{\sin (\pi/h)(z - b - i n k)}{\sin (\pi/h)(z + b - i n k)} \tag{3.2}$$

The signs develop from the fact that the file of order n is inverse to that of order zero when n is odd and similar when n is even. In consequence, the potential function due to the double infinite series of vortices comprising also the vortices of the wing itself, will be obtained by making the summation $\sum_{-\infty}^{+\infty} F_{(n)}$

n taking all entire values positive and negative including $n = 0$. We shall then have

$$F = \sum_{-\infty}^{+\infty} F_{(n)} = \frac{\Gamma}{2 i \pi} \sum_{-\infty}^{+\infty} (\pm 1) \log \frac{\sin (\pi/h)(z - b - i n k)}{\sin (\pi/h)(z + b - i n k)} \tag{3.3}$$

The corresponding complex velocity will be given by

$$w = \frac{dF}{dz} = \frac{\Gamma}{2 i h} \sum_{-\infty}^{+\infty} (-1)^n \left[\cot \frac{\pi}{h}(z - b - i n k) - \cot \frac{\pi}{h}(z + b - i n k)\right]$$

which after some simple transformation becomes

$$w = \frac{\Gamma}{i h} \sum_{-\infty}^{+\infty} \left[(-1)^n \frac{\sin 2 \pi b/h}{\cos \frac{2 \pi b}{h} - \cos \left(\frac{2 \pi z}{h} - i 2 \pi n \frac{k}{h}\right)}\right] \tag{3.4}$$

Thus at the point $z = 0$ for sections sufficiently far downstream, the induced velocity has the value

$$w_{(z=0)} = -\frac{i \Gamma}{h} \sum_{-\infty}^{+\infty} \left[(-1)^n \frac{\sin 2 \pi b/h}{\cos \frac{2 \pi b}{h} - \cosh 2 \pi n \frac{k}{h}}\right] \tag{3.5}$$

In order to obtain the velocity due to the vortex images alone (velocity due to the complementary potential φ_c) it is necessary to subtract from w the induced velocity due to the vortices representing the wing itself. This has, at any point z, the value

$$w_1 = \frac{-i \Gamma}{2 \pi} \left(\frac{1}{z - b} - \frac{1}{z + b}\right) \tag{3.6}$$

and at the point $z = 0$, this gives

$$w_{1\,(z=0)} = \frac{i \Gamma}{\pi b}$$

It results that the velocity due to the complementary potential φ_c, has, at a great distance downstream and for $z = 0$, the value

$$w_{c\,(z=0)} = (w - w_1)_{(z=0)} = \frac{-i\Gamma}{h}\left[\sum_{-\infty}^{+\infty} \frac{(-1)^n \sin 2\pi b/h}{\cos \frac{2\pi b}{h} - \cosh 2\pi n \frac{k}{h}} + \frac{h}{\pi b}\right] \tag{3.7}$$

At the wing $A_0 B_0$, the mean value of this disturbing velocity will have one half this value. Furthermore, in accordance with relation (3.7) this velocity becomes reduced to the single vertical component v_c and we have $w_{c\,(z=0)} = (u_c - i\,v_c)_{z=0} = -(i\,v_c)_{z=0}$

whence
$$v_{c\,(z=0)} = \frac{\Gamma}{2h}\left[\sum_{-\infty}^{+\infty} \frac{(-1)^n \sin 2\pi b/h}{\cos 2\pi b/h - \cosh 2\pi n\, k/h} + \frac{h}{2b}\right] \tag{3.8}$$

It thus appears that the constraint due to rectangular walls of area $S_0 = hk$ is equivalent to the action of a flow producing a change in the angle of incidence which in general will have the value

$$\triangle\alpha = \frac{v_{c\,(z=0)}}{U_0} = \frac{\Gamma}{2h} A = C_L \frac{S}{8 S_0} \frac{k}{b} A = \varepsilon C_L \frac{S}{S_0} \tag{3.9}$$

where
$$A = \sum_{-\infty}^{+\infty} \frac{(-1)^n \sin 2\pi b/h}{\cos \frac{2\pi b}{h} - \cosh 2\pi n\, k/h} + \frac{h}{\pi b} \tag{3.10}$$

The correlative induced resistance will then be

$$\triangle D \cong L \triangle \alpha$$

whence
$$\triangle C_D = \frac{C_L^2 S}{8 S_0} \frac{k}{b} A = \varepsilon \frac{C_L^2 S}{S_0} \tag{3.11}$$

The calculation of the factor A can easily be made by writing the relation (3.10) in the form

$$A = \left[\frac{h}{\pi b} - \cot \frac{\pi b}{h} + 2 \sin \frac{\pi b}{h} \sum_{1}^{\infty} \frac{(-1)^n}{\cos \frac{\pi b}{h} - \cosh 2\pi n \frac{k}{h}}\right] \tag{3.12}$$

in which the term $\cot \pi b/h$ corresponds to $n = 0$. Usual values of k/h may be comprised between 1/4 and 4. It may thus be noted that it is sufficient to make $n = \pm 1, \pm 2$ (and perhaps ± 3 for small values of the ratio k/h) for the series under the sign k/h is rapidly convergent. The velocity v_c is ascending and the corrections $\triangle\alpha$ and $\triangle C_L$ must therefore be added to the experimental values.

M. Carafoli[1] has calculated the values of $\varepsilon = A k/8b$ for various ratios k/h and for the two values $h/2b = 2$ and $h/2b = 1.5$. The following table gives the numerical results.

---

[1] Carafoli, E., Contribution à la théorie de la sustentation en aérodynamique, Thèse de Doctorat ès-sciences, Paris, 1928.

These results may be compared with those obtained by Glauert[1] and Terazawa[2].

Making use of an approximative method in which the influence of the ratios $k/2b$ and $h/2b$ was neglected, Glauert calculated the values of a factor $(A k)/8b$ which permits the calculation of the corrections $\triangle \alpha$ and $\triangle C_D$ by means of (3.9) and (3.11). For values of $k/h \geqslant 1$, the values found by Glauert are very near those found by Carafoli with $h/2b = 2$, but the differences between the two sets of results become more notable for small values of k/h. For the value $k/h = 1$ (square section), the values of Carafoli and Glauert are very near together.

By way of experimental verification, the following may be noted.

Glauert[3] applied the corrections (calculated by the approximate method) to the results of tests on a monoplane wing and on a biplane cellule carried out in two wind tunnels, of square section, 7 ft. and 4 ft. respectively on the side, and found a very satisfactory agreement between the results thus corrected.

Terazawa applied Villat's method to the case of a wing tested in a rectangular tunnel in a constrained stream. He preferred to represent the complementary disturbing velocity at the wing by the mean value of the vertical induced velocity, instead of by the value corresponding to the mid section.

The factor $\varepsilon = A\,k/8b$ is then expressed by the equation

$$\varepsilon = \frac{1}{4\pi}\cdot\frac{h}{2b}\cdot\frac{k}{2b}\,K\left(\frac{b}{h}\,e^{-\pi k/h}\right) \qquad (3.13)$$

where K is an elliptic function, somewhat complicated, but permitting of simplification for the usual cases where $e^{-\pi k/h}$ is very small.

Terazawa has given only the following results

$\frac{h}{b} =$	1	1
$\frac{b}{L} =$	1.33	2.33
$\varepsilon = \frac{A}{8}\cdot\frac{K}{b} =$	0.171	0.143

Equation (3.12) applied to these two cases would have given respectively, $\varepsilon = 0.147$ and 0.138. With the usual proportions of $h/2b$ between 1.33 and 2.00, it may be assumed that the corrections calculated by the formulas (3.9), (3.10), and (3.11) will be quite sufficiently close.

Rosenhead (*loc. cit.*) has also dealt with the case of a constrained stream of rectangular section, taking his departure from a potential defined by the elliptic functions of $(y \pm b)/h$ of periods 1 and $\tau = i\,2k/h$.

[1] Glauert, M., The Interference of Wind Channel Walls on the Aerodynamic Characteristics of an Aerofoil, British A.R.C., R. and M. 867, March 1923.

[2] Terazawa, M. K., On the Interference of Wind Tunnel Walls., Report No. 44 Ae. R. I., Tokyo, 1929.

[3] Glauert, M., British A.R.C., R. and M. 889, 1923—24.

The calculation leads to corrections $\triangle C_D$ and $\triangle \alpha$ of which the values given by Glauert are the first approximations.

Rosenhead considers also that it is sufficient in practice, to limit the approximation to the first few terms of the convergent series, taking the first power of b/h in the developments. In such case the corrections for an elliptical distribution of Γ are expressed by simplified formulas:

$$\triangle \alpha = \varepsilon \frac{S}{S_0} C_L$$

$$\triangle C_D = \varepsilon \frac{S}{S_0} C_L^2$$

in which

$$\varepsilon \cong \frac{\pi k}{2h}\left[\frac{1}{12} + 2\sum_{p=1}^{\infty} \frac{p}{1+e^{\frac{2p\pi k}{h}}}\right]$$

By the use of the same general principles of vortex images, but carried out with slightly different detail, we may obtain a more general view of the corrections for an air-stream of rectangular section under various conditions of restraint or freedom at the experimental section.

We state again the general assumptions made and the program followed in the development of these results.

a) The assumption, as hitherto, of a uniform distribution of circulation along the span of the wing.

b) The replacement of the actual physical limitation of the stream by an infinite system of vortex images of the marginal vortices of the wing itself.

c) The calculation of mean velocity v_m due to these images by the relation:

$$v_m = \frac{1}{2b}\int_{-b}^{+b} v\, d\, y \tag{3.14}$$

where v is the velocity produced at a point on the wing between $-b$ and $+b$ ($2b$ = span of wing).

d) The calculation of the corrections by the formulas:

$$\triangle \alpha = \frac{v_m}{V} = C_L \frac{S}{S_0}\varepsilon \tag{3.15}$$

$$\triangle C_D = C_L^2 \frac{S}{S_0}\varepsilon \tag{3.16}$$

in which ε is the same factor as in (3.9) and (3.11).

In all cases, the wing is assumed to be supported in the center of the section of the stream.

Case I. Rectangular Section, Half Constrained by Horizontal Walls (Parallel to the Span)[1]. The general scheme of the system of vortex

[1] TOUSSAINT, A., Sur les corrections à apporter aux caractéristiques aérodynamiques d'une aile sustentatrice expérimentée dans une soufflerie à veine rectangulaire semi-guidée par des parois parallèles à l'envergure de l'aile et à la vitesse du vent., Comptes rendus de l'Académie des Sciences, Feb. 20, 1934.

images is shown in Fig. 25. It comprises vertical files of vortices with alternate signs as determined from the two initial wing-tip vortices.

The complex potential $f(z)$ is then calculated for a vertical file of vortices and there follows directly the velocity $1/2\ (df/dz)$ at any point z due to a single vortex file.

These results are then combined for double vertical files. The diagram shows that the double vertical files are similar to the initial file for $y = \pm m h$ when m is even, and inverse when m is odd.

For the initial double file ($y = \pm b$) the mean velocity due to the wing tip vortices of the wing itself is subtracted from the mean velocity calculated from (3.14). The influence of the images of this double initial file is thus given by the expression:

$$v_m(0) = -\frac{\Gamma}{\pi b} \log \frac{\tanh \pi b/k}{\pi b/k} \tag{3.17}$$

The images comprising the remaining double files $x = \pm mb$ give, in the same way,

$$v_m(\pm m) = \frac{-\Gamma}{4\pi b}(-1)^m \sum_1^\infty \log \frac{\tanh(\pi/2k)(mh + 2b)\tanh(\pi/2k)(mh - 2b)}{\tanh^2(\pi m h/2k)}$$

Finally replacing Γ by $(C_L/2)\, U_0\, S/L$, the correction $\triangle\alpha$ is found from (3.15) with a value of ε given by:

$$\left.\begin{aligned} \varepsilon_1 = \frac{-1}{4\pi (h/k)(2b/h)^2} \Big[& \log \frac{\tanh \pi b/k}{\pi b/k} + \\ & + (-1)^m \log \frac{\tanh(\pi/2k)(mh + 2b)\tanh(\pi/2k)(mh - 2b)}{\tanh^2(\pi m h/2k)} \end{aligned}\right\} \tag{3.18}$$

The computation of ε_1 is simple, for it is only necessary to carry it out for values of $m = 1$, and 2 for normal values of the ratio h/k.

Fig. 25 gives the results for various values of h/k and of $2b/h$.

It is readily seen that for values of $2b/h$ lying between 0.5 and 0.7 and for values of h/k between 1.15 and 1.36, the corrections are zero or negligibly small. For smaller values of h/k the corrections are negative, the same as for a free stream, and inversely for larger values. However, for the usual values of $2b/h$, the positive corrections are always somewhat smaller than the negative.

A stream of air under half constraint, in this manner, has then certain advantages with regard to these corrections, aside from any matters of experimental convenience which the presence of a "floor" and a "ceiling" might furnish. It would appear, therefore, that this type of constraint for a rectangular air stream is to be recommended, especially for private industrial laboratories, where the facility of construction may be added to the advantages already noted.

For the limiting case (purely theoretical) where $2b/h = 0$, the results become those found by Theodorsen[1].

Case II. Rectangular Section, Half Constrained by Vertical Walls (At Right Angles to the Span)[2]. The general scheme of the vortex images is shown in Fig. 26. It comprises vertical files of vortices of the same sign as those at the wing tips, (+) or (—) according to the file considered.

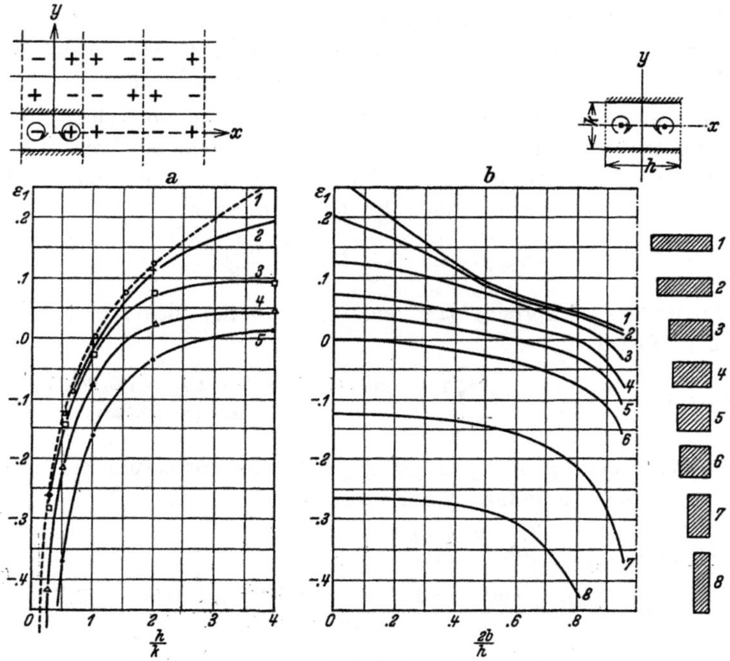

Fig. 25.

The association of double vertical files shows that they are all similar to the initial double file. The latter gives, by the action of its images, the mean vertical velocity

$$v_m(0) = \frac{-\Gamma}{4\pi b} \log \frac{\sinh(2\pi b/k)}{2\pi b/k} \tag{3.19}$$

In a similar manner the images composing the other double files give

$$v_m(\pm m h) = -\frac{\Gamma}{4\pi b} \sum_{+1}^{+\infty} \log\left[1 - \frac{\sinh^2 2\pi b/k}{\sinh^2 \pi m h/k}\right] \tag{3.20}$$

---

[1] THEODORSEN, T., The Theory of Wind Tunnel Wall Interference, U.S. N.A.C.A. Report No. 410, 1931.

[2] TOUSSAINT, A., Des corrections à apporter aux caractéristiques aérodynamiques d'une aile sustentatrice expérimentée dans une soufflerie à veine rectangulaire semi-guidée par des parois latérales perpendiculaires à l'envergure de l'aile, Comptes rendus de l'Académie des Sciences, Feb. 27, 1934.

and finally we have

$$\varepsilon_2 = \frac{-1}{4\pi\ (h/k)\ (2\,b/h)^2}\left[\log\frac{\sinh\,(2\,\pi\,b/k)}{2\,\pi\,b/k} + \sum_1^{\infty}\log\left(1 - \frac{\sinh^2 2\,\pi\,b/k}{\sinh^2 \pi\,m\,h/k}\right)\right] \tag{3.21}$$

The calculation of ε_2 is made with $m = 1$ and 2 for the usual values of h/k. Fig. 26 gives the results of these calculations. As may be seen,

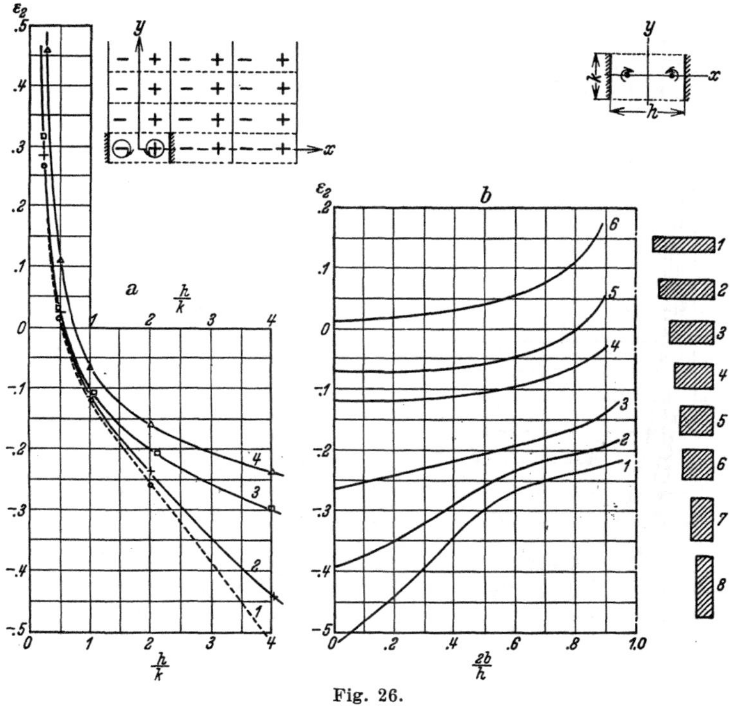

Fig. 26.

for $2b/h$ between 0.5 and 0.8, and for h/k between 0.6 and 0.75, the corrections are zero or negligibly small.

For the case of $2b/h = 0$, Theodorsen's results[1] are again produced.

Case III. Rectangular Stream Fully Constrained. This case is, of course, the same as that already considered in the earlier part of the present section. Repeating, however, to a slight degree and in the form similar to that employed for the two preceding cases, we have the general

[1] *loc. cit.*

scheme of vortex images shown in Fig. 27. The vertical files associated in pairs give double files similar to the initial double file. The calculation of ε_3 is then reduced to the calculation of ε_1, the factor $(-1)^m$ being omitted. This gives

$$\varepsilon_3 = \frac{-1}{4\pi\,(h/k)\,(2\,b\,h)^2}\left[\log\frac{\tanh \pi\, b/k}{\pi\, b/k} + \sum_1^{\infty} \log \frac{\tanh\,(\pi/2\,k)\,(m\,h+2\,b)\,\tanh\,(\pi/2\,k)\,(m\,h-2\,b)}{\tanh^2\,(\pi\,m\,h/2\,k)}\right] \tag{3.22}$$

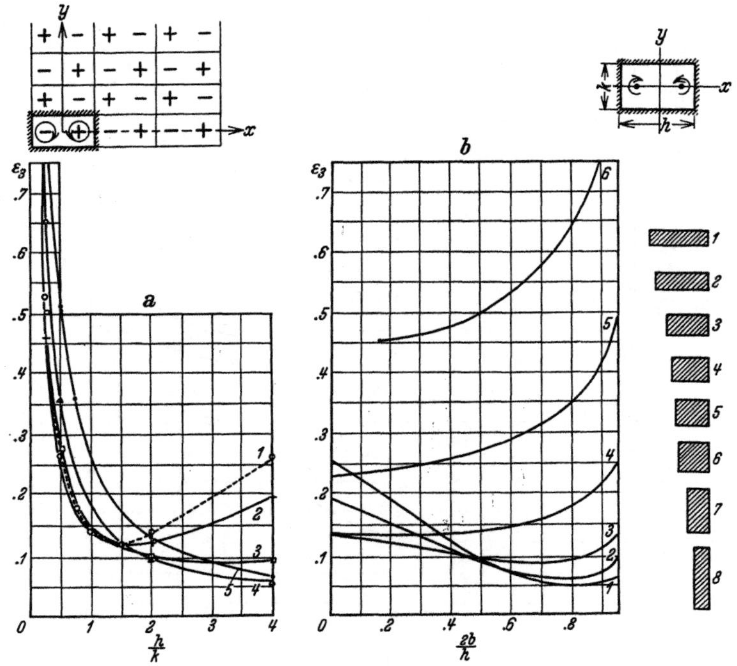

Fig. 27.

Fig. 27 gives the results of this calculation and it is seen that the corrections are always positive. For the case of $2b/h = 0$, Theodorsen's results[1] are again produced.

Case IV. Rectangular Stream, Free at Test Section. The general scheme of images is shown in Fig. 28. The association of the vertical files give double files similar or inverse to the initial double file. The calculation of ε_4 is reduced therefore to the calculation of ε_2, with, however, the introduction of the factor $(-1)^m$. This gives:

[1] *loc. cit.*

$$\varepsilon_4 = \frac{-1}{4\pi(h/k)(2b/h)^2}\left\{\left[\log\frac{\sinh 2\pi b/k}{2\pi b/k} + \right.\right.$$
$$\left.\left. + (-1)^m \sum_1^\infty \log\left(1 - \frac{\sinh^2 2\pi b/k}{\sinh^2 \pi m h/k}\right)\right]\right\} \qquad (3.23)$$

Fig. 28 gives the results of this calculation, and it is seen that the corrections are always negative.

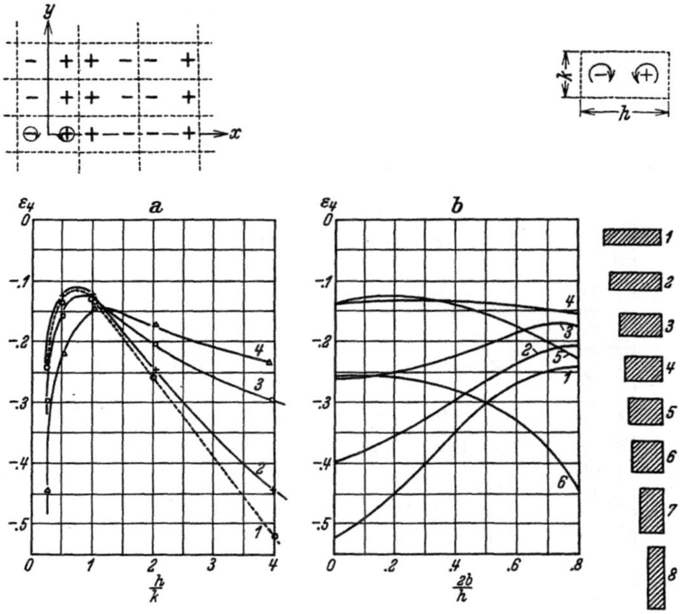

Fig. 28.

4. Tunnel with Rectangular Section with Two-Dimensional Flow. In this case the wing extends to the limits of the stream so that there are no wing tip vortices trailing into the flow and the distribution of the circulation is uniform. This arrangement is used for the purpose of approaching as closely as possible to the conditions corresponding to two-dimensional flow. However, the stream being limited above and below the wing, there is reason to examine the constraints due to this limitation. As in the preceding case, we shall consider the condition with the air stream free or constrained.

a) Constrained Flow. The vortex system representative of the wing is reduced to a lifting vortex of circulation Γ, located, for example, at a distance k_1 from the bottom of the tunnel. Let k be the total height of the air stream. See Fig. 29.

The method of images leads to the replacement of the constraints due to the rigid walls by an indefinite series of vortex images, constituting a vertical file of vortices of alternate circulation $\pm\Gamma$. The distance between the vortices of the same circulation is $2k$. The initial vortices will have as coordinates $z_0 = 0$ and $z'_0 = -i\,2k_1$. The potential function due to this vertical file will then have the value

$$F = \frac{\Gamma}{2i\pi} \log \frac{\sinh \frac{\pi z}{2k}}{\sinh \frac{\pi}{2k}(z + 2ik_1)} \tag{4.1}$$

The complex velocity w at any point z will then be

$$w = \frac{dF}{dz} = \frac{-i\Gamma}{4k}\left[\coth \frac{\pi z}{2k} - \coth \frac{\pi}{2k}(z + 2ik_1)\right] \tag{4.2}$$

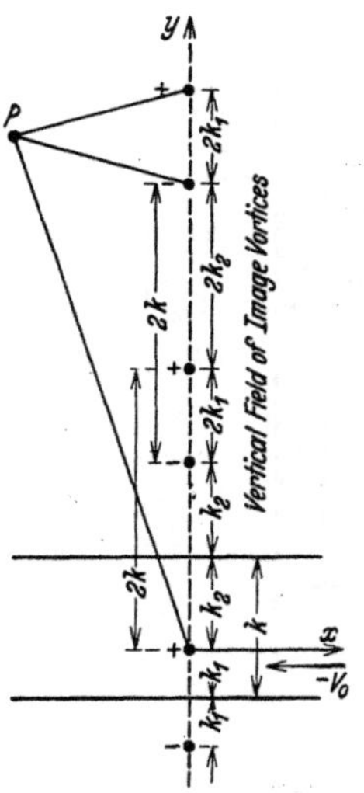

Fig. 29.

It is readily seen that on the walls [*i. e.* for $z = -ik$ and for $z = i(k - k_1)$] this velocity has no normal component.

At the point $z = 0$ the velocity w becomes reduced to

$$w_{z=0} = \frac{i\Gamma}{4k} \coth \frac{i\pi k_1}{k} = \frac{\Gamma}{4k} \coth \frac{\pi k_1}{k} = u_{c\,(z=0)} \tag{4.3}$$

The lifting vortex itself having no influence at the wing, this expression represents the complementary velocity representative of the constraint due to the walls of the tunnel. It is seen that this constraint is equivalent to a decrease in the velocity of flow $-U_0$.

However, if the wing is placed in the middle of the flow ($k_1 = k/2$), this complementary velocity cancels itself. Nevertheless, and aside from this possible constraint, the stream at the wing may be curved by reason of the fact that upstream and downstream, *i. e.* $z \neq 0$, the complementary velocity has a vertical component which may be computed by subtracting from the expression for w the velocity due to the vortex proper of the wing, and calculating the expression

$$-iv_c = p.\,im.\,(w - w_1) = p.\,im.\left(w - \frac{i\Gamma}{2\pi z}\right) \tag{4.4}$$

We shall carry out this calculation for the points situated on the axis ox ($z = x$), that is,

$$(-iv_a)_{y=0} = p\ im.\ \frac{-i\Gamma}{4k} \coth \frac{\pi z}{2k} - \coth \frac{\pi}{2k}(z + 2ik_1) - \frac{2k}{\pi z}$$

A simple computation then gives

$$v_{c\,(y=0)} = \frac{\Gamma}{4\,k}\,\frac{\coth\frac{\pi z}{2\,k}\left(\coth\frac{2\,z}{2\,k}-1\right)}{\coth\pi\frac{k_1}{k}+\coth^2\frac{\pi z}{2\,k}} - \frac{2\,k}{\pi z} \tag{4.5}$$

It is seen immediately that for points respectively upstream and downstream, $v_{c\,(y=0)}$ changes sign with x. The stream-lines are then curved at the wing, with concavity directed upward.

According to Carafoli[1], the influence of this curving of the stream-lines is evidenced by an increase of circulation corresponding to an increase of the angle α by an amount equal to $c/4\,R_1$ where c is the chord of the wing and R_1 is radius of curvature of the stream-line. The value of R_1 at the wing will be given by the equation

$$\frac{1}{R_1} = -\frac{1}{U_0}\left(\frac{\partial v_c}{\partial x}\right)_{z=0} = \frac{1}{U_0}\,p.\,im.\left(\frac{d\,(w-w_1)}{d\,z}\right)_{z=0} \tag{4.6}$$

and for the general case this gives

$$\frac{1}{R_1} = \frac{-\pi\Gamma}{12\,U_0\,k^2}\left(\frac{1}{3}-\mathrm{cosech}^2\frac{\pi k_1}{k}\right) \tag{4.7}$$

For the particular case where $k_1 = k/2$, this gives

$$\frac{1}{R_1} = \frac{\pi\Gamma}{12\,U_0\,k^2}$$

The angle $\triangle\alpha$ corresponding to this effect of increased circulation will then be

$$\triangle\alpha = \frac{1}{4}\cdot\frac{c}{R_1} = \frac{\pi\Gamma c}{48\,U_0\,k^2} = C_L\,\frac{\pi}{96}\,\frac{c^2}{k^2} \tag{4.8}$$

The constraint due to the walls limiting such a two-dimensional flow is therefore equivalent to an induced angle $\triangle\alpha$ which must be added to the experimental values of the incidence for a given lift.

In other words, for a given angle of incidence, the coefficients of lift in a two-dimensional flow limited as in a tunnel or unlimited as in the open air, are in the ratio

$$\frac{C_{L\infty}}{C_{L\,(lim.)}} \cong \frac{1}{1+\frac{\varkappa\pi}{48}\,\frac{c^2}{k^2}}$$

where $\varkappa = \pi\,(1+\varepsilon)$ is the ratio $C_L/2\alpha\,(\infty)$.

For the usual forms of wing profiles, we may take

$\varkappa \cong 3.14 \times 1.1 = 3.45$

and there results

$$\frac{C_{L\,\infty}}{C_{L\,(lim.)}} = \frac{1}{1+0.225\,c^2/k^2}$$

The table below gives the values of this ratio for various values of c/k

c/k	$=0.1$	0.2	0.3	0.4	0.5
$\frac{C_{L\,\infty}}{C_{L\,(lim.)}}$	$=0.98$	0.957	0.937	0.917	0.90

[1] Carafoli, E., Thèse *loc. cit.*

By a method of calculation utilizing conformal transformation and Villat's[1] formula for Dirichlet's problem in a circular ring, T. Sasaki[2] has found that the sustentation in plane flow limited by rigid walls is slightly decreased. The following table gives values of the ratio $C_{L(\infty)}/C_{L(lim.)}$, coefficients of lift for plane flow, unlimited and limited, for various values of the ratio c/k, c being the chord of the wing (located at the center of the flow) and k the height of the flow between the limiting walls above and below.

$\frac{c}{k} =$	0.10	0.20	0.30
$\frac{C_{L(\infty)}}{C_{L(lim.)}} =$	1.02	1.052	1.10

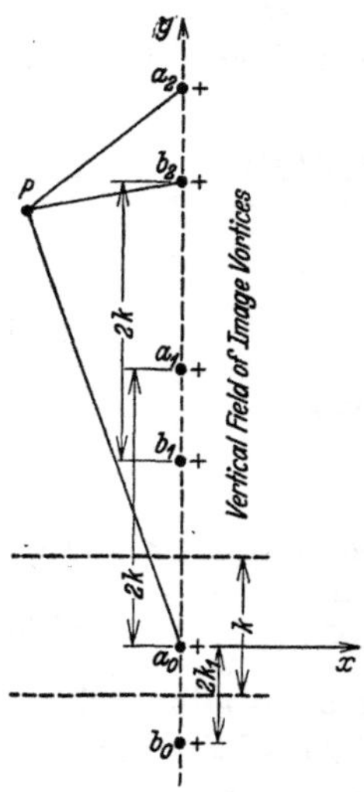

Fig. 30.

Sasaki notes further, that there is no complementary induced resistance accompanying this reduction of lift. These results are not in accord with those indicated by Carafoli, and it would appear that there is need of further investigation in order to settle the question. It would be sufficient to test, in the same tunnel, wings of the same profile but with varying values of the ratio c/k.

The comparison of the coefficients for the same angle of incidence would then show the influence of the ratio c/k.

b) Free Stream. The method of images leads to a replacement of the constraints due to a limitation of the plane flow above and below the wing, by an infinite series of vortex images constituting a file of vortices of the same circulation (Fig. 30).

Then taking the initial vortices at $z_0 = 0$ and $z_0' = -i\,2k_1$ the corresponding potential function is expressed in the form

$$F = \frac{\Gamma}{2i\pi}\left[\sum_{-\infty}^{+\infty} \log(z - 2ink) + \sum_{-\infty}^{+\infty} \log(z + 2ik_1 - 2ink)\right] \tag{4.9}$$

The complex velocity at any point z will be

$$w = \frac{dF}{dz} = \frac{\Gamma}{2i\pi}\left[\sum_{-\infty}^{+\infty} \frac{1}{z - 2ink} + \sum_{-\infty}^{+\infty} \frac{1}{z + 2ik_1 - 2ink}\right. \tag{4.10}$$

[1] VILLAT, H., Leçons sur l'hydrodynamique, Chapter II, Paris, 1929.

[2] SASAKI, T., On the Effect of the Wall of a Wind Tunnel Upon the Lift Coefficient of a Model, Report No. 46 Ae. R. I., Tokyo.

This expression may be written in the form

$$w = \frac{\Gamma}{2\,i\,\pi}\left[\frac{1}{z} + \sum_{1}^{\infty}\frac{2\,z}{z^2+4\,n^2k^2} + \frac{1}{z+2\,i\,k_1} + \sum_{1}^{\infty}\frac{2\,(z+2\,i\,k_1)}{(z+2\,i\,k_1)^2+4\,n^2k^2}\right]$$

and making use of the relation[1]

$$coth\, x = \frac{1}{x} + \sum_{1}^{\infty}\frac{2\,x}{x^2+n^2\pi^2}$$

and putting $x = \pi\, z/2\,k$, the value of w takes the form

$$w = -\frac{i\,\Gamma}{4\,k}\left[coth\,\frac{\pi\,z}{2\,k} + coth\,\frac{\pi}{2\,k}\,(z+2\,i\,k_1)\right] \qquad (4.11)$$

Aside from the sign of the second term, this expression is identical with that found for the value of w in the case of constrained plane flow.

We shall then have at the wing a complementary velocity

$$w_{(z=0)} = u_c = -\frac{\Gamma}{4\,k}\,cot\,\frac{\pi\,k_1}{k} \qquad (4.12)$$

which is to be added to $-U_0$ and which becomes zero in the particular case where $k_1 = k/2$.

Furthermore, the stream is curved at the wing with the concavity downward and the corresponding radius of curvature is given by

$$\frac{1}{R_1} = -\frac{\pi\,\Gamma}{8\,U_0\,k^2}\left(\frac{1}{3} + cosech^2\,\frac{\pi\,k_1}{k}\right) \qquad (4.13)$$

However, in this case of the free stream, the curvature is extended both upstream and downstream from the wing. Thus for the usual case where $k_1 = k/2$, the value of the complementary velocity at ∞ upstream and at ∞ downstream is given by

$$w_{(z=\pm\infty)} = \frac{\Gamma}{2\,i\,k}\left(coth\,\frac{\pi\,z}{k}\right)_{z=\pm\infty} = \mp\frac{i\,\Gamma}{2\,k} = (-i\,u_c)_{z=\pm\infty}$$

From this follows,

$$u_{c\,(z=\pm\infty)} = \pm\frac{\Gamma}{2\,k} = \pm\frac{C_L}{4}\,U_0\,\frac{c}{k} \qquad (4.14)$$

At a finite distance upstream (or downstream) the curvature of the stream is characterized by the value of w for $z = x \pm i\,k/2$ and always for $k_1 = k/2$ we have

$$w = \frac{-i\,\Gamma}{2\,k}\,coth\,\frac{\pi}{k}\left(x \pm \frac{i\,k}{2}\right) = -\frac{i\,\Gamma}{2\,k}\,tanh\,\frac{\pi\,x}{k} \qquad (4.15)$$

The inclination of the stream at the distance x upstream will then be characterized by the angle

$$\triangle\,\alpha\,(x) = \frac{u_{c\,(x)}}{U_0} = \frac{\Gamma}{2\,k\,U_0}\,tanh\,\frac{\pi\,x}{k} = \frac{C_L}{4}\,\frac{c}{k}\,tanh\,\frac{\pi\,x}{k} \qquad (4.16)$$

[1] This may be derived from the corresponding expression for *cot* x by putting $i\,x$ for x. For the latter formula see Theory of Infinite Processes, Smail pp. 229 237, or Villat *loc. cit.* p. 58.

Practically the two-dimensional (plane) stream emerges from a constrained section $A_0\, B_0$ located upstream from the wing. The curvature of the stream at the wing will then be measured by the angle $\triangle\, \alpha\, (x)$ and the aerodynamic resultant $L\, \infty$ will be normal to the local relative wind which makes this angle with the direction of U_0 at the section of emergence.

The experimental measurements made under these conditions will then comprise these components

$$\left. \begin{aligned} L_c &= L_\infty \cos \triangle\, \alpha\, (x) \cong L_\infty \\ D_c &= L_\infty \sin \triangle\, \alpha\, (x) = L_c\, \triangle\, \alpha\, (x) \end{aligned} \right\} \tag{4.17}$$

with an effective incidence

$$\alpha_\infty = \alpha_e - \triangle\, \alpha\, (x) \tag{4.18}$$

Thus the influence of the free plane flow is reduced to the following corrections to be applied to the experimental values.

1) For a given coefficient C_L, the experimental values of the incidence α_e must be decreased by an angle $\triangle\, \alpha$ given by the equation

$$\triangle\, \alpha_{(x)} = \frac{C_L}{4} \frac{c}{k} \cdot \tanh \frac{\pi\, x}{k} \tag{4.19}$$

2) For this same value of C_L, the experimental values of the drag must be decreased by a quantity $\triangle\, C_D$ given by the equation

$$\triangle\, C_{D\,(x)} = \frac{C_L^2}{4} \frac{c}{k} \tanh \frac{\pi\, x}{k} \tag{4.20}$$

The following table gives the values of $\tanh\, (\pi\, x/h)$ as a function of (x/k)

$x/k =$	0	0.10	0.20	0.30	0.40	0.50	0.60
$\tanh \frac{\pi\, x}{k} =$	0	0.304	0.557	0.736	0.85	0.917	0.955

Thus when the distance between the wing and the section of emergence is greater than 0.6 h, the corrections due to the curvature of the plane flow tend toward a limit value,

$$\triangle\, \alpha_{(x = \infty)} = \frac{C_L}{4} \frac{c}{k}$$

The preceding corrections may be expressed *for a given value of the angle of incidence* by the ratio $C_{L\,(\infty)}/C_{L\,(c)}$ and by the value of $\triangle\, C_D$ corresponding to $C_{L\,(\infty)}$. Thus we have from (4.19), where $\varkappa = C_{L\,\infty}/2\,\alpha = \pi\,(1 + \varepsilon)$ [1],

$$\frac{C_{L\,(\infty)}}{C_{L\,(c)}} = \frac{1}{1 - \triangle\, \alpha/\alpha} = \frac{1}{1 - \varkappa \frac{c}{2\,k} \tanh \pi\, x/k} \tag{4.21}$$

and $$\triangle\, C_{D\,(\alpha)} = C_{L\,(\infty)}^2 \frac{c}{4\,k} \tanh \frac{\pi\, x}{k} = C_{L\,(c)}^2 \frac{c/4\,k \tanh \pi\, x/k}{1 - \varkappa\, (c/2\,k) \tanh \pi\, x/k}$$

[1] This comes directly from the well known formula $C_{L\,\infty} = 2\,\pi\,\alpha\,(1 + \varepsilon)$, see Division E II (8.20).

The following table gives the values of $C_{L(\infty)}/C_{L(c)}$ and of $\triangle C_{D(\alpha)}/C^2_{L\infty}$ for a practical case where $x/k = 0.30$ and for various values of c/k.

$c/k =$	0.10	0.20	0.30	0.40
$C_{L(\infty)}/C_{L(c)} =$	1.15	1.34	1.62	2.05
$\triangle C_{D(\alpha)}/C^2_{L(\infty)} =$	0.0635	0.1275	0.190	0.255

It may be here noted that the values of the ratio $C_{L(\infty)}/C_{L(c)}$ as given in the table above are in good agreement with the corresponding values calculated by Sasaki (*loc. cit.*) who has treated the problem in a somewhat different manner as previously noted. It must be remembered, however, that Sasaki's values relate to the theoretical case of a free stream indefinitely extended upstream and downstream. For this limit case, the values of the ratio furnished by (4.21) are somewhat larger, notably for large values of c/k. Here again a carefully planned series of tests would furnish a more satisfactory knowledge of these correction factors for the case of a free two-dimensional (plane) flow. However, by way of experimental verification, reference may be made to the tests carried out in the aerodynamic laboratory at Göttingen[1].

The plane flow and the model made use of had the following characteristics: $c = 0.60$, $k \cong 2m\,15$ whence $c/k = 0.28$

$x \cong 1.m\,34$, $x/k = 0.62$.

The corrections applied correspond to those of (4.19) and (4.20) for the limit case where $x = \infty$. The results as corrected were in very good agreement with those obtained by the usual tests with small models, for which, moreover, the corrections for limitation of flow are much smaller and better known.

But in this case we should have

$$tanh\,\frac{\pi x}{k} = 0.960$$

and in consequence the application of (4.19) and (4.20) would have given, for this case, satisfactory values.

It seems logical that the formulas for correction should take account of the distance x between the wing and the section of emergence of the free stream.

It should be further noted that in spite of the uncertainty regarding the corrections to be applied in the case of a constrained plane flow, these corrections are much smaller for the same value of the ratio c/k, than those to be applied in the case of a free plane flow. Furthermore in the first case there is no correction for the drag. For tests which should approach as nearly as possible to the conditions of an unlimited plane flow, the constrained plane flow is then clearly preferable to the free plane flow. In any case, however, there should be added to the corrections previously discussed, those relating to the influence of the

[1] Ergebnisse der Aerodynamischen Versuchsanstalt zu Göttingen, 1. Lieferung, p. 54.

gradient of static pressure and, in case of need, those relating to the effect of obstruction.

5. Body in Plane Flow of which the Wake is Composed of Alternate Vortices. It is known that for an indefinite flow about a body, the train of alternate vortices forms a stable configuration when the width h of the train and the interval a between two vortices of the same sign satisfy the relation, (see Fig. 31)

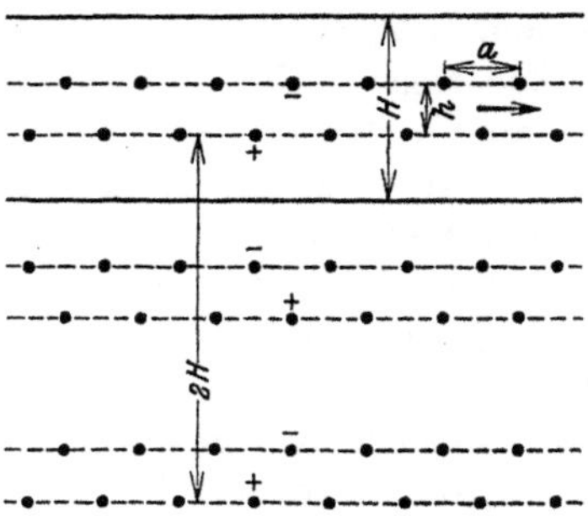

Fig. 31.

$$\sinh \frac{\pi h}{a} = 1 \text{ whence } \frac{h}{a} = 0.2806$$

Under these conditions the proper velocity of the vortices is $(-U_0 + u_0)$ where $u_0 = \frac{\Gamma}{2a} \tanh \frac{\pi h}{a}$

and from which, $\Gamma = 2\sqrt{2}\, a u_0$.

The aerodynamic resistance of a body (of transverse dimension b) is expressed by the drag coefficient

$$C_{D(\infty)} = \frac{D}{b(\varrho/2)U_0^2} = \left(5.656 - 2.24\frac{u_0}{U_0}\right)\frac{u_0 h}{U_0 b}$$

In order to calculate this resistance it is necessary to know, either theoretically or experimentally, the values of u_0/U_0 and of $u_0 h/U_0 b$ for cylindrical bodies of various forms of cross-section.

In order to determine the influence of limiting walls on a guided stream of finite width $H > h$, recourse may be had to the method of

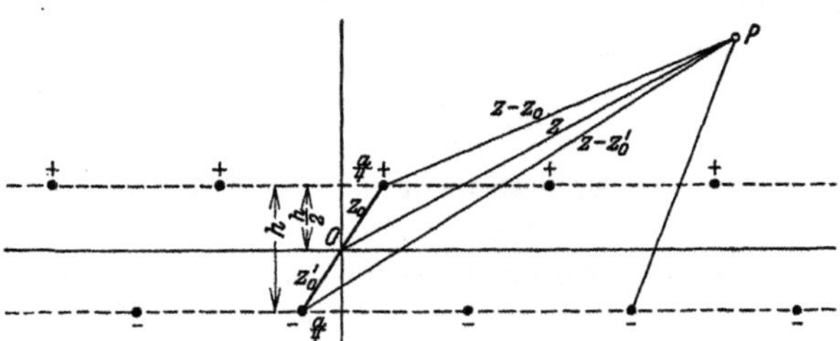

Fig. 32.

images, and the actual walls thus replaced by an infinite series of files or trains of vortex images.

For the file of the real vortices, the potential function referred to axes x and y is expressed for the region downstream sufficiently far away by the equation[1] [z = complex $(x + iy)$]

$$F = \frac{\Gamma}{2i\pi} \log \frac{\sin(\pi/a)(z - z_0)}{\sin(\pi/a)(z - z_0')} \tag{5.1}$$

with $z_0 = a/4 + i\,h/2$ and $z_0' = -z_0$. See Fig. 32.

[1] See VILLAT, Leçons sur la Théorie des Tourbillons, Chap. IV, Paris 1930.

The potential for a file of image vortices of order n will have the same expression with

$$\left.\begin{aligned} z_0 &= \frac{3a}{4} + \frac{ih}{2} \pm i\,n\,H \\ z_0' &= \frac{a}{4} - \frac{ih}{2} \pm i\,n\,H \end{aligned}\right\} \text{for } n \text{ odd}$$

and

$$\left.\begin{aligned} z_0 &= \frac{a}{4} + \frac{ih}{2} \pm i\,n\,H \\ z_0' &= -\frac{a}{4} - \frac{ih}{2} \pm i\,n\,H \end{aligned}\right\} \text{for } n \text{ even}$$

We consider first the case of n odd. This will give

$$F_{(n)} = \frac{\Gamma}{2\,i\,\pi} \log \frac{\sin \frac{\pi}{a}\left(z - \frac{3a}{4} - \frac{ih}{2} \mp i\,n\,H\right)}{\sin \frac{\pi}{a}\left(z - \frac{a}{4} + \frac{ih}{2} \mp i\,n\,H\right)}$$

and the corresponding complex velocity will be

$$\left.\begin{aligned} w_{(n)} = \frac{dF}{dz}_{(n)} = \frac{\Gamma}{2\,i\,a}\Big[\cot \frac{\pi}{a}\left(z - \frac{3a}{4} - \frac{ih}{2} \mp i\,n\,H\right) - \\ - \cot \frac{\pi}{a}\left(z - \frac{a}{4} + \frac{ih}{2} \mp i\,n\,H\right)\Big] \end{aligned}\right\} \tag{5.2}$$

This expression after a few elementary transformations becomes

$$w_{(n)} = \frac{\Gamma}{i\,a} \cdot \frac{\cosh \pi\,h/a}{i \sinh \pi\,h/a + \cos 2\,\pi\,z/a \cosh 2\,\pi\,n\,H/a \pm i \sin 2\,\pi\,z/a \sinh 2\,\pi\,n\,H/a} \tag{5.3}$$

For practical applications, $2\pi\,n H/a$ is much greater than $\pi\,h/a$ for in general $H/h > 6$. In these conditions if we put $2\,\pi\,H/a = \sigma$ with $\sigma = 12\,\pi\,h/a$ and $h/a = 0.2806$, we shall have $e^{\sigma} \geqslant 40{,}000$.

We may thus, with Glauert[1], simplify the expression of w_n, even for $n = 1$, by neglecting $e^{-\sigma}$ in comparison with e^{σ} in the values of $\cosh\, n\,\sigma$ and of $\sinh\, n\,\sigma$ and by neglecting $\sinh \pi\,h/a \leq \sinh \sigma/12$ in comparison with e^{σ}.

We thus find

$$w_{(n)} = \frac{2\,\Gamma}{i\,a} e^{-n\sigma} \cosh \frac{\pi\,h}{a} \left(\cos 2\pi\,z/a \mp i \sin 2\pi\,z/a\right) \tag{5.4}$$

Associating the values of n odd, positive and negative, and limiting the values of n to ± 1, we have finally

$$w \cong \frac{-4\,i\,\Gamma}{a} e^{\sigma} \cosh \frac{\pi h}{a} \cos \frac{2\,\pi\,z}{a}$$

A similar calculation can be made for the case where n is even, but with the preceding approximation, we may neglect the values of w for $n \geqslant 2$.

[1] Glauert, H., The Characteristics of a Kármán Vortex Street in a Channel of finite Breadth. British A.R.C., R. and M. 1151, 1928—29.

The expression above, then, represents the complementary velocity w_c (at a point z) due to the action of the trains of image vortices (aside from the velocity due to the train of real vortices).

Calculating the value of this velocity at the "initial vortex" $z = z_0 = a/4 + ih/2$, we find

$$w_{c\,(z=z_0)} = (u_{c\,0} - i\,v_{c\,0}) = -\frac{4\,\Gamma}{a} \cdot e^{-\sigma} \cosh \frac{\pi h}{a} \sinh \frac{\pi h}{a}$$

The velocity u_0 produced at the same point by the action of the real vortices will have the value

$$u_0 = \frac{\Gamma}{2a} \tanh \frac{\pi h}{a}$$

and in consequence the resultant velocity on any vortex in a guided flow will be

$$u = U_0 + u_{c\,0} = \left(1 - 8e \cosh^2 \frac{\pi h}{a}\right) u_0 \qquad (5.5)$$

The correction $(u - u_0)$ due to the influence of the walls is in the neighborhood of 0.04%. It is then negligible and it can be assumed that the criterion of stability of the vortex train remains the same as in an unlimited stream.

The complete potential function for a guided flow is then

$$F = F_1 + F_2 = \varphi + i\psi$$

where

$$F_1 = -\frac{i\,\Gamma}{2\pi} \log \frac{\sin(\pi/a)(z - z_0)}{\sin(\pi/a)(z + z_0)}$$

and

$$F_2 = -\frac{2\,i\,\Gamma}{\pi} e^{-\sigma} \cosh \frac{\pi h}{a} \sin \frac{2\pi z}{a}$$

Glauert, from this, derives for the stream function along the walls in constrained flow the expression

$$\varphi_{(wall)} = \pm \frac{\Gamma h}{2a} = \text{const.}$$

whence results the existence of a flow directed upstream of velocity $\Gamma h/aH$ at all the points of a downstream section such as BB'.

In order to satisfy the equation of continuity, the rate of the flow across the section downstream BB' must be equal to that across the section upstream AA', which is $-U_0 H$. In consequence of the flow upstream of velocity $\Gamma h/aH$ across BB', the condition of continuity imposes downstream, the existence of a general velocity of flow $-W$ greater than $-U_0$, *i. e.*,

$$-W = -U_0 - \frac{\Gamma h}{a\,H}$$

It results that the general velocity of the real vortices downstream will be $-W + u_0$.

In these conditions the calculation of the drag of the body may be effected anew, following Kármán's method by considering the motion of the fluid relative to the vortices in the wake downstream. Thus the fluid has the relative velocity u_0 with reference to the vortices, the body has the relative velocity $-(W - u_0)$ with reference to the fluid and the upstream velocity of the fluid is $(-U_0 + W - u_0)$.

Glauert thus finds that the drag coefficient for a body forming a wake of alternate vortices has the value

$$C'_D = \frac{R'_x}{b\,(\varrho/2)\,U_0^2} = \left(5.656 - 2.24\,\frac{u'_0}{U_0}\right)\frac{u'_0\,h'}{V_0\,b} + 24\left(\frac{u'_0\,h'}{V_0\,b}\right)^2 \cdot \frac{b}{H} \qquad (5.6)$$

This expression is similar to that of Kármán for the case of a plane unlimited flow.

In order to use this expression it is necessary to know the values of u'_0/U_0 and of h'/b for a constrained flow; for, *a priori*, it cannot be assumed that these ratios preserve values applicable to the case of unlimited flow.

Glauert has given a theoretical approximate solution for this problem and the results of his study are in agreement with experimental results for bodies of abrupt form such as a plane plate normal to the flow but excluding cylindrical bodies with section oval or fusiform.

From this study it results that the drag coefficient in constrained flow is given by the equation

$$C_{Dc} = C_{D\infty} + 32\left(\frac{u_0\,h}{U_0\,b}\right)^2 \frac{b}{H}$$

u_0/U_0 and h/b being the values relative to a plane unlimited flow. For a plane plate normal to the velocity, Heisenberg's[1] values may be adopted, *i. e.* $\dfrac{u_0}{U_0} = 0.2295$ and $\dfrac{u_0\,h}{U_0\,b} = 0.3535$

There results for the case of a plane plate,

$$C_{Dc} = C_{D\infty} + 4\,\frac{b}{H}$$

or again
$$\triangle C_D = C_{D\infty} - C_{Dc} = -4\,\frac{b}{H}$$

The coefficient C_D of a plane plate of infinite extent being near to 2, it is seen that a ratio $b/H = 1/20$ leads to a correction $\triangle C_D$ of about 10% of the value of C_D.

6. Influence of Rigid Walls Limiting a Plane Constrained Flow on the Resistance of Cylinders with Sections Elongated in the Direction of the Flow. In this case the search for the complementary potential corresponding to the constraints of the walls limiting a plane flow must be made by assuming a uniform plane flow without circulation around the obstacle. This problem has been solved by Sir Richard Glazebrook[2] for the case of Rankine ovals. In practice the profile of the body under

[1] Phys. Zeitschr., Vol. 23, p. 363, 1922.

[2] Glazebrook, R., Trans. Inst. Naval Architects, p. 155, 1909.

test is frequently derived from the circle by conformal transformation and it therefore becomes of interest to treat the case of the circular cylinder.

a) Case of Circular Cylinder (Radius R). The constraints due to the limiting walls may be replaced by an infinite series of images situated at the distances $\pm i\,nk$, k being the height of the plane flow.

The cylinder itself and its images may be considered as resulting from the superposition of a flow of translation of velocity $-U_0$ parallel to O_x with the flow due to a doublet located at the center of each of the circles. The potential function for an image of order n, in these conditions will be[1]

$$F_{(n)} = M\frac{1}{z \pm i\,n\,k} - U_0 z$$

and the total potential for the collectivity of images and real cylinder will be

$$F = -U_0 z + M\left[\frac{1}{z} + \sum_1^\infty\left(\frac{1}{z - i\,n\,k} + \frac{1}{z + i\,n\,k}\right)\right] \tag{6.1}$$

which with $M_1 = M\,\pi/h$ becomes

$$F = -U_0 z + M_1\left[\frac{1}{\pi\,z/k} + \frac{2z}{\pi\,k}\sum_1^\infty \frac{1}{z^2/k^2 + n^2}\right] \tag{6.2}$$

or again $F = -U_0 z + M_1 \coth\dfrac{\pi z}{k}$ (6.3)

The factor M_1 is found by the condition that for $z = \pm R = x$, the velocity $dF/dz = d\varphi/dx = 0$. We find then successively,

$$F_{(z = \pm x)} = \mp U_0 x \pm M_1 \coth\frac{\pi x}{k} = \varphi_0$$

$$\frac{dF}{dz}{}_{(z = \pm R)} = \mp U_0 \mp M_1 \frac{\pi/k}{\sinh^2 \pi R/k} = 0$$

whence $M_1 = -(U_0 k/\pi)\,\sin h^2\,\pi\,R/k$.

Finally the total potential takes the form

$$F = -U_0 z - \frac{-U_0 k}{\pi}\sinh^2\frac{\pi R}{k}\coth\frac{\pi z}{k} \tag{6.4}$$

whence

$$w = \frac{dF}{dz} = u - i\,v = -U_0\left(1 - \frac{\sinh^2 \pi R/k}{\sinh^2 \pi z/k}\right) \tag{6.5}$$

On the walls where $z = x \pm i\,k/2$, it is easily seen that the normal component is zero and the velocity reduces to

$$w_1 = u_1 = -U_0\left(1 + \frac{\sinh^2 \pi R/k}{\cosh^2 \pi x/k}\right) \tag{6.6}$$

It is desirable also to show that the deformation of the circular contour resulting from this flow is negligible. To this end it is sufficient to take

---

[1] The potential $F_{(0)}$ due to a doublet located at the origin 0 is $F_{(0)} = M/z$ where $M = -2a\mu$ in the notation of Division B IV 10.

the value of F at a point $z = ib$ and then find the relation between R and b. This gives

$$F_{(z=ib)} = -U_0 i b + \frac{i U_0 k}{\pi} \sinh^2 \frac{\pi R}{k} \cot \frac{\pi b}{k} = (\varphi + i \psi)_{z=ib}$$

whence $$\psi_{(z=ib)} = -U_0 b + \frac{U_0 k}{\pi} \sinh^2 \frac{\pi R}{k} \cot \frac{\pi b}{k} = 0 \tag{6.7}$$

and $$\sinh \frac{\pi R}{k} = \sqrt{\frac{\pi b}{k} \tan \frac{\pi b}{k}}$$

From this it is readily seen that for the usual values of R/k, the ratio R/b is sensibly 1, at least within a few thousandths.

We can now calculate the velocity w at any point whatever of the circle of radius R. To this end the expression for w may be transformed in the manner suggested by Hans Ermisch[1] *i. e.*

$$w = u - i w = -U_0 \left(1 - \sinh^2 \frac{\pi R}{k} q e^{i\delta}\right) \tag{6.8}$$

where $$q = \frac{1}{\cosh^2 \pi x/k - \cos^2 \pi y/k}$$

and $$\tan \delta = 2 \frac{\tanh \pi x/k \tan \pi y/k}{\tan^2 \pi y/k - \tanh^2 \pi x/k}$$

This will give

$$\frac{w^2}{U_0^2} = [1 - 2 \sinh^2 \pi R/k \, q \cos \delta + \sinh^4 \pi R/k \, q^2] \tag{6.9}$$

In a plane unlimited flow, the velocity at a point $z = Re^{i\theta}$ on the circle is given by the equation

$$\frac{V_\infty}{U_0} = 2 \sin \theta$$

If V and V_∞ are calculated for various points on the circle (remembering that V and V_∞ denote respectively the constrained and the unconstrained flow at any point on the circle), and the ratio $m = V/V_\infty$ is found, it will appear that m varies but little over the entire contour of the circle. It is then sufficient to determine the value at the point $z = iR$ ($\theta = \pi/2$). For this point we find,

$$m_1 = \frac{1}{2}\left(1 + \frac{\sinh^2 \pi R/k}{\sin^2 \pi R/k}\right) \tag{6.10}$$

The following table gives the values of m_1 for various values of the ratio $k/2R$

$k/2R =$ 8.5	5.66	4.25	3.40	2.83	2.25
$m_1 =$ 1.01	1.025	1.0475	1.075	1.115	1.195
$m_1^2 =$ 1.02	1.05	1.097	1.156	1.243	1.430

[1] Ermisch, H., Strömungsverlauf und Druckverteilung an Widerstandskörpern in Abhängigkeit von der Reynolds'schen Zahl. Abhandlungen aus dem Aerodynamischen Institut Aachen, Heft 6, 1927.

And the following table gives the values of m at various points on the contour for the ratio $k/2R = 2.25$.

θ	0	23.5°	33.2°	48°	60°	70.5°	80.5°	90°
m		1.175	1.18	1.181	1.19	1.192	1.192	1.195

It is seen that for this particularly severe case, m varies but slightly except near the point of velocity zero upstream and down.

Finally then, the influence of the walls limiting a plane guided flow, appears as an increase of velocities on the circular contour in the ratio $m = m_1$. It is therefore equal to the action of a flow of velocity $m\,U_0$ instead of U_0.

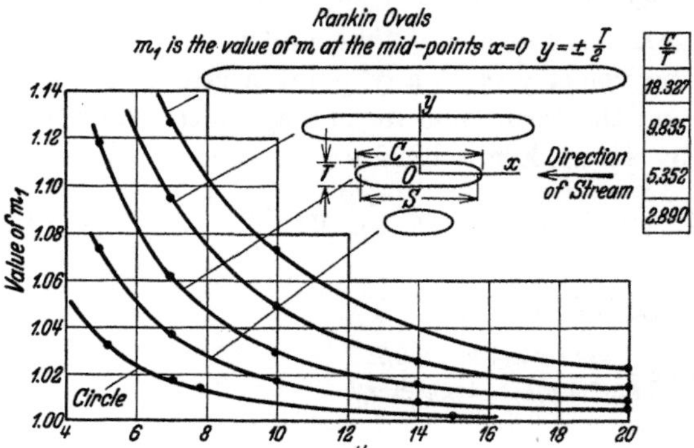

Fig. 33.

b) Case of Cylinders of which the Profile is a Rankine Oval or a Biconvex. Fage[1] has found an analogous result in calculating the ratio m at various points of contours defined by five Rankine ovals for which he had adopted Glazebrook's stream function.

These sections were characterized by the ratios l/t of the major to the minor axis. The value of the ratio $m = V/V_\infty$ was calculated for points corresponding to the relative ordinates $y/t = 0.5$, 0.4, 0.25 for the influence due to a plane flow of which the dimensions were characterized by the ratio k/t from 28 to 5. The calculation shows that the ratio m is almost constant over the entire contour except for locations near the point of stagnation at the extremities of the major axis. At the same time when the ratio k/t decreases notably, it is found that the value of m suffers a continued decrease from the value m_1 at the principal section ($y/t = 1$).

The curves of Fig. 33 represent the values of m_1 for the various ovals and for the various values of k/t.

Fage considers that the modifications of form of the various ovals as a function of the ratio k/t may be neglected, as indeed we have seen above for the circular cylinder of which the section becomes an ellipse in the total flow including the images of the real doublet.

In this most unfavorable case, the error in the value of m_1 does not reach 1%.

[1] Fage, A., On the Two-Dimensional Flow Past a Body of Symmetrical Cross-Section Mounted in a Channel of Finite Breadth, British A.R.C., R. and M. 1223, 1928—29.

This same author carried out later, systematic tests on the influence of rigid walls with plane flow on the drag of bodies such as wings with symmetric profile and profiles in the form of Rankine ovals. He found that the relative increase in the drag coefficient in free plane flow could be represented in the form

$$\frac{\triangle C_D}{C_{D\,\infty}} = \frac{C_{D\,c} - C_{D\,\infty}}{C_{D\,\infty}} = k_1\,(m_1^2 - 1)$$

or again

$$\frac{C_{D\,(c)}}{C_{D\,(\infty)}} = 1 + k_1\,(m_1^2 - 1) \qquad (6.11)$$

where m_1 is the theoretical ratio of the velocities on the profile in unlimited plane velocity, and k_1 is an empirical coefficient which depends chiefly on the form of the profile and which allows for the difference between the actual and the theoretical velocity.

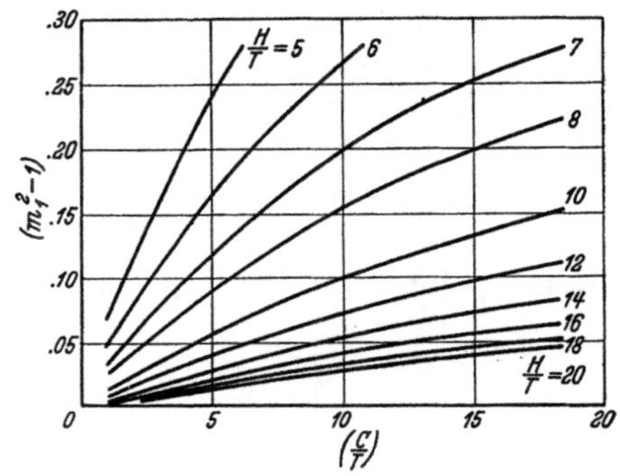

Fig. 34.

It results from these tests that the factor k_1 depends chiefly on the coefficient $C_{D\,\infty}$. The following numerical values are drawn from Fage's mean curve,

k_1	0.5	1.0	1.25	1.5	1.75	2.33
$C_{D\,\infty}$	0.005	0.010	0.015	0.02	0.03	1.20
		(elongated forms)				(circle)

where C_D represents the drag coefficient for motion along the direction of the major axis.

The factor $(m_1^2 - 1)$ depends only on the ratios k/t and c/t. The curves of Fig. 34 give the values of $(m_1^2 - 1)$ for various values of these ratios.

It is of interest to note with Fage, that the formula proposed is applicable even in the case of a circular cylinder, but that the coefficient of drag corresponding is much greater than that for cylinders with elongated sections. Furthermore in his tests on the cylinder, Fage reached a value of the Reynolds number of about 2.65×10^4. ($U_0 D = 4.14\, f^2/s$).

The corresponding coefficient C_D has a value of 1.2 (for an infinite extension) and the real flow must comprise a wake with alternate vortex trains. It follows that the influence of the limitation of the plane flow calculated by the equation

$$\frac{\triangle C_D}{C_{D\,\infty}} = 2.33\,(m_1^2 - 1) \qquad (6.12)$$

with

$$m_1 = \frac{1}{2}\left(1 + \frac{\sinh^2 \pi\, R/k}{\sin^2 \pi\, R/k}\right) \qquad (6.13)$$

comprises implicitly the influence of the alternate vortices on the drag.

These results seem to admit of extension to the case of profiles of lifting wings. For ordinary tests the dimensions of the models are relatively small and the preceding corrections are negligible. However, for special tests, such for example as the drag of the profile as a function of its surface covering, it might be found advantageous to employ models of large dimensions and in such case these corrections would become applicable.

Lock's Approximate Method[1]. The method of images has been employed by Lock, in dealing with this problem, with some simplification in the calculation of the complementary velocities due to the images. Thus for the case of the circular cylinder, the complementary velocity due to the real doublet is given [as may be seen from (6.1)] by

$$w = -M \sum_{1}^{\infty} \left[\frac{1}{(z - i\,n\,k)^2} + \frac{1}{(z + i\,n\,k)^2} \right] \tag{6.14}$$

At the point $z = 0$ we shall have

$$w_{(z=0)} = u_1 = 2\,M \sum_{1}^{\infty} \frac{1}{n^2\,k^2} = \frac{\pi^2}{3}\,\frac{M}{k^2} \tag{6.15}$$

If we neglect the slight deformation, relative to the circle, of the closed contour resulting from the unlimited plane flow, completed by the potential of the velocities due to the images of the real doublet, we may take for M the value corresponding to the real doublet in unlimited plane flow of velocity U_0, that is,

$$M = M_0 = U_0\,R^2$$

Under these conditions, for the cylinder of radius R in a plane flow of height h, we shall have

$$\frac{u_1}{U_0} \sim \frac{\pi^2\,R^2}{3\,k^2} = \frac{\pi^2}{12}\left(\frac{2\,R}{12}\right)^2 = 0.822\left(\frac{2\,R}{k}\right)^2 \tag{6.16}$$

This approximate value of u_1/U_0 is to be compared with the values of $(m_1 - 1)$ previously calculated. The following table gives the result of this comparison.

$k/2R$	= 8.5	5.66	4.25	3.40	2.83	2.225
$2R/k$	= 0.1175	0.1765	0.235	0.294	0.354	0.450
u_1/U_0	= 0.0113	0.0257	0.0455	0.0712	0.1027	0.166
$(m_1 - 1)$	= 0.010	0.025	0.0475	0.075	0.115	0.195

For usual values and even for exceptional values of $2\,R/k$, the agreement is very satisfactory.

Lock has also applied this method to the case of cylinders of which the directrix is:

[1] Lock, M., The Interference of a Wind Tunnel on a Symmetrical Body, British A.R.C., R. and M. 1275, 1929.

1) An ellipse.

2) A Rankine oval.

3) A Joukowski symmetrical profile generalized by the Kármán-Trefftz method to give a finite angle at the trailing edge.

4) A Joukowski symmetrical profile (angle zero at the trailing edge).

It is known that the Joukowski transformation $z = \zeta + c^2/\zeta$ may be generalized in the form[1]

$$\frac{z - q\lambda}{z + q\lambda} = \frac{(\zeta - \lambda)q}{(\zeta + \lambda)q}$$

and that this latter form may be expressed by the expansion

$$z = \zeta + \sum_{n=1}^{n} \frac{n\,\varkappa_n}{\zeta^n} \tag{6.17}$$

in which

$$\varkappa_1 = \frac{q^2 - 1}{3} c^2 = \lambda^2$$

and for the Joukowski transformation, $q = 2$ and $\lambda = c$.

We assume first a plane uniform flow transformed about a circle and the latter then transformed around a Joukowski profile. The complex velocities at any homologous points are in the ratio

$$\frac{w_\zeta}{w_z} = \frac{dz}{d\zeta} = 1 - \sum_{n=1}^{n} \frac{n\,\varkappa_n}{\zeta^{n+1}} \tag{6.18}$$

For the first transformation we have $\lambda = c = R$ and the velocity w_ζ $(n = 1)$ in the unlimited flow about the circle is given by

$$w_\zeta = w_z \frac{dz}{d\zeta} = -U_0\left(1 - \frac{R^2}{\zeta^2}\right) \tag{6.19}$$

(see the Joukowski form above).

For the calculation in the present case, the function of transformation for the Joukowski profile may be limited to the first two terms $(\varkappa = 1)$ thus,

$$z \cong \zeta + \left(\frac{q^2 - 1}{3}\right)\frac{c^2}{\zeta} \tag{6.20}$$

with $q = 2$ in the case of the Joukowski profiles and $q < 2$ for the Kármán-Trefftz profiles. Under these conditions we shall have, with analogous approximations suitable especially for large values of ζ

$$\left.\begin{aligned}
\frac{dz}{d\zeta} &= 1 - \left(\frac{q^2-1}{3}\right)\frac{c^2}{\zeta^2}\\
w_{(\zeta)} = w_{(z)}\frac{d\zeta}{dz} &\cong -U_0\left[1 - \frac{R^2}{\zeta^2}\right]\left[1 + \left(\frac{q^2-1}{3}\right)\frac{c^2}{\zeta^2}\right]\\
&\cong -U_0\left[1 - \frac{R^2}{\zeta^2} + \frac{q^2-1}{3}\,\frac{c^2}{\zeta^2} + \cdots\right]\\
&\cong -U_0\left[1 - \frac{1}{\zeta^2}\left\{R^2 - \left(\frac{q^2-1}{3}\right)c^2\right\}\right]
\end{aligned}\right\} \tag{6.21}$$

[1] Toussaint, A. and Carafoli, E., Théorie et Tracés des Profils d'ailes sustentatrices, pp. 25, 50., Paris, 1927.

For points located on the axis, we shall have

$$w_{(z=iy)} \cong -U_0 - \frac{U_0 R^2}{y^2}\left[1 - \left(\frac{q_2-1}{3}\right)\frac{c^2}{R^2}\right] \tag{6.22}$$

Thus in the case of a profile in an unlimited plane flow, the velocity due to the system of sources and sinks equivalent to this profile is equal to the velocity $(U_0 R^2/\zeta^2)$ due to a doublet, multiplied by the factor

$$\left[1 - \frac{q^2-1}{3}\frac{c^2}{R^2}\right]$$

The limited plane flow around such a profile may be represented by a plane flow around a system composed of the real profile and an infinite series of profile images; such system resulting from the application of the function of transformation to each of the circles of the unlimited system composed of the doublet images and the real doublet.

In consequence, at the real profile, the complementary velocity due to the profile images will be equal to the complementary velocity due to the doublet images multiplied by the factor $\left(1 - \frac{q^2-1}{3}\frac{c^2}{R^2}\right)$, that is,

$$u_{1\,(z)} = \frac{\pi^2}{3}\frac{U_0 R^2}{k^2}\left(1 - \frac{q^2-1}{3}\frac{c^2}{R^2}\right)$$

[see (6.14), (6.11), (6.21)] with $y = nk$ and $-\infty \ldots n \ldots +\infty$.

For the usual profiles we have

$$\frac{R}{c} = 1 + \frac{x\,t}{l} + \frac{x^2\,t^2}{l^2}$$

With $x = 0.77$ for ordinary Joukowski profiles,
and $x = 0.62$ for usual Kármán-Trefftz profiles,
and $c \cong l/4$.

Lock introduces a coefficient λ[1] which depends on the form of the contour and the ratio of its two axes l/t. This coefficient is defined by:

$$u = U_0\lambda\frac{t^2}{4\,y^2} = \frac{U_0\lambda}{4\,n^2}\left(\frac{t}{k}\right)^2$$

where u is the added image velocity for a single image and the summation is represented by $\sum_{-\infty}^{+\infty} 1/n^2 = \pi^2/3$. (Note that here $t =$ twice the value of Lock's t.)

Neglecting terms of higher order, we find here as the value of this coefficient

$$\lambda \cong \frac{(4-q^2)}{12}\cdot\left(\frac{l}{t}\right)^2 + \frac{x}{2}\left(\frac{l}{t}\right) + \frac{3}{4}x^2$$

This, with $q = 2$ (Joukowski profiles), gives

$$\lambda_{(J)} = \frac{x}{2}\frac{l}{t} + \frac{3}{4}x^2$$

[1] Lock, *loc. cit.*, p. 621, and diagram 4 for values of λ.

and for Kármán-Trefftz profiles with

$$q = 2.5 - 0.5\frac{R}{c} \qquad \text{and} \qquad x = 0.62$$

$$\lambda_{(K.\,T.)} = 0.385 + 0.415\frac{l}{t}$$

For an elliptical profile of which the axes are l and t, Lock found

$$\lambda_{(E)} = 0.5 + 0.5\frac{l}{t}$$

Under these conditions it may be assumed, as previously, that the increase of drag due to the action of the walls in limiting the plane flow, will be given by

$$\left.\begin{aligned}\frac{C_{D\,(k)} - C_{D\,(\infty)}}{C_{D\,(\infty)}} = \frac{\triangle C_D}{C_{D\,(\infty)}} &= k_2\,(m_1^2 - 1) = k_2\left[\left(1 + \frac{u_1}{U_0}\right)^2 - 1\right] = \\ &= k_2\,\lambda\frac{t^2}{k^2}\left[1.644 + 0.675\,\frac{t^2}{k^2}\right]\end{aligned}\right\}$$

which for small values of t/k may be reduced to

$$k_2\,(m_1^2 - 1) \cong 1.644\,k_2\,\lambda\frac{t^2}{k^2}$$

If these results are compared with those calculated or found experimentally by Fage[1], the following conclusions may be formulated.

1) The values of u_1/U_0 are in very good agreement with the theoretical values of $(m_1 - 1)$ for all contours for which $l/t \leqslant 6$, that is, for $t/l \geqslant 0.17$.

2) For contours more extended, even to $l/t = 18.5$ the agreement is also very good so long as the chord of the profile does not notably exceed the height of the flow ($l \cong k$).

3) Comparison with the experimental results shows that the influence of the dimensions of the flow is of order $(t/k)^2$ and the values of the coefficient k_2 are comprised between 1 and 1.9 for all the usual contours (especially those of wings) where the drag coefficient $C_D \leqslant 0.03$.

In general then, the influence of a limited plane flow on the resistance of cylindrical bodies of which the directrix is a contour of elongated form (similar to the profile of wings) is expressed by an increase of the drag coefficient which may be expressed by the formula

$$\triangle C_D = C_{D\,(\infty)}\left[1.644\,k_2\,\lambda\,\frac{t^2}{k^2}\right]$$

λ being the coefficient noted above, and k_2 being an empirical coefficient which depends on the form of the contour and on its resistance in an unlimited flow.

For a greatly elongated contour of which the chord l is large in comparison with the height k of the flow, we may still evaluate the complementary velocity u_1 by assuming that the velocity $U_0 + u$ is constant

[1] Fage, A., British A.R.C., R. and M. 1223.

in the entire space comprised between the maximum thickness of the contour and the walls limiting the flow. Under these conditions, the equation of continuity gives

$$U_0 k = (U_0 + u)(k - t)$$

whence
$$u = \frac{t}{k-t} U_0$$

If now we consider the velocity $(U_0 + u_0)$ which is observed in an unlimited plane flow in the zone of maximum thickness, the complementary velocity u_1 may be defined by the equation

$$\frac{u_1}{U_0} = \frac{u}{U_0 + u_0} = \frac{t}{k-t} \frac{U_0}{U_0 + u_0}$$

The values of $U_0 + u_0$ as above defined may be calculated theoretically from the form of the contour.

If this procedure is applied to the Rankine ovals of length-breadth ratio 9.8 and 18.3. Lock finds that the values of u_1/U_0, thus calculated, agree to an approximation of 1% with the theoretical values, if the length of the contour is at least equal to the height h of the flow.

By a combination of the two approximate methods for the calculation of u_1/U_0, the mean value adopted will be very near the theoretical values when $t/k \leqslant 0.2$.

7. Influence of the Walls Limiting the Flow on the Drag of a Body of Revolution. Lock (*loc. cit.*) has proposed to extend to the case of flow in three dimensions, the results previously found for plane flow with cylindrical obstacles.

To this end he considers that the complementary velocity, due to the images of the sources and sinks which define the body of revolution, will be of the form $\frac{u_1}{U_0} = \tau \lambda \left(\frac{s}{S}\right)^{3/2}$ in which

τ is a constant coefficient the value of which depends only on the form of the flow.

λ is a coefficient which depends only on the form of the body with reference to its relative elongation. For a sphere $\lambda = 1$.

s/S is the ratio of the area of maximum section of the body to that of the flow.

The values of τ, to a first approximation are as follows:

Square section	0.81
Circular section	0.80
Rectangular section (2/1)	1.03

The values of λ have been calculated for the Rankine "ovoids" and for spheroids.

Lock found for the case of a free flow, that to a first approximation, values are given by the equation

$$\frac{u_1}{U_0} = \tau \lambda \left(\frac{s}{S}\right)^{3/2}$$

at which the values are now

Free stream, square section, $\tau = -0.24$.

Free stream, circular section $\tau = -0.20$.

It is to be noted that the value of u_1/U_0 is about 1/4 that corresponding to the case of a limited flow and that the correction $\triangle C_D$ is notably less and of the opposite sign. To these approximate theoretical corrections, Lock proposes the application of an empirical factor k_3, analogous or equal to the factor k_2 found for the case of plane flow.

EXPERIMENTAL METHODS—WIND TUNNELS PART 2

By

Eastman Jacobs,

N.A.C.A Langley Field, Virginia

CHAPTER I

SCALE EFFECT

1. Introductory. The fact that wind-tunnel model tests are almost indispensable to the development of new types of aircraft and to the study of the aerodynamic properties of bodies is generally recognized. Practically all that is known about the aerodynamic properties of bodies, that is, how the air acts upon them, has been learned as a result of wind-tunnel tests. A large number of airplane builders conduct model tests as a matter of routine, several in their own wind tunnels. It is at once evident, therefore, that a proper understanding of the results and limitations of wind-tunnel model tests is necessary to the intelligent study and practice of aerodynamics and aeronautic engineering.

The fact, however, that the results of wind-tunnel tests on small models are often not directly applicable to full-scale conditions is not so generally recognized. The air in a wind tunnel does not necessarily act on the small model in a manner similar to that of the free air on the similar full-scale object. A part of any difference between them is due to different basic conditions; the model is placed in a limited stream of relatively turbulent air, whereas the full-scale object passes through a very large volume of air, the turbulent condition of which is problematical. Differences in the action of the air on two similar bodies may likewise exist apart from any dissimilarity in the surrounding conditions when the two similar objects are of different size.

Naturally the larger body may be expected to experience larger air forces because of its greater size; and, if its speed is higher, because of its greater speed. Such differences are not ordinarily referred to as scale effect and disappear if the forces are expressed in suitable coefficient form; but any change in the value of the coefficient with scale, or in other words, any variation from the expected change of force with size and speed is referred to as *scale effect*.

When a large range of velocity or size is involved, there will exist in general a scale effect; that is, the value of the coefficients will not remain constant. In certain special cases the coefficients may change as much as several hundred per cent and they may change so rapidly with the scale that the familiar speed square and area laws of the force variation with velocity and size do not apply even approximately. An example might be cited of two geometrically similar bodies of different size the larger of which would experience the lower resistance. It is evident, therefore, that the question of scale effect is one which must be dealt with as long as the results of model tests are to be used.

2. Fundamental Conceptions. Before proceeding to a more detailed discussion of scale effect, we may note that the general ground for the application of the laws of kinematic similitude (see Division A IV) to two systems comprising solid bodies and a fluid in relative motion, is based on the assumption of geometrical similarity as between the solid bodies and kinematic similarity as between the relative motions. The former may be closely realized but we can, by no means, be so sure of the latter.

However, let us consider for the time being two systems which are assumed to be fully similar: a full-scale flow, the characteristics of which will be denoted by symbols with the subscript (1), and a model flow, the characteristics of which will be denoted by symbols with the subscript (2). If the flows are steady, and the velocities, V, are sufficiently low so that changes in the density, ϱ, may be neglected, then the velocities V_1 and V_2 will maintain the constant ratio V_1/V_2 at corresponding points of the flows and if stable flow is to be maintained by the action of pressures alone, the pressure differences, P, between pairs of corresponding points must satisfy the condition

$$\frac{P_1}{P_2} = \frac{\varrho_1 V_1^2}{\varrho_2 V_2^2} \tag{2.1}$$

Conversely, if the pressure forces are the only ones acting, (2.1) will be satisfied at all points if the boundary conditions are similar. The ratio of the air forces on corresponding areas is then,

$$\frac{F_1}{F_2} = \frac{\varrho_1 V_1^2 l_1^2}{\varrho_2 V_2^2 l_2^2} \tag{2.2}$$

where l is a characteristic length. The above equations will be seen to express a balance between the pressure forces and the inertial forces in the flow. If a new set of forces is brought into play, this balance can no longer be maintained, unless the new set of forces, F', (the viscous forces) are in the same ratio:

$$\frac{F'_1}{F'_2} = \frac{\varrho_1 V_1^2 l_1^2}{\varrho_2 V_2^2 l_2^2} \tag{2.3}$$

but the viscous forces acting on a given area are proportional to the coefficient of viscosity, μ; the velocity gradient, which for similar flows is proportional to V/l; and the area, l^2; therefore

$$\frac{F'_1}{F'_2} = \frac{\mu_1 V_1/l_1}{\mu_2 V_2/l_2} \cdot \frac{l_1^2}{l_2^2} = \frac{\mu_1 V_1 l_1}{\mu_2 V_2 l_2} \tag{2.4}$$

Combining (2.3) and (2.4) we find the necessary condition for flow similarity:

$$\frac{\varrho_1 V_1^2 l_1^2}{\varrho_2 V_2^2 l_2^2} = \frac{\mu_1 V_1 l_1}{\mu_2 V_2 l_2}$$

or

$$\frac{\varrho_1 V_1 l_1}{\mu_1} = \frac{\varrho_2 V_2 l_2}{\mu_2} \tag{2.5}$$

A model and a full-scale flow can not be similar, if the forces resulting from the viscosity of the air are appreciable, unless (2.5) is satisfied. If (2.5) is satisfied, the forces may be compared by (2.2); but if the flows are not similar, the model test can give no direct indication of the full-scale forces.

The non-dimensional ratio $\varrho V l/\mu$ has been given the name "*Reynolds number*" after the English physicist, Osborne Reynolds, who first recognized its importance in connection with the conditions under which fluid movement in pipes changes from laminar to turbulent flow. The Reynolds number is a measure of the dynamic scale of the flow, and two flows having equal values of the Reynolds number are said to be at the same scale. It is important also to note that the Reynolds number is a measure of the relative importance of pressure and viscous forces in the flows; under otherwise similar conditions, the viscous forces at any point tend to become smaller relative to the pressure forces as the Reynolds number is increased.

3. Conditions for Flow Similarity. As shown above, when the pressure and viscous forces are both important in a flow, one condition for flow similarity is that the Reynolds numbers must be equal; there may be other forces and other conditions, however. Aside from the obvious condition that the objects must be geometrically similar, the motions of the two objects must also be similar, if the motion is other than rectilinear. For example, a model propeller must be tested at the same effective pitch ratio, V/nD, as that for which the characteristics of the full-scale propeller are desired. The ratio of the tangential to the forward velocity of the blade elements is then the same in both and the motions are similar.

If gravitational forces are important in a flow, as they are in the testing of ship models, it is necessary that the velocities be in the ratio of the square roots of similar linear dimensions—a condition known as Froude's law.

If the speeds are sufficiently high, the pressure differences between different points become so large that density changes within the fluid may no longer be neglected. A third condition must then be satisfied if the model and full-scale flows are to be alike. It may be shown by the application of the theory of dimensions[1], that in consequence of this condition, the ratios of the velocities of the bodies to the velocities of sound in the two fluid media must be equal in the model and full-scale flows.

When the required conditions for flow similarity are not exactly satisfied, experiments must be relied upon to determine whether or not the model tests are of value. In the usual form of aerodynamic model testing, the value of the Reynolds number is much lower than the full-scale value. Consequently, the viscous forces, as compared with the pressure or mass forces, are relatively much larger in the model flow; and, strictly speaking, the model and full-scale flows are not similar. The coefficient of viscosity of air is very small and the viscous forces in a flow are usually small as compared with the pressure forces. Experiments have repeatedly shown, however, that no slipping takes place between the fluid and the solid surface of an immersed object. The fluid immediately adjacent to the solid surface is at rest with respect to the surface. It is evident, therefore, that however small the viscosity of the fluid, the viscous forces near a solid surface in the flow become sufficiently large to predominate and to bring to rest the fluid actually in contact with the surface.

The layer of fluid adjacent to the surface in which the viscous forces are important is referred to as the boundary layer. Outside of this thin layer the viscous forces, at values of the Reynolds number common to aeronautical work, are probably so small that their immediate effect is negligible. The question may then naturally arise: How can the value of the Reynolds number, which determines the relative importance of the viscous forces, determine the characteristics of a flow about an object if these forces are important only in a very thin layer adjacent to the object? It is known that the character of the flow in the boundary layer may determine the general character of the entire flow, either through direct action by controlling the shearing forces transmitted to

[1] See Division A IV, also Division H I 3.

the surface of the object, or through indirect action affecting the entire flow.

It is not surprising to find that the intensity of the shearing forces transmitted to the surface of an object depends on the value of the Reynolds number. A discussion of this subject will be found in 4. Only a small portion of the air forces most frequently encountered in practice may be attributed directly to skin friction. Usually, therefore, the indirect effect of the boundary layer upon the entire flow is the more important.

As an example of the indirect effect of the character of the flow in the boundary layer, the characteristics of the flow about a bluff body may be considered. The explanation of the flow about a sphere, due to Prandtl, may be taken as typical. In the early stages of the flow, soon after the motion has begun, the distribution of the pressures and velocities in the flow is approximately that which would be expected if the effect of viscosity were entirely absent. This type of flow is symmetrical in front and behind the sphere and gives rise to no drag force. Air particles in passing along the surface of the sphere must be accelerated in passing from the front pole to the equator and then decelerated in passing from the equator to the rear pole. The potential head, represented by the impact pressure at the front pole, must be transformed into velocity head. Then, in order to maintain this type of flow, this velocity head must be transformed back into the same potential head when the particles again come to rest at the rear pole. During the acceleration period, however, the particles have lost a part of their energy as a result of the action of viscous forces. Consequently, when at the equator they meet with an adverse pressure gradient, their kinetic energy is insufficient to carry them against this adverse pressure on to the rear pole. The particles therefore tend to come to rest and to accumulate behind the sphere before reaching the rear pole. These stationary particles, under the action of the adverse pressure gradient, then tend to flow back toward the low pressure region at the equator. The result is that a secondary return flow is started, underrunning the main flow, separating it from the surface of the sphere and thus forming a turbulent dead air region behind the sphere.

The point of separation moves forward along the surface until a position is reached where the momentum carried into the inner layers by the viscous forces from the overrunning flow is sufficient to prevent a further progression of the underrunning return flow. The position of this equilibrium point will determine the size of the turbulent wake and the characteristics of the entire flow. In this way, it is possible for the viscous forces to determine the characteristics of an entire flow in spite of the fact that the viscosity of the fluid may be small and its immediate effects confined to a very thin layer.

4. Scale-Effect on Certain Bodies—Flat Plates. When a flat plate is exposed to an airstream with its faces parallel to the direction of flow, no pressure forces can act upon it in the direction of motion, if the plate is sufficiently smooth and thin. The drag forces on such a plate must then be entirely the result of viscous forces. As we have already seen, it has been experimentally established that there is no slipping between the plate and the air immediately adjacent to it, and hence, with respect to the plate, the velocity of the air immediately adjacent to it is zero.

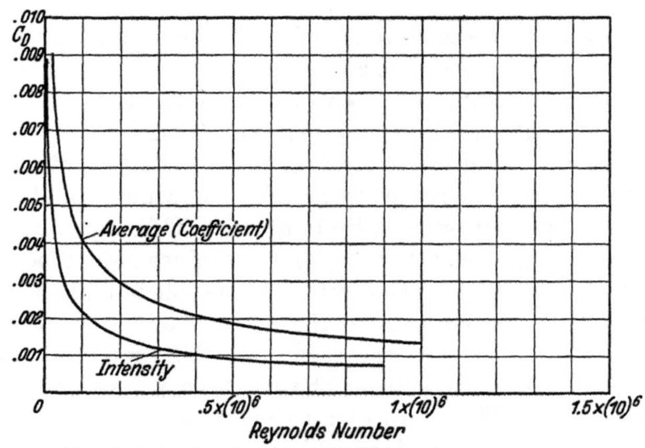

Fig. 1. Skin Friction on Flat Plate, Laminar Flow.

The intensity of the viscous drag on the surface of the plate may, therefore, be written

$$F = \mu \left(\frac{d u}{d y}\right)_0$$

where u is the velocity, y the distance taken ⊥ to the surface and the subscript 0 implies the ratio du/dy at the surface of the plate.

The scale effect may then be predicted in a general way from this equation. As we move from the leading edge of the plate toward the trailing edge, we may think of the Reynolds Number of the flow at the point increasing as the distance x from the nose is increased.

Near the nose of the plate, the velocity gradient at the surface (du/dy) will be large and consequently the intensity of the viscous drag will be large. As we increase the Reynolds number by passing downstream along the plate, the air retarded by the plate tends to accumulate and this layer of retarded air becomes thicker with the result that the velocity gradients become smaller. Accordingly, as we increase the Reynolds number, the intensity of the skin friction decreases[1] as shown in Fig. 1.

[1] The exact solution of the laminar flow over a thin flat plate has been given by H. Blasius: Zeitschrift für Math. u. Physik, Vol. 56, p. 1, 1908. His solution leads to the result: $C_D = 1.34\, R^{-1/2}$.

As we pass backward along the plate, the boundary layer continues to thicken and the shearing stress on the plate to decrease, but it has been shown experimentally that this type of flow is unstable if the Reynolds number exceeds a value known as the lower critical value. The laminar flow which we have been discussing may, however, continue beyond this value of the Reynolds number if the inflow on the plate is sufficiently steady[1]. However, if we continue downstream along the plate, we shall eventually reach a point where the flow in the boundary

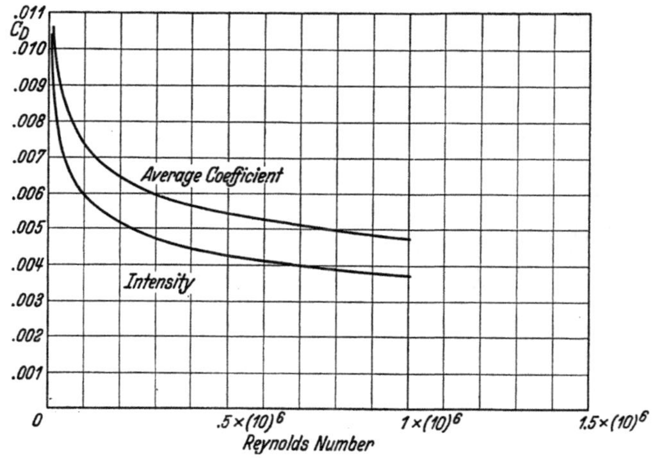

Fig. 2. Skin Friction on Flat Plate, Turbulent Flow.

layer changes from laminar to turbulent. The transition of the flow to the turbulent condition brings about drastic changes in the air forces. Whereas, previously the shearing stresses were transmitted through the boundary layer only by virtue of the fluid viscosity, the stresses are now transmitted by the scouring action of the turbulent air. High velocity air is carried very close to the surface of the plate by transverse flow in the eddies. If the flow in the boundary layer had been turbulent from the first, the intensity of the skin friction would have been something like that shown by the curves of Fig. 2. We may expect to find a more or less abrupt rise in the intensity of the skin friction at the transition point. In Fig. 3, the intensity is shown by the dotted line to rise to the turbulent skin friction intensity and then to fall off along a curve as though the flow had been turbulent from the beginning.

Now passing on downstream along the plate, with further increase in the Reynolds number, the turbulent boundary layer increases in thickness, as did the laminar boundary layer, but the intensity of the skin friction decreases more slowly. The boundary layer now consists

---

[1] U.S. N.A.C.A. Technical Report No. 342.

of an outer layer of air in the turbulent state, underlaid by an inner layer adjacent to the surface in which the forces continue to be transmitted to the surface through the action of viscosity.

The skin friction drag coefficient may be obtained by integrating the intensity. The scale effect on the drag of a flat plate may then be shown by a curve such as that marked "average (coefficient)" in Fig. 3. The Blasius law of skin friction with laminar flow may be represented by a straight line plot of the drag coefficient against the Reynolds

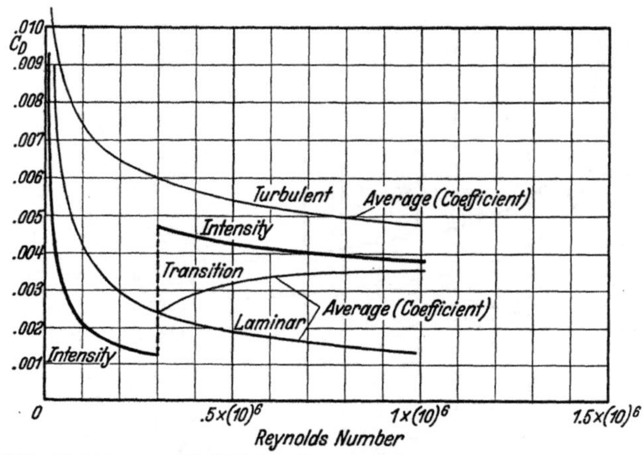

Fig. 3. Skin Friction on Flat Plate, Combined Laminar and Turbulent Flow.

number on logarithmic scales. Prandtl's semi-empirical relation for turbulent flow in the boundary layer[1] is

$$C_D = .074 R^{-1/5}$$

and is also represented as a straight line on logarithmic scales. These curves are represented in Fig. 4 together with several curves that represent the drag coefficient corresponding to conditions with part laminar and part turbulent flow in the boundary layer. The different curves were obtained by assuming different values of the critical Reynolds number; that is, the Reynolds number at which the transition from laminar to turbulent flow in the boundary layer takes place.

Airships. The drag of an airship is due mainly to the skin friction on its surface. Measurements of the distribution of the normal pressures over the surface of airship forms have repeatedly shown that the drag due to the pressure forces alone is small. It is not surprising, therefore, to find that the scale effect on the drag of airships resembles somewhat the scale effect on the skin friction drag of flat plates. In Fig. 5 are shown the results of drag tests of the same model airship in several

[1] Ergebnisse der Aerodynamischen Versuchsanstalt zu Göttingen, III. Lieferung, Berlin, 1927.

different wind tunnels. Lines indicating the skin friction on a rectangular flat plate having the same length and area have been computed from the relations given in the previous section and are plotted on the same sheet for comparison.

It will be noted that the drag coefficient curves plotted against the Reynolds number resemble somewhat the transition curves for flat plates. Aside from the existence of some pressure drag, the following considerations suggest caution regarding an undue extension of the analogy

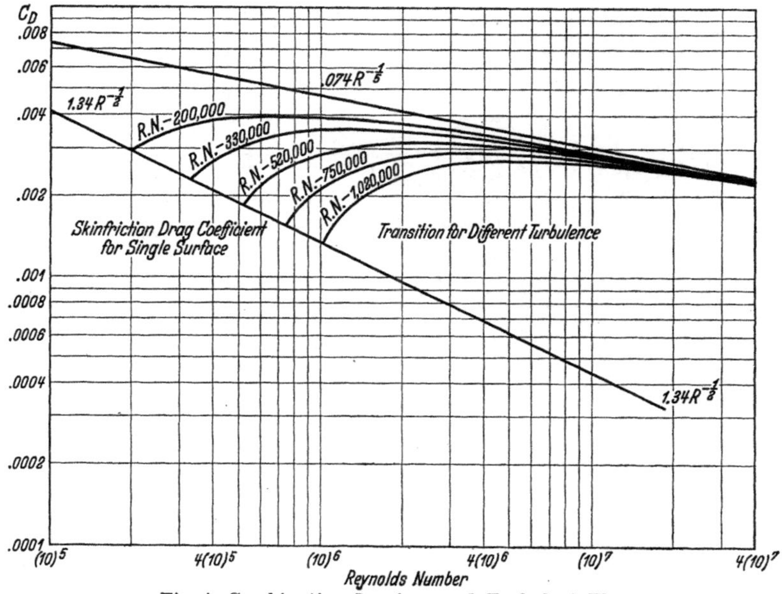

Fig. 4. Combination Laminar and Turbulent Flow.

between the drag of airships and the drag of flat plates. Over the forward portion of the airship hull a pressure gradient exists which opposes the accumulation of air in the boundary layer. Furthermore, the boundary layer tends to remain thin because, as a result of the increasing circumference of the body, a given volume of reduced-energy air forms a thinner layer as it moves aft. Over the after portion of the hull, opposite conditions obtain[1].

Dryden[2], in connection with a study of wind tunnel turbulence, has attempted a quantitative study of the variation of drag coefficient of an airship with the Reynolds number. His studies indicate that the different results obtained from different wind tunnels may be attributed to differences in the turbulence of the airstreams. Referring to Fig. 5,

[1] These effects have subsequently been analyzed by Clark Millikan, A.S.M.E. Trans., Applied Mechanics, Vol. 54, No. 2, pp. 29—43, January 30, 1932.

[2] U.S. N.A.C.A. Technical Report No. 342.

the drag coefficients as obtained from different tunnels are seen to tend to converge as the Reynolds number is increased, toward the curve obtained from the Variable Density Wind Tunnel[1] (curve marked A). This curve was obtained from tests in the original Variable Density Wind Tunnel. Airship tests made since the tunnel was rebuilt, have given results at the lower values of the Reynolds number more like those from other tunnels because the turbulence has been reduced. The above discussion suffices to show the futility of an attempt to predict

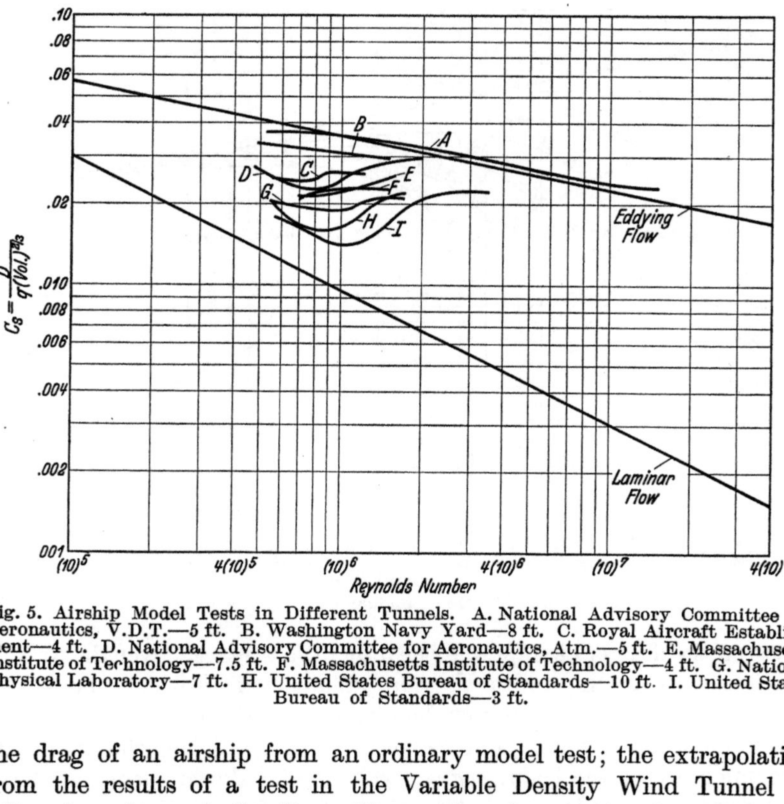

Fig. 5. Airship Model Tests in Different Tunnels. A. National Advisory Committee for Aeronautics, V.D.T.—5 ft. B. Washington Navy Yard—8 ft. C. Royal Aircraft Establishment—4 ft. D. National Advisory Committee for Aeronautics, Atm.—5 ft. E. Massachusetts Institute of Technology—7.5 ft. F. Massachusetts Institute of Technology—4 ft. G. National Physical Laboratory—7 ft. H. United States Bureau of Standards—10 ft. I. United States Bureau of Standards—3 ft.

the drag of an airship from an ordinary model test; the extrapolation from the results of a test in the Variable Density Wind Tunnel to full-scale values of the Reynolds number for airships is sufficiently hazardous.

Spheres. The scale effect on the drag of spheres may be taken as typical of that on bluff bodies that have no definite point at which the flow leaves the surface. The drag of such bodies is due mainly to the action of pressure forces rather than to viscous forces. The scale effect that is observed must therefore be attributed to the indirect action of the viscous forces as pointed out under the general discussion of scale effect.

[1] See Chapter II. Also U.S. N.A.C.A. Technical Report No. 227.

The curve in Fig. 6 represents the variation of the drag coefficient of a sphere with the Reynolds number. The curve is characterized by a very steep portion indicating a rapid reduction in the drag coefficient with increasing values of the Reynolds number in the region of the value known as the critical Reynolds number.

In 3, the manner in which a separation point of the flow is established on the surface of the sphere was discussed. It will be remembered that at the separation point the further progression of the returning or underrunning flow from the dead air region behind the sphere is prevented by the shearing stresses from the overrunning flow. We may now explain the rapid reduction of the drag coefficient as the Reynolds number

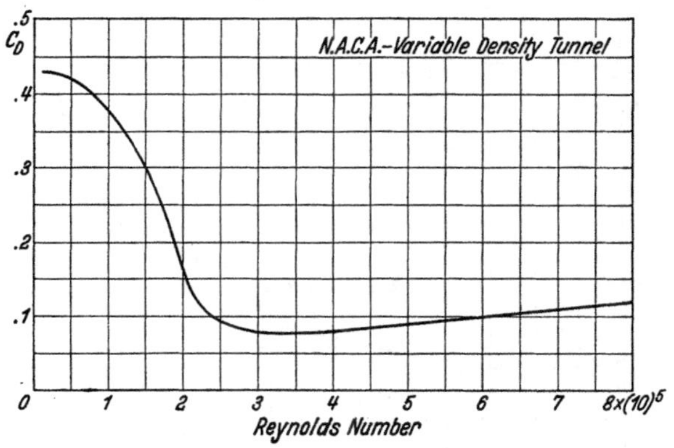

Fig. 6. Variation of Sphere Drag Coefficient with Scale.

is increased. Experiments[1] have shown that below the critical Reynolds number the separation point is forward of the equator. The turbulent wake is therefore large and the drag coefficient is comparatively high. However, as the Reynolds number is increased, the boundary layer flow in the neighborhood of the separation point becomes turbulent and the separation point is forced backward by the scouring action of the turbulent air until the turbulent wake becomes comparatively small and the flow approaches more closely the symmetrical potential flow which gives rise to no drag. If the sphere is tested in an airstream having a greater degree of turbulence, we may expect to find that the transition to turbulent flow in the boundary layer takes place sooner and that the value of the critical Reynolds number is reduced[2].

The nature of the conditions giving rise to the scale effect in the range of the Reynolds number above the critical value is not so well

[1] U.S. N.A.C.A. Technical Memorandum No. 475.

[2] For a more complete discussion of the aerodynamic characteristics of the sphere see U.S. N.A.C.A. Technical Reports Nos. 185 and 342, Technical Note No. 312 and Technical Memorandum No. 475.

understood. However, the results of sphere tests in the Variable Density Wind Tunnel have shown that as the Reynolds number is increased above the critical value, the drag coefficient falls to a minimum and then increases slowly. From this fact, it may be inferred that after the turbulent condition of the boundary layer has been fully established, the scouring action of the turbulent layer tends to decrease with increasing values of the Reynolds number[1].

Airfoils. Although a thorough understanding of the scale effect on the aerodynamic characteristics of airfoils would be very desirable, the subject is much more complicated than those already discussed, and

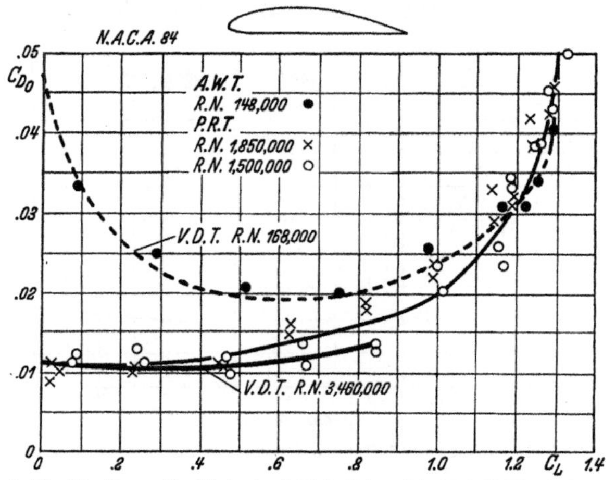

Fig. 7. Profile Drag Coefficient of N.A.C.A.—84 Airfoil Plotted on C_L.

it is not as well understood. In general, the lift of an airfoil, within the range of angles of attack in which it is usually employed, is not subject to important scale effect corrections. The same applies to the induced drag. Except for a thin boundary layer and narrow wake, the flows that give rise to the lift and induced drag are of the potential type and are essentially independent of the viscosity. Nevertheless, two of the most important aerodynamic characteristics of an airfoil vary with the Reynolds number: the profile drag and the useful angular range. The latter tends to determine the maximum lift coefficient of the airfoil. No attempt will be here made to give a full discussion of the manner in which these characteristics vary with the Reynolds number. However,

[1] Some theoretical and experimental substantiation of this theory as applied to other conditions may be found in the following papers:

Prandtl, L., On the Rôle of Turbulence in Technical Hydrodynamics, Lecture (Paper No. 504), World Engineering Congress, Tokyo, Trans. 1930.

Nikuradse, J., Versuche über die Strömung von Wasser durch divergente und konvergente Kanäle, Heft 289, Forschungsarb. d. Vereins Deutscher Ingenieure, 1929.

a considerable number of airfoils have been tested at large values of the Reynolds number in the Variable Density Wind Tunnel of the National Advisory Committee for Aeronautics[1] and of these a certain number of results may be noted. Presumably the best way to evaluate the scale effect correction for any airfoil is to compare its characteristics with those of similar airfoils from tests in a Variable Density Wind Tunnel.

The profile drag coefficient usually tends to decrease with increasing values of the Reynolds number, and with most airfoils the angular range of low profile drag tends to increase with the Reynolds number. Fig. 7 is a plot of the profile drag coefficient, C_{D0}, of the N.A.C.A.–84 airfoil. The results of tests, one in the Variable Density Tunnel at a comparatively low value of the Reynolds number, one in the Propeller Research Tunnel of the National Advisory Committee for Aeronautics and one in the Variable Density Tunnel under pressure, are given together. The marked drop of the profile drag coefficient, for this airfoil near zero lift, in passing to large values of the Reynolds number, may be attributed to an extension of the angular range of approximately potential flow to lower angles. This fact is shown by pressure distribution measurements on the same airfoil in the Variable Density Tunnel[2].

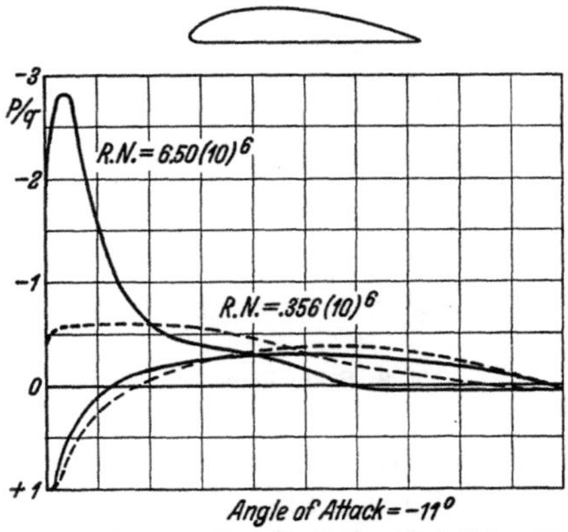

Fig. 8. Pressure Distribution on N.A.C.A.–84 Airfoil.

The results of these measurements near zero lift at two widely different values of the Reynolds number are shown together in Fig. 8. It will be noted that the distribution of the pressures from the low scale test differs considerably from the distribution to be expected from a potential flow.

Thin symmetrical airfoils may be compared with flat plates as regards their minimum drag scale effect. In Figs. 9 and 10, the results of measurements of the minimum drag coefficient of such an airfoil[3] are compared with the drag curves for a flat plate. It will be noted that the minimum drag scale effect curve for the airfoil (Fig. 9) resembles one of the corresponding transition curves for a flat plate (Fig. 4). When the turbulence of the airstream was increased by the introduction of a screen, the drag coefficients obtained (Fig. 10) followed the skin friction drag for a flat plate with turbulent flow in the boundary layer.

[1] U.S. N.A.C.A. Technical Report No. 352, also Technical Report No. 460.
[2] U.S. N.A.C.A. Technical Report No. 353.
[3] U.S. N.A.C.A. Technical Note No. 364.

The scale effect on the maximum lift of an airfoil may be compared with the scale effect on the drag of a sphere. The flow over the upper surface of the airfoil encounters an adverse pressure gradient as does

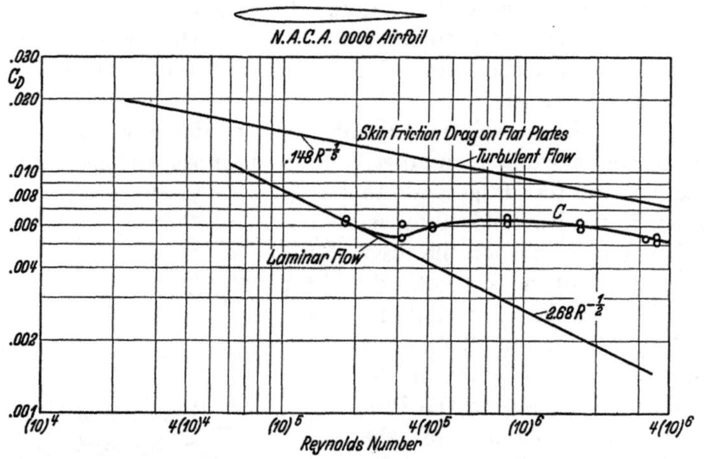

Fig. 9. Scale Effect on Minimum Drag of Thin Symmetrical Airfoil.

the flow over the sphere. Separation of the flow and the consequent loss of lift are controlled by the boundary layer conditions. The flow in the boundary layer becomes turbulent more readily at large values

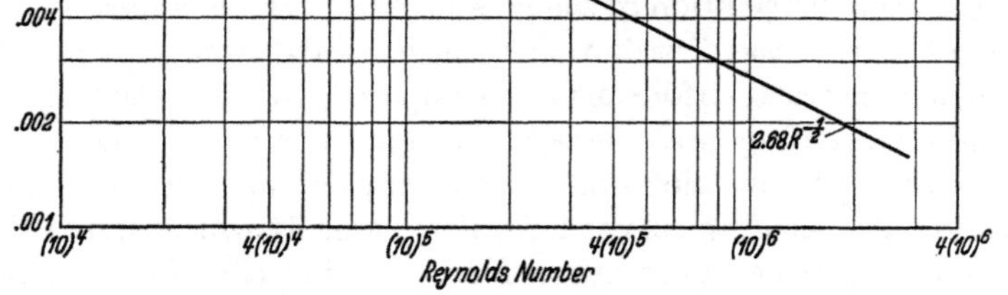

Fig. 10. Scale Effect on Minimum Drag of Thin Symmetrical Airfoil.

of the Reynolds number, and the scouring action of the turbulent air tends to delay the formation of a separation point or of a broad turbulent region over the upper surface until higher angles of attack are reached. Therefore we may expect to find that the maximum lift coefficients of most airfoils increase with increasing values of the Reynolds number.

Certain airfoils, however, fail to show an increasing maximum lift coefficient with Reynolds number within the range in which we are interested. Two examples are shown in Figs. 11 and 12. Thin airfoils

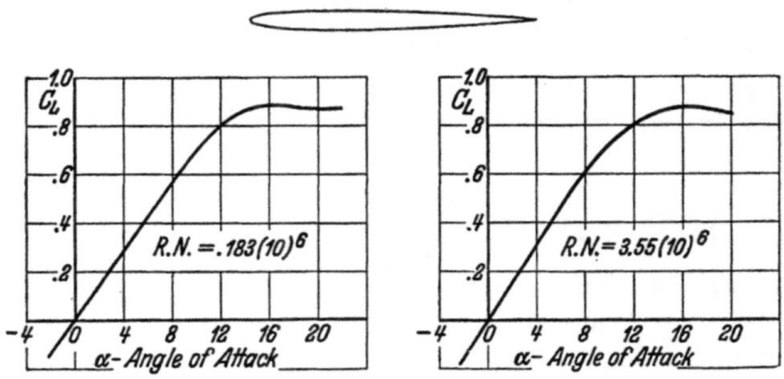

Fig. 11. Scale Effect on Thin Symmetrical Airfoil.

having relatively sharp leading edges have shown little change in maximum lift with Reynolds number. This fact might be explained by considering that the Reynolds number formed by using twice the leading edge

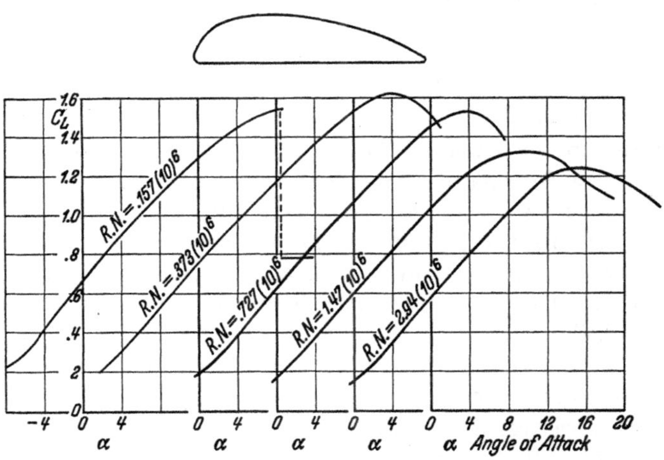

Fig. 12. Scale Effect on Thick "High Lift" Airfoil.

radius as the characteristic length has not reached the critical value for a cylinder, even when the Reynolds number formed by using the chord of the airfoil has reached 3,550,000 (Fig. 11).

On the other hand, there is a class of thick highly cambered airfoils which display high maximum lift coefficients when tested in ordinary atmospheric tunnels but give lower maximum lift coefficients at values of the Reynolds number approaching those of flight. Fig. 12 illustrates the scale effect on such an airfoil.

This type of effect might be explained in the same manner as for the increasing drag of a sphere above the critical value of the Reynolds number. A suitable Reynolds number, formed from a length comparable with the radius of curvature of the upper surface in the neighborhood of the maximum adverse pressure gradient, would be larger than the critical Reynolds number. Further experiments in a Variable Density Tunnel will be required to properly develop these problems.

CHAPTER II

THE VARIABLE DENSITY TUNNEL

1. General Description. The Variable Density Wind Tunnel of the National Advisory Committee for Aeronautics, located at Langley Field, Virginia, was especially designed to reach high values of the Reynolds number in aerodynamic model testing. This is accomplished by utilizing compressed air as a working fluid instead of air of normal density. The Reynolds number,

$$R = \frac{\varrho V l}{\mu} \quad \text{or} \quad \frac{V l}{\nu},$$

may be increased by increasing the density of the fluid, ϱ. As a matter of fact, the Reynolds number under otherwise similar conditions is almost directly proportional to the pressure or density of the air, because the coefficient of viscosity, μ, is but little influenced by the density or pressure. The effect of increasing the pressure is therefore to reduce the kinematic viscosity.

Increase in the value of the Reynolds number through the use of a fluid of low kinematic viscosity is thus naturally suggested, and it is not easy to say who first may have suggested compressed air for this purpose. However, Munk in 1920 considered the feasibility of a wind tunnel utilizing compressed air as a working fluid, and following this, the design of the tunnel was begun by the Committee's staff at Langley Field in 1921.

The Variable Density Tunnel (Fig. 13) is enclosed in a riveted steel tank 15 feet in diameter and 34 feet 6 inches long, capable of withstanding a working pressure of somewhat more than twenty atmospheres. An elongated toroidal structure inside divides the interior into three parts: a central throat, part of which forms the test section; an annular return passage formed between the inner structure and the outer shell; and a dead air space within the toroid, which provides space for the balance and other operating equipment[1]. The various operations required within the tunnel are performed by small electric motors. Observations are made by viewing the instruments in the dead air space through peep holes in the side of the tank and windows in the inner wall of the return passage.

[1] U.S. N.A.C.A. Technical Report No. 227, also subsequent paper describing the rebuilt tunnel, Technical Report No. 416.

The inner part of the tunnel was originally built of wood fastened to a steel structure. After several years of operation, a fire, probably electrical in origin, destroyed the interior of the tunnel. In rebuilding, an all metal construction was employed. In order to reduce the airstream turbulence, to increase the accuracy of measurements and to facilitate working in the tunnel, various other small changes in detail were also made.

The air in the tunnel is now circulated by a three-bladed propeller driven by a 425 horsepower direct-current motor. The large scale tests are usually made under the following approximate conditions:

Pressure: 20 atmospheres.
Dynamic pressure:
600 Kg/m² (122.88 lb. per sq. ft.).
Velocity: 75.8 ft. per sec.
Temperature:
30 to 65 degrees C.
Reynolds number: 3,300,000 (Airfoil model, chord 5 inches).
Propeller speed:
800 to 900 r. p. m.
Power input: 200 h. p.

The balance consists essentially of a U-shaped cradle, surrounding the lower part of the airstream, to which the model is rigidly attached. The cradle is suspended from four knife-edges, which are in effect carried on the overhanging ends of the lift and moment balance beams, so that forces acting on the cradle are counteracted by a movable counterpoise on the beams. In order to counteract

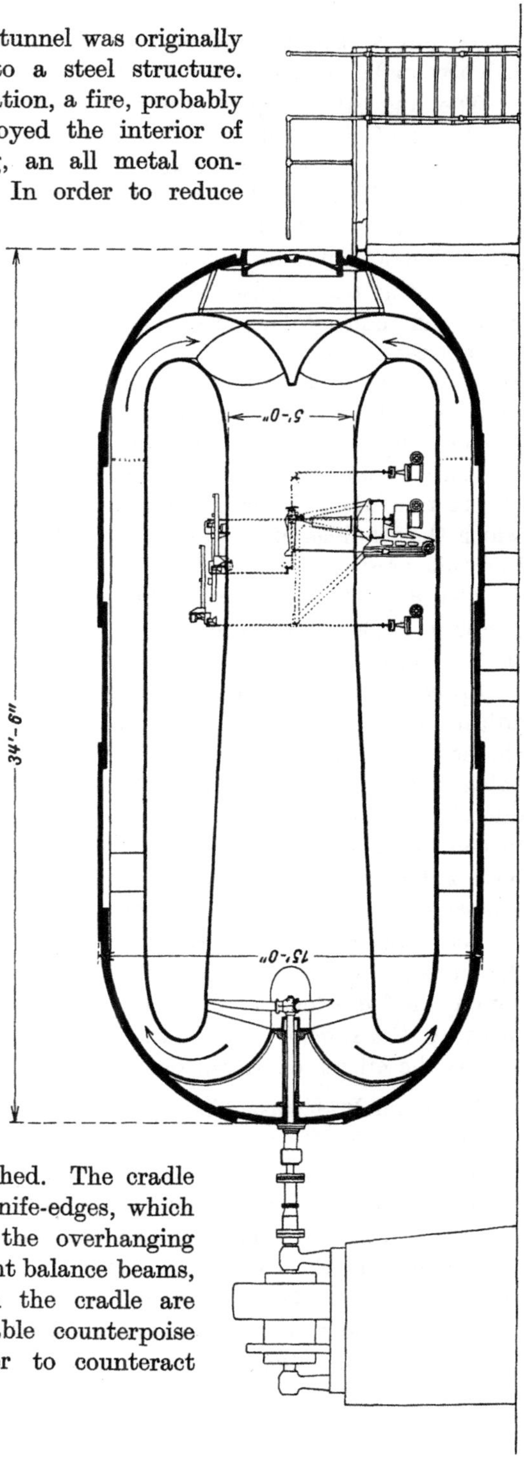

Fig. 13.

larger forces, counterweights may be placed on bridges under the balance cradle by means of motor driven camshafts.

The drag balance operates in a similar manner, except that the forces are taken off horizontally and transmitted to the drag balance beam through bell cranks. The model under test is supported on the cradle by means of two fixed struts and is free to pivot about the upper end of the struts. A third strut, which is moved up and down by means of an electric motor to vary the angle of attack, completes the support.

A routine test consists of the measurement of the forces on the lift, drag, and moment balances. The dynamic pressure is indicated by a manometer connected between static pressure orifices, one set of four being located in the small diameter 28 inches ahead of the position of the model, and the other set of four at the forward end of the return passage. By comparing the indications of this manometer with those of another connected to a standard Pitot-tube arranged to traverse the test section when the model is not in place, a factor is determined by which the readings of this manometer are multiplied to obtain the true dynamic pressure.

The Reynolds number is obtained from the air temperature and pressure, dynamic pressure, and the chord or characteristic length of the model, through the use of a nomographic chart.

Space does not permit a detailed consideration of the calibrations, adjustments, and corrections which must be made. The balance is checked for alignment periodically in order to ensure that the drag balance measures only horizontal forces. The deviation of the airflow direction in the tunnel from the horizontal is determined periodically by conducting a test of an airfoil in both the erect and inverted positions. Such a test determines what proportion of the lift is to be added or subtracted from the drag in order to correct for an inclined airflow.

The support drag is normally obtained by conducting a test with the supports in place, connected inside of, but not touching, a hollow dummy airfoil which is supported independently of the balance. In this way, the support drag in the presence of the model is obtained as well as the effects of extraneous air currents on the balance.

2. Special Problems. The operation of the Variable Density Wind Tunnel is more difficult than the operation of an ordinary wind tunnel and presents a number of special problems. There are three fundamental reasons for the greater difficulty of operating this type of wind tunnel:

1. The necessity of providing remote control for all apparatus inside the tunnel, and the time and expense involved in entering the tunnel to make adjustments or changes.

2. The necessity of providing sufficiently accurate models. All parts of the model, including small details and its surface condition as well,

must be made so carefully that they may be considered, in the highest possible degree, as geometrically similar to the full-size object.

3. The necessity of dealing with the relatively large forces experienced by the model and the balance.

As an example of a problem of the first class, that of obtaining the support drag at various angles of attack may be considered. It is necessary either to let out the compressed air after a test at one angle of attack, enter the tunnel, alter the dummy airfoil supports to correspond to a new angle of attack and then compress the air in the tunnel again, or it is necessary to construct a complicated device controllable from outside the tunnel to adjust the attitude of the dummy airfoil in such a way as to be independent of the balance and, at the same time, maintain a small clearance between it and the supports which must be mounted to move with the angle of attack mechanism on the balance.

Very little is known regarding the effects of small protuberances from the surface of bodies, except that they may, under some conditions, become very important. Small wires protruding from the surface of a sphere have been shown[1] to greatly influence its drag. Recent tests in the Variable Density Wind Tunnel have shown that the maximum lift coefficient of a moderately thick symmetrical airfoil is not seriously affected by coating its surface with No. 180 carborundum if the Reynolds number is within the range covered by atmospheric tunnels. In passing to full scale, however, the maximum lift coefficient of the polished airfoil increases, whereas that of the coated airfoil remains unchanged.

As another example of the increased importance of model details in testing at large Reynolds numbers, the results of tests[2] of a model "Sperry Messenger Airplane" may be considered. The model airplane as originally tested was constructed in the same manner as wind-tunnel models generally, omitting completely or forming approximately many of the small parts of the original airplane. The drag coefficients, as determined from the tests of this model were low as compared with values obtained from flight tests of the full-size airplane. The model was then rebuilt to include many small parts, such as control horns, which had been omitted. The results of these tests are compared in Fig. 14. It will be noted that while the effect of the details on the drag is small at low values of the Reynolds number, the effect becomes very important when the Reynolds number approaches its full-scale value. These results point to the conclusion that complicated objects, such as complete airplanes, are extremely difficult to model with sufficient accuracy to insure reliable results when tested at high Reynolds numbers.

[1] U.S. N.A.C.A. Technical Report No. 185.
[2] U.S. N.A.C.A. Technical Report No. 225.

The third consideration, that of dealing with the comparatively large air forces on the model, dictates the type of balance and support that is used. The air loads on the models may be as large as 300 pounds per square foot; but, because the air speed is comparatively low, the forces encountered are not as large as they would be if the same Reynolds number had been obtained by increasing the velocity or the size of the model rather than by increasing the density of the air. If the value of the Reynolds number is increased by changing the velocity or the size of the model, forces equal to the full-scale forces are encountered when the full-scale value of the Reynolds number is reached. Although the forces on the models used in the Variable Density Tunnel are much less than the full-scale forces, they frequently must be considered in the design of models that are to be tested.

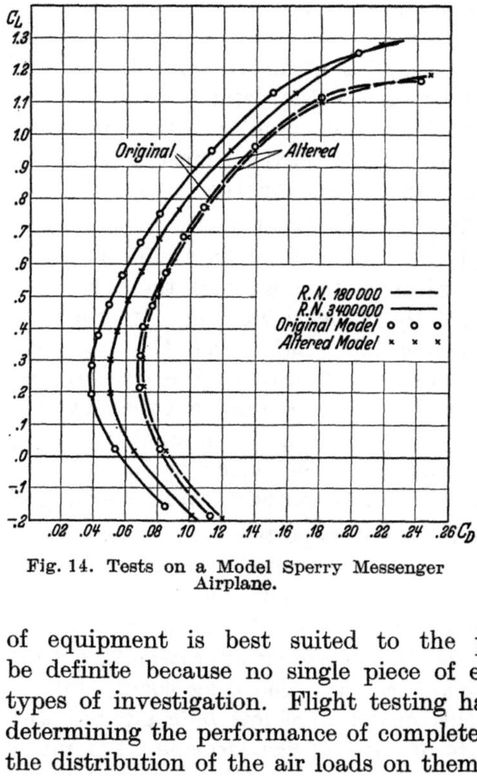

Fig. 14. Tests on a Model Sperry Messenger Airplane.

3. Advantages of a Variable Density Wind Tunnel. Granting that special equipment is required with which to carry out investigations at large values of the Reynolds number, the question arises as to what kind of equipment is best suited to the purpose. The answer cannot be definite because no single piece of equipment is best suited to all types of investigation. Flight testing has proved very satisfactory for determining the performance of complete airplanes and for investigating the distribution of the air loads on them, but this method is not suited for the determination of the characteristics of the component parts. Large tunnels have proved efficient for investigating propellers and parts of airplanes, but the large tunnel is expensive to build and to operate; most of the models are expensive, unless existing airplanes are used; and the tests are made less rapidly than in a small tunnel.

The Variable Density Wind Tunnel combines some of the advantages of both the small and the large tunnel. A tunnel of this type may be of such a size that models for it are quickly and cheaply constructed or altered. The models are not difficult to handle and the power required to operate the tunnel is only a little more than that used by an ordinary wind tunnel. The Variable Density Tunnel, therefore, provides the most satisfactory means of carrying out a certain class of investigations at full-scale value of the Reynolds number. In addition, it provides by far the best means of investigating the effects of varying the Reynolds number. When the Reynolds number is changed by varying the density, the air forces vary directly as the Reynolds number, and to investigate a twenty-fold range of the Reynolds number, it is necessary only to deal with a twenty-fold range of air forces. However, if the Reynolds number were varied throughout the same range by varying V or l, it would be necessary to deal with a four-hundred-fold range of air forces. Moreover, the extent to which V may be increased is limited by compressibility effects which may become important at speeds greater than three to five hundred feet per second. To change l involves the construction of a new model which, in all probability, will not be geometrically similar to the original. If a larger model is built and tested in the same tunnel, the model will occupy a greater proportion of the tunnel and be subject to different interference corrections. On the other hand, if a larger tunnel is employed, differences in the results due to differences in the characteristics of the air flow in the two tunnels may mask any difference due to true scale effect. It may be therefore concluded that the Variable Density Wind Tunnel provides the best possible means of investigating scale effect because all test conditions remain as nearly similar as possible throughout a very large range of the Reynolds number.

CHAPTER III

EXPERIMENTAL METHODS FOR THE INVESTIGATION OF AERODYNAMIC PHENOMENA AT HIGH SPEEDS

1. Introductory. When the velocity involved in an air flow is of the order of 100 miles per hour or less, it has been found experimentally and theoretically that the character of the flow is little affected by the compressibility of the fluid. While it is true that when the fluid involved is a gas it is by no means incompressible, still the pressure differences within the flow at these low speeds are so small as compared with the initial pressure that density changes resulting from pressure variation within the flow are for practical purposes negligible. Theoretical investigations have shown[1] that when the velocities involved are not small as compared with the velocity of sound in the gas, density variations within the flow become of importance and the flow is said to be subject to compressibility effects.

Compressibility effects were first encountered in connection with the study of ballistics. As regards the practical aerodynamics of airplanes, compressibility effects were considered negligible until operating conditions were altered by the increasing power and revolution speeds of

[1] See Division H.

motors. The fact that the motors were usually coupled directly with their propellers resulted in the use of propellers, the blade tip sections of which had speeds comparable with the velocity of sound. Recent refinements in airplane design have so increased the speed of airplanes that engineers must now deal with air flows about airplane parts that attain velocities exceeding one half that of sound. Even at moderate airplane speeds the effects of compressibility should not be disregarded if accurate measurements are desired. The importance of experimental investigations of airplane parts at high speeds, including airfoil sections for propellers and airplane wings, is therefore apparent. This chapter considers the experimental methods employed for this purpose; *i. e.*, the investigation of the aerodynamic characteristics of bodies as affected by the compressibility encountered at high speeds.

2. Early Experiments. The first experimental data throwing light on aerodynamic phenomena at very high speeds were obtained through tests on projectiles. These tests indicated that as the velocity increases the drag increases more rapidly than the square of the velocity, and very much more rapidly on approaching the velocity of sound. At speeds above the velocity of sound, the drag was found to increase more rapidly than the square of the velocity, but not as rapidly as the rate found for speeds in the neighborhood of the velocity of sound. At speeds above the velocity of sound, spark photographs showed the existence of shock waves from the projectile. These shock waves were considered theoretically and the theoretical treatment also showed that the importance of compressibility effects in a gas flow is determined by the ratio V/V_c, where V is the flow velocity and V_c is the velocity of sound in the gas. In other words, two otherwise similar flows are equally affected by compressibility when the ratio V/V_c has the same value for one as for the other.

Later, the realization that the aerodynamics of propeller blade sections might be subject to compressibility effects led to the development at McCook Field Dayton, Ohio, of the first equipment suitable for the aerodynamic investigation of airplane parts at high speed. Caldwell and Fales[1] in 1920 reported wind tunnel studies dealing with aerodynamic phenomena at high speeds. They describe the design and construction of a 14-inch diameter open-return type tunnel, driven by a 200-horsepower motor, and capable of speeds up to 650 feet per second, in which airfoil sections suitable for use as propeller blade sections were investigated. While these investigations demonstrated the existence of some compressibility effects, the value of the work is limited because lift measurements only were made and because the highest speeds attained were low as compared with the velocity of sound.

[1] Caldwell, F. W. and Fales, E. N., Wind Tunnel Studies in Aerodynamic Phenomena at High Speed, U.S. N.A.C.A. Report No. 83, 1920.

The first aerodynamic tests at speeds approaching that of sound were reported by G. P. Douglas and R. McKinnon Wood[1] in 1922. Their investigations consisted of propeller tests in an ordinary wind tunnel employing a model airplane propeller driven at tip speeds above that of sound. This excellent investigation demonstrated conclusively the existence of marked adverse compressibility effects when tip speeds reach a value in excess of approximately 850 feet per second. These experiments, although of immediate practical value, are not well suited for a basic study of compressibility phenomena, and other efforts to provide more suitable equipment for obtaining such data have since been made.

Briggs, Hull and Dryden[2] reported in 1925 a series of airfoil tests at high speeds. Speeds approaching that of sound were attained in the jet from an orifice 12.24 inches in diameter. The air was supplied by means of a 5,000-horsepower air compressor which was under test at the Lynn Plant of the General Electric Company at the time. Several 3-inch chord propeller sections, mounted so as to extend across the open jet, were tested with this equipment. The results, however, were not altogether satisfactory, partly because the test conditions were not entirely under the control of the investigators and partly because the airfoil set-up involved large and uncertain aspect ratio correction and unknown influences of the jet-boundary and model-end conditions.

More recently, several investigations have been made, but all in relatively small tunnels. Briggs and Dryden[3] in 1927 reported tests made on 1-inch chord airfoils in a 2-inch diameter open jet, at speeds up to and slightly above the speed of sound. This investigation included pressure-distribution measurements, but was subject to the same limitations as their previous investigation.

Stanton's[4] experiments, reported in 1928, should be mentioned because they gave the first results obtained from tests made under conditions which permitted the determination of airfoil-section characteristics. He employed a 3-inch diameter closed-throat tunnel capable of speeds up to 1.7 V_c. The air for the tunnel was supplied from a 530-horsepower compresser. The value of the results was limited, however, by the fact that relatively large models were employed in relation to the diameter of the jet; even so, the dynamic scale or Reynolds number of the model tests was very small.

---

[1] Douglas, G. P. and Wood, R. McKinnon, The Effects of the Tip Speed on Airscrew Performance, British A.R.C. R. and M. No. 884, 1924.

[2] Briggs, L. J., Hull, G. F. and Dryden, H. L., Aerodynamic Characteristics of Airfoils at High Speeds, U.S. N.A.C.A. Report No. 207, 1925.

[3] Briggs, L. J. and Dryden, H. L., Pressure Distribution over Airfoils at High Speeds, U.S. N.A.C.A. Report No. 255, 1927.

[4] Stanton, T. E., A High Speed Wind Tunnel for Tests on Aerofoils, British A.R.C. R. and M. No. 1130, 1928.

Busemann[1], at Göttingen a little later, developed a rather novel form of high-speed tunnel. In order to avoid the use of a high-power compressor, he employed a large tank (10 cubic meters) which is evacuated by means of a relatively small compressor. The tunnel then draws air from the room and discharges it into the evacuated tank through a variable-area restriction in which the speed reaches that of sound. The tank thus permits a run of approximately 10 seconds' duration at high speed. The air for the tunnel is taken from the room through a honeycomb and bell-mouth entrance cone, from which it forms an open jet passing across the test chamber. From the test chamber the air passes through the velocity-controlling restriction into the evacuated tank. It is evident that as long as the tank pressure is sufficiently low to maintain the speed of sound in the restriction, the tunnel velocity depends only on the cross-sectional area of the variable restriction and will therefore remain constant and independent of the tank pressure. The equipment, however, is small and is subject to many of the same limitations as those noted previously.

3. High Speed Tunnel of the U.S. National Advisory Committee for Aeronautics. *a) Development.* In 1927 the National Advisory Committee for Aeronautics began the development of more suitable equipment for investigating compressibility effects. The principal part of the equipment consisted of a comparatively large tunnel of a novel type using compressed air from the N.A.C.A. variable-density wind tunnel as the source of motive power. The high-pressure air from the variable-density tunnel was ultilized as an induction-jet, similar in principle to that employed with steam injectors.

The application of the induction principle to a high-speed tunnel was the outgrowth of experimental investigations of thrust augmentors for jet propulsion made by Shoemaker and the author[2] who, at the suggestion of G. W. Lewis, applied the result to the development of an induction-jet tunnel. A model tunnel having a throat diameter of approximately 1 inch was first built and a series of tests conducted. As with the thrust augmentors, high velocity compressed-air jets were employed to induce the flow. The tunnel air stream was composed wholly of the air drawn in from the room and which passed through the tunnel before reaching the compressed-air injector jet. These model tests were used as a guide in determining the general proportions of the high-speed tunnel, the results having indicated that speeds in excess of the velocity of sound were attainable.

---

[1] Busemann, A., Profilmessungen bei Geschwindigkeiten nahe der Schallgeschwindigkeit (im Hinblick auf Luftschrauben). Jahrbuch der Wissenschaftlichen Gesellschaft für Luftfahrt, pp. 95—99, 1928.

[2] Jacobs, Eastman, N. and Shoemaker, James M., Tests on Thrust Augmentors for Jet Propulsion, U.S. N.A.C.A. Technical Note No. 431, 1932.

Following the model tests, a 12-inch diameter high-speed tunnel was constructed in the variable-density wind tunnel building in 1928. Further developments were carried out with this tunnel, from which the present form was evolved. The later part of the development is the work of Stack; parts of the following material descriptive of this equipment and of the test methods employed are taken from his description of the tunnel in U.S. N.A.C.A. Technical Report No. 463.

Incidentally, it should be noted that rather extensive model investigations having do to with induction-jet tunnels have subsequently been carried out in England. Professor G. I. Taylor became interested in the N.A.C.A. high-speed tunnel during a visit to the Committee's laboratory in 1929, and, upon his return to England and at his suggestion, the late Sir Thomas Stanton began a series of model tests[1] with a view toward the development of a high-speed induction-jet wind tunnel to be used with the British Aeronautical Research Committee's projected compressed-air tunnel. This work was carried on after his death by Bailey and Wood[2], who have contributed much to the present knowledge of the behavior of the induction jet.

b) Description and Arrangement. The N.A.C.A. high-speed tunnel has several advantages over other high-speed tunnels. The diameter of the tunnel is large as compared with most of the tunnels in which high-speed tests have been made. The flow past the model is relatively non-turbulent, since the air stream in the tunnel throat is composed of the air taken directly from the relatively quiet air in the building in which the tunnel is housed. Moreover, the model airfoils employed for tests extend through the walls and are supported outside the air stream, thus eliminating uncertain end effects and support interference, and permitting the direct measurement of airfoil-section characteristics.

The general structural arrangement of the tunnel without balance and mountings is shown in Fig. 15. Compressed air from the variable-density wind tunnel is piped into the high-pressure chamber and discharged through the annular nozzle *A A* as shown in the Figure. The jet from this nozzle induces a flow of air from the room through the lower portion of the tunnel, where the model is carried on a photo-recording balance suitably installed in the space *B*. The air in the tunnel stream mixes with the high-pressure air in the diffuser, which conducts the air through the roof of the building. Six vanes are employed at the mouth of the entrance cone to prevent twist developing in the air stream on entering

[1] Stanton, T. E., The Development of a High-Speed Induction Wind Tunnel. Confidential Aeronautical Research Committee Report T-3094, 1931.

[2] Bailey A., and Wood, S. A., Development of a High Speed Induced Wind Tunnel, British A.R.C. R. and M. 1468, 1932.

Bailey, A. and Wood, S. A., Principles of the Air Injector, British A.R.C. R. and M. 1545, 1933.

the entrance cone. The test section is approximately 11 inches in diameter and 7 inches long, and is made slightly divergent in order to reduce the axial static-pressure gradient. The included angle between the walls of the exit cone is 4.6°; this portion of the tunnel ends in an abrupt step just below the annular high-pressure nozzle. The diffuser, in which the mixing takes place above the high-pressure nozzle, is 19 feet 10 inches long and the included angle between diametrically opposite elements is 4.8°.

c) The Motive Power. At the end of a test at high Reynolds number in the variable-density tunnel, a relatively large supply of air at an initial pressure of 300 pounds per square inch is available. This air is discharged from the high-pressure air chamber through the converging-diverging annular nozzle shown in the figure. This nozzle has a minimum annular opening of 0.06 inch and a divergent portion with an angle of divergence of 11.1°. Approximately 20 minutes are required to completely discharge the tank of compressed air through the high-speed tunnel. Only the first few minutes, however, are available for tests at very high speeds. With the arrangement shown, speeds closely approaching the velocity of sound can be attained.

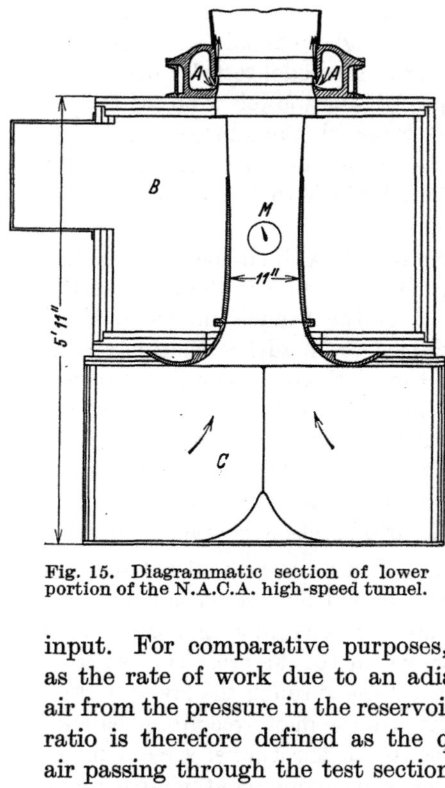

Fig. 15. Diagrammatic section of lower portion of the N.A.C.A. high-speed tunnel.

d) Energy Ratio. The energy ratio is difficult to determine for this type of tunnel because of the uncertainty of the value of the power input. For comparative purposes, however, the power input is taken as the rate of work due to an adiabatic expansion of the high-pressure air from the pressure in the reservoir to atmospheric pressure. The energy ratio is therefore defined as the quotient of the kinetic energy of the air passing through the test section in a unit time divided by the power due to adiabatic expansion of the high-pressure air. The value for the tunnel as operated varies considerably over the speed range but at a speed of 0.5 V_c the value is 1.6.

e) Air-Flow Characteristics. The characteristics of the air flow in the tunnel are excellent, principally because the air is taken through the entrance cone with very low losses from the comparatively quiet air in the building in which the tunnel is housed. The variation in dynamic pressure across the test section, except for a very thin boundary layer near the walls, is less than 0.5 per cent. The directional variation of the air flow is believed to be less than 1/4°, and the turbulence is believed to be very small.

4. Description of Balance. The balance must measure the large range of forces resulting from the wide speed range over which tests are made. Because of the limited time available for observations, it must be automatically recording. Reference may be made to the original description for details. The balance measures the lift, drag, and pitching moment by multiplying and recording the deflection of steel springs (cantilever-beam type) to which the forces are transmitted. The essential parts consist of a cast-iron cradle mounting a yoke to which the model is attached, the linkages necessary to transmit the forces to the steel springs and a camera for magnifying and recording the deflections of the springs. A part of the yoke to which the model airfoil passing across the test section is secured, is rotatable, so that the angle of attack of the model may be varied. The linkages, springs, and recording systems are designed to record photographically the deflections of the three beams from which the lift, drag, and pitching moment of the airfoil are determined. The record shows also the pressure difference recorded by a standard N.A.C.A. pressure cell on the same film, from which the dynamic pressure, q, corresponding to each test may be obtained.

The airfoil models usually employed are constructed of steel, by means of a special airfoil-generating machine. Models of 2-inch chord are ordinarily used; they extend across the tunnel and through holes in the tunnel-wall. In order to reduce the flow between the tunnel and the dead-air space, and the resulting interference, these holes are covered with specially designed end plates. The end plates are made of brass and are circular in form, fitting into recesses cut in the tunnel walls. The plates are sufficiently flexible so that they can bend and rotate as the angle of attack is changed, thus maintaining the contour of the walls, and at the same time, a small clearance around the hole in the plate through which the model airfoil passes.

5. Dynamic-Pressure and Velocity Determination. The dynamic-pressure and velocity are computed from Bernoulli's equation for a compressible fluid. The equation is

$$P_a = P_s\left[1 + \frac{k-1}{2}\left(\frac{V_s}{V_c}\right)^2\right]^{\frac{k}{k-1}}$$

where P gives the pressure in the fluid, the subscript a denoting atmospheric conditions, the subscript s denoting conditions at the test section of the tunnel; V_c is the velocity of sound in the fluid for the conditions at the test section, and k is the ratio of the specific heats for air, numerically equal to 1.4. The formula is derived by substituting the pressure-density relation for adiabatic flow in the general form of the Bernoulli equation. A more convenient form of the foregoing equation for use when V_s/V_c is less than unity is obtained by expanding in the series

$$P_a = P_s + \frac{1}{2}\varrho_s V_s^2 \left[1 + \frac{1}{4}\left(\frac{V_s}{V_c}\right)^2 + \frac{1}{40}\left(\frac{V_s}{V_c}\right)^4 + \frac{1}{1600}\left(\frac{V_s}{V_c}\right)^6 \ldots\right]$$

This is the form used to compute the dynamic pressure, which is therefore

$$q = \frac{1}{2}\varrho_s V_s^2 = \frac{P_a - P_s}{1 + \frac{1}{4}\left(\frac{V_s}{V_c}\right)^2 + \frac{1}{40}\left(\frac{V_s}{V_c}\right)^4 + \frac{1}{1600}\left(\frac{V_s}{V_c}\right)^6 \ldots}$$

The values of V_s/V_c are computed from the first form of the equation. Solving this equation for V_s/V_c gives

$$\frac{V_s}{V_c} = \sqrt{\frac{2}{k-1}\left[\left(\frac{P_a}{P_s}\right)^{\frac{k-1}{k}} - 1\right]}$$

The derivation of these formulas is dependent primarily upon the assumption of true adiabatic expansion and the absence of losses due to friction. The errors induced by the first of these assumptions are probably small because of the rapidity with which the heat energy of the air is converted into kinetic energy. In order to check the validity of the formulas, the total pressure at the test section has been measured by comparing the pitot impact pressure with the room pressure; it was found that its value differed from the room pressure by less than 0.02 per cent of the dynamic pressure, except for a very small core at the center and a relatively thin layer adjacent to the wall.

The difference between the static pressure at the test section and the atmospheric pressure cannot be reliably determined from a direct measurement during a test because of the presence of the model. Accordingly, calibrated static-pressure orifices are used for this purpose. Four small holes in the tunnel wall 10 inches below the location of the quarter-chord axis of the model are connected to a manifold which is in turn connected to pressure-measuring devices. The calibration factor is determined from simultaneous measurements of the quantities (P_a—P_s) and (P_a—P_{sp}), where the subscript sp denotes conditions at the static-pressure orifices. The static pressure at the test section is taken as the value registered by four holes in a tube located along the axis of the tunnel. The holes in the tube are 90^0 apart and are in the horizontal plane which passes through the quarter-chord position of the airfoil model. In order to avoid end interferences, the tube extends from the relatively low velocity region at the mouth of the entrance cone to a point 20-1/4 inches above the model location. The calibration factor as computed from these measurements is

$$F = \frac{q}{P_a - P_{sp}}$$

For use in computing test results, both this factor and V/V_c are plotted against $P_a—P_{sp}$. The variation of these quantities with atmospheric pressure for constant values of $P_a—P_{sp}$ is negligible except at speeds in the immediate neighborhood of the velocity of sound.

The only pressure measured during a test is the difference between the atmospheric pressure and that at the static-pressure orifices. A single-tube mercury manometer is mounted outside the tunnel to provide the operator with means for adjusting the speed and measuring the pressure

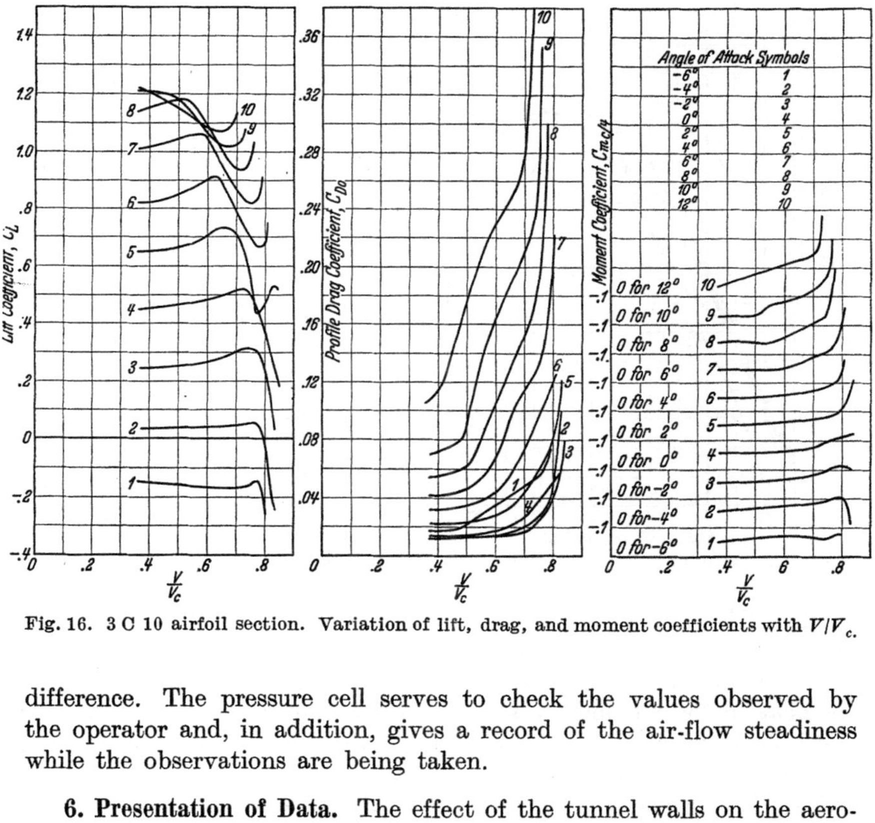

Fig. 16. 3 C 10 airfoil section. Variation of lift, drag, and moment coefficients with V/V_c.

difference. The pressure cell serves to check the values observed by the operator and, in addition, gives a record of the air-flow steadiness while the observations are being taken.

6. Presentation of Data. The effect of the tunnel walls on the aerodynamic characteristics of airfoils tested in this tunnel is not definitely known or understood. The effect of variations in the form and clearance of the end plates is known to be critical and, accordingly, the end plates are very carefully adjusted. Some preliminary tests with airfoils having different chords indicate that no correction need be applied to these data to obtain characteristics for infinite aspect ratio. In other words, it is believed that the data may be directly applied in practical design problems as airfoil-section data.

The data are presented in two graphic forms. The first form, which shows directly the effects of compressibility, consists of plots of the various coefficients for a given angle of attack against V/V_c. The other form consists of plots of the lift coefficient and the drag coefficient against angle of attack for several speeds.

Samples of typical airfoil test data obtained from the tunnel are shown in Figs. 16 and 17. The characteristics shown are for the 3 C10 airfoil section, a section of 10 per cent thickness derived from the Clark Y

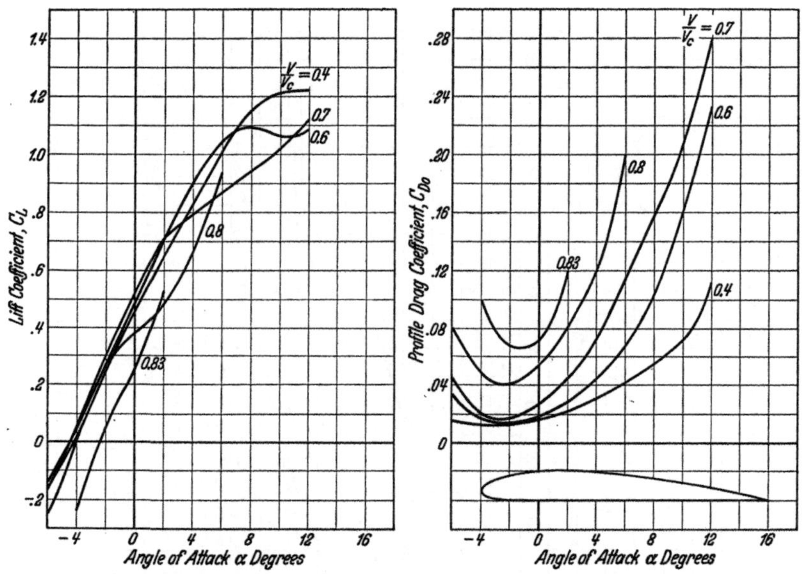

Fig. 17. 3 C 10 airfoil section. Aerodynamic characteristics at various values of V/V_c.

section. The results shown in Fig. 17 are representative airfoil characteristics showing typical compressibility effects. It will be noted that as the speed of the air flowing past the airfoil is increased, the lift, drag, and moment coefficient undergo a small numerical increase which continues until a compressibility burble occurs. As the speed is increased further, the breakdown of the flow corresponding to the compressibility burble is evidenced by a marked decrease in the lift coefficient, a rapid increase in the drag coefficient, and a movement of the angle of zero lift toward 0^0.

The results of the airfoil investigations[1] that have been made with the N.A.C.A. high-speed tunnel equipment cannot be considered in detail here. It is sufficient to point out, however, that marked compressibility effects are observed at speeds well below that of sound and within the speed range of practical aerodynamics.

[1] Stack, John and von Doenhoff, Albert E., Tests of 16 Related Airfoils at High Speeds, U.S. N.A.C.A. Technical Report No. 492, 1934.

Bibliography.

In addition to the references given throughout the text of the Division and which relate more especially to the theoretical aspects of tunnel wall interference, selected references are given as below to articles dealing specifically with descriptions of existing wind tunnels in various parts of the world and with their instrumental equipment.

American.

New Type of Wind Tunnel, N.A.C.A. Technical Note No. 60, 1921.

High Velocity Wind Tunnels, N.A.C.A. Technical Memorandum No. 318, p. 29, 1925.

High Speed Wind Tunnels, Scientific American, Vol. 133, pp. 275—276, October, 1925, New York.

Variable Density Wind Tunnel, N.A.C.A. Technical Report No. 227, 1925.

Wire Suspensions in Wind Tunnels, N.A.C.A. Technical Memorandum No. 342, p. 14, 1925.

The New York University Wind Tunnel, Aviation, Vol. 20, No. 9, pp. 293—295, March 1, 1926, New York.

Study of Open Jet Wind Tunnel Cones, N.A.C.A. Technical Note, No. 260, p. 15, 1927.

The New Aerodynamic Laboratory of the University of Toronto, Journal of the Engineering Institute of Canada, Vol. 10, No. 8, pp. 390—399, August, 1927, Montreal.

An Automatic Speed Control for Wind Tunnels, N.A.C.A. Technical Note, No. 278, p. 15, 1928.

On Improvement of Air Flow in Wind Tunnels, N.A.C.A. Technical Memorandum No. 470, p. 9, 1928.

Twenty Foot Propeller Research Wind Tunnel, N.A.C.A. Technical Report No. 300 1928.

Investigation of Air Flow in Open Throat Wind Tunnels, N.A.C.A. Technical Report No. 322, 1929.

Motor Control for Wind Tunnel. Precision Speed Regulation for the Wind Tunnel Motor at California Institute of Technology, Journal A.I.E.E., Vol. 48, No. 9, pp. 686—691, September, 1929, New York.

Berliner-Joyce Wind Tunnel, Aero Digest, Vol. 17, No. 5, p. 86, November, 1930 New York.

Calibration Constant of Wright Field 5-Foot Wind Tunnel, Air Corps Information Circular, Vol. 7, No. 643, p. 17, March 1, 1930, Washington. Air Corps Technical Report No. 3082.

Constant Speed Wind Tunnel Control, Aviation, Vol. 29, No. 4, p. 247, October, 1930, New York.

The New American Wind Tunnels, Journal Royal Aeronautical Society, Vol. 34, No. 235, pp. 559—576, July, 1930, London.

Vertical Wind Tunnel, N.A.C.A. Technical Report No. 387, 1931.

The 7 by 10. Foot Wind Tunnel, N.A.C.A. Technical Report No. 412, 1932.

Variable Density Wind Tunnel, N.A.C.A. Technical Report No. 416, 1932.

Full Scale Wind Tunnel, N.A.C.A. Technical Report No. 459, 1933.

English.

Experiments on Models of a Duplex Wind Channel, British A.R.C., R. and M. 552, 1917—18.

Reports on Tests of a Model of the Proposed No. 2, 7-Foot Wind Channel, At the R.A.E., British A.R.C., R. and M. 574, 1918—19.

The No. 2 7-Foot Wind Channel at the R.A.E., British A.R.C., R. and M. 847, 1922—23.

The "Duplex" Wind Tunnel of the National Physical Laboratory, British A.R.C., R. and M. 879, 1923—24.

Wind Tunnel, Journal Royal Aeronautical Society, Vol. 31, No. 200, pp. 799—806, August, 1927, London.
A High Speed Wind Channel for Tests on Aerofoils, British A.R.C., R. and M. 1130, 1927—28.
The One-Foot Wind Tunnel at the National Physical Laboratory; Including Particulars of Calibration Made with a Pitot Tube and Vane Anemometer at Low Speeds, British A.R.C., R. and M. 1103, 1927—28.
Variable Density Wind Tunnel. Report of the Scale Effect Panel, British A.R.C., R. and M. 1149, 1927—28.
The Armstrong-Whitworth Wind Tunnels, Aircraft Engineering Supplement, Flight, Vol. 21, No. 13, pp. 258d—258g, March, 1929, London.
Experiments on Models of a Compressed Air Wind Tunnel, British A.R.C., R. and M. 1355, 1930—31.
The 5-Foot Open Jet Wind Tunnel, British A.R.C., R. and M. 1364, 1930—31.

Continental.

Nouvelles Recherches sur la Résistance de l'air et l'aviation faites au laboratoire d'Auteuil, par G. Eiffel, p. 1, 1914.
Göttingen Wind Tunnel for Testing Aircraft Models, U.S. N.A.C.A. Technical Note No. 66, 1920.
See also: Berichte und Abhandlungen der Wissenschaftlichen Gesellschaft für Luftfahrt, a supplement to Zeitschrift für Flugtechnik und Motorluftschiffahrt, September, 1920.
Wind Tunnel of the Zeppelin Airships, U.S. N.A.C.A. Technical Memorandum No. 202, 1923.
See also: Zeitschrift für Flugtechnik und Motorluftschiffahrt, January 31 and February 15, 1921.
French Wind Tunnel Methods, Journal Royal Aeronautical Society, Vol. 33, No. 226, pp. 838—877, October, 1929, London.
Bulletins du Service Technique de l'Aéronautique, No. 19, 1924 and No. 66, 1930.
The Electrodynamometric Balance of the Small Wind Tunnel of the French Service of Aerodynamic Research, U.S. N.A.C.A. Technical Memorandum No. 556, p. 10, 1930.
Ein einfaches Sechskomponenten-Meßgerät der neuen Windkanalanlage am Kyffhäuser-Technikum, Bad Frankenhausen, Zeitschrift für Flugtechnik und Motorluftschiffahrt, p. 393, August 18, 1930.
Versuche an einem Windkanalmodell, Zeitschrift für Flugtechnik und Motorluftschiffahrt, p. 107, February 28, 1931.
Gilles, A., Beschreibung des Aerodynamischen Institutes der Technischen Hochschule Aachen und seiner Versuchseinrichtungen, Abhandlungen aus dem Aerodynamischen Institut an der Technischen Hochschule, Aachen, Heft 10, 1931.
Wind Tunnel of the Bucharest Polytechnic Institute, U.S. N.A.C.A. Technical Memorandum No. 651, 1931.
See also: L'Aéronautique, No. 157, p. 193, June, 1932.
Il Laboratorio Aerodinamico Della Direzione Superiore Studi Ed Esperienze Del Ministero Del-L'Aeronautica, L'Aerotecnica, Vol. XII, No. 2, p. 143, February, 1932.
Publications scientifiques et Techniques du ministère de l'Air, Service des Recherches de l'aéronautique, No. 5, 1932.
Der Windkanal der Focke-Wulf Flugzeugbau A.G., Zeitschrift für Flugtechnik und Motorluftschiffahrt, p. 305, June 14, 1932.
Annales du conservatoire National des Arts et Metiers, Série IV, Spécial No., p. 75, 1933.
Der Kleine Windkanal der D.V.L., Zeitschrift für Flugtechnik und Motorluftschiffahrt, p. 559, October 28, 1933.
Rendiconti sperimentali del Regio Istituto Superiore di Ingegneria di Torino, L'Aerotecnica, Vol. XIV, No. 4, 1934.

INDEX

GPSR Compliance

The European Union's (EU) General Product Safety Regulation (GPSR) is a set of rules that requires consumer products to be safe and our obligations to ensure this.

If you have any concerns about our products, you can contact us on ProductSafety@springernature.com

In case Publisher is established outside the EU, the EU authorized representative is:

Springer Nature Customer Service Center GmbH
Europaplatz 3
69115 Heidelberg, Germany

Batch number: 10682323

Printed by Printforce, the Netherlands